Salters-Nuffield Advanced Biology for Edexcel A2 Biology

STUDENT BOOK

A PEARSON COMPANY

Pearson Education Limited
Edinburgh Gate
Harlow
Essex
CM20 2JE
United Kingdom

and Associated Companies throughout the world

www.pearson.com

First published 2006
Published as trial edition 2003
This edition published 2009
15 14 13
10 9 8 7 6

ISBN 978-1-4082-0591-4

SNAB project editor: Anne Scott
Edited by Kate Redmond
Designed and illustrated by Pantek Arts Ltd, Maidstone, Kent
Picture research by Kay Altwegg and Charlotte Lippmann
Index by John Holmes

Printed in Malaysia, KHL CTP

Contents

Contributors

Many people from schools, colleges, universities, industries and the professions have contributed to the Salters-Nuffield Advanced Biology project. They include the following.

Central team

Angela Hall	Nuffield Curriculum Centre
Michael Reiss	Institute of Education, University of London
Anne Scott	University of York Science Education Group
Sarah Codrington	Nuffield Curriculum Centre

Authors

Angela Hall	Nuffield Curriculum Centre
Gill Hickman	Ringwood School
Sue Howarth	Tettenhall College
Sarah Middlewick	Ringwood School
Nick Owens	
Michael Reiss	Institute of Education, University of London
Anne Scott	University of York Science Education Group
Nicola Wilberforce	Esher College

Editor

Anne Scott

Acknowledgements

We would also like to thank the following for their advice and assistance.
Teachers, technicians and students at schools and colleges running the Salters-Nuffield Advanced Biology course.

Shona Fletcher	University of York
Martin Hall	Natural History Museum
Professor Robin Millar	University of York
Gary Skinner	
Iain Thyne	University of York
Sandra Wilmott	University of York Science Education Group

Sponsors

We are grateful for sponsorship from the Salters' Institute and the Nuffield Foundation, who have continued to support the Salters-Nuffield Advanced Biology project after its initial development and who have enabled the production of these materials.

Authors of the previous editions

This revised edition of the Salters-Nuffield Advanced Biology course materials draws heavily on the initial project development and the work of previous authors.

Jamie Copsey	Jersey Zoo
Nan Davis	
Jon Duveen	City & Islington College
Brian Ford	Colchester Sixth-Form College
Richard Fosbery	The Skinners' School
David Greenwood	Greenhead College, Huddersfield
Ginny Hales	Cambridge Regional College
Steve Hall	King Edward VI School, Southampton
Gill Hickman	Ringwood School
Malcolm Ingram	
Liz Jackson	King James School
Christine Knight	
Pauline Lowrie	Sir John Deane's College
Nick Owens	Oundle School, Peterborough
Jenny Owens	Rye St Antony School, Oxford
Ron Pickering	Sir John Deane's College
Jacquie Punter	Brighton & Hove Sixth Form College
Cathy Rowell	University of York
David Slingsby	
Mark Smith	Leeds Grammar School
Jane Wilson	Eggbuckland Community College, Plymouth
Mark Winterbottom	King Edward VI School, Bury St Edmunds

Advisory Committee for the initial development

Professor R McNeill Alexander FRS	University of Leeds
Dr Roger Barker	University of Cambridge
Dr Allan Baxter	GlaxoSmithKline
Professor Sir Tom Blundell FRS (Chair)	University of Cambridge
Professor Kay Davies CBE FRS	University of Oxford
Professor Sir John Krebs FRS	Food Standards Agency
Professor John Lawton FRS	Natural Environment Research Council
Professor Peter Lillford CBE	University of York
Dr Roger Lock	University of Birmingham
Professor Angela McFarlane	University of Bristol
Dr Alan Munro	University of Cambridge
Professor Lord Robert Winston	Imperial College of Science, Technology and Medicine

Please cite this publication as: Salter-Nuffield Advanced Biology A2 Student Book, Edexcel Pearson, London, 2009

About the course

Welcome to the second year of Salters-Nuffield Advanced Biology (SNAB). SNAB is much more than just another A-level specification. It is a complete course with its own distinctive philosophy. The course is supported by a comprehensive set of teaching, learning and support materials which embrace a student centred approach. SNAB combines the key concepts underpinning biology today, combined with the opportunity to gain the wider skills that biologists now need.

A context-led approach

All eight topics use a context-led approach; a storyline or contemporary issue is presented, and the relevant biological principles are introduced when required to aid understanding of the context.

In each topic you will study more than one area of biology. In later topics, you will meet many of the ideas again and develop them further. For example, immunity appears in Topic 6 Infection immunity and forensics, and is revisited in Topic 7 Run for your life, in the context of overtraining.

Building knowledge through the course

In SNAB there is not, for example, a topic labelled 'biochemistry' containing everything you might need to know on carbohydrates, fats, nucleic acids and proteins. In SNAB you study the biochemistry of these large molecules bit by bit throughout the course when you need to know the relevant information for a particular topic. In this way information is presented in manageable chunks and builds on existing knowledge.

Ideas introduced at AS are revisited in A2, for example, the control of heart rate in Toic 7 extends the work done on heart and circulation in the first AS topic.

Activities as an integral part of the learning process

SNAB encourages an active approach to learning. Throughout this book you will find references to a wide variety of activities. Through these, you will learn both content and experimental techniques. In addition, you will develop a wide range of skills, including data analysis, critical evaluation of information, communication and collaborative work.

Within the electronic resources you will find animations on such things as photosynthesis, muscle contraction and the nerve impulse. These animations are designed to help you understand the more difficult bits of biology. The support sections should be useful if you need help with biochemistry, mathematics, ICT, study skills, the examination or coursework.

SNAB and ethical debate

With rapid developments in biological science, we are faced with an increasing number of challenging decisions. The use of drugs in sport is one ethical dilemma considered in the A2 course.

In SNAB you develop the ability to discuss and debate these types of biological issues. There is rarely a right or wrong answer; rather you learn to justify your own decisions using ethical frameworks.

Exams and coursework

Edexcel examines SNAB A2 as the context-led approach within the Edexcel A2 level Biology specification. The Edexcel exams reward your ability to reason scientifically and to use what you have learned in new contexts, rather than merely being able to regurgitate huge amounts of information you have learnt off by heart. Most of the exam questions are structured ones, but you will also write more extended answers. You can find out more about the coursework and examinations within the electronic resources and in the specification.

We feel that SNAB is the most exciting and up-to-date advanced biology course around. Whatever your interests are – whether you want to study a biological subject at University, or use the qualification in other ways – we hope you enjoy the course.

Any questions?

If you have any questions or comments about the materials you can let us know via the website or write to us at:

The Salters-Nuffield Advanced Biology Project
Science Curriculum Centre
University of York
Heslington
York YO10 5DD

Email uyseg-snab@york.ac.uk
www.advancedbiology.org

How to use this book

There are a number of features in the student books that will help your learning and help you find your way around the course.

This A2 book covers the four A2 topics. These are shown in the contents list, which also shows you the page numbers for the main sections within each topic. There is an index at the back to help you find what you are looking for.

Main text

Key terms in the text are shown in **bold type**. These terms are defined in the interactive glossary that can be found on the software using the 'search glossary' feature.

There is an introduction at the start of each topic and this provides a guide to the sort of things you will be studying in the topic.

There is an '**Overview**' box on the first spread of each topic, so you know which biological principles will be covered.

Overview of the biological principles covered in this topic

The topic starts by looking at examples of ecosystems, building on the ideas met in Topic 4. You will see how both biotic and abiotic factors affect ecosystems, and understand more about how ecosystems are continuously changing. Changes to ecosystems are partly a result of natural processes such as succession.

Occasionally in the topics there are also '**Key biological principle**' boxes where a fundamental biological principle is highlighted.

Key biological principle: Homeostasis

The need for homeostasis

In humans, if cells are to function properly, the body's internal conditions must be maintained within a narrow range of the cells' optimum conditions. The maintenance of a stable internal environment is called **homeostasis**.

This is partly achieved by maintaining stable conditions within the blood, which in turn gives rise to the tissue fluid that bathes the body's cells. In the blood the concentration of glucose, ions and carbon dioxide must be kept within narrow limits. In addition, the water potential (determined by the concentration of solutes in the blood),

'**Did you know?**' boxes contain material that will not be examined, but we hope you will find it interesting.

? Did you know?

Those who influenced the development of Darwin's theory

Darwin's own observations of organisms were key to the development of his ideas on evolution but other scientists influenced his thinking, e.g. Thomas Malthus's ideas about the factors influencing the size of populations. In 1798, Malthus wrote an essay about the potential of the human population to grow exponentially. He argued that this would lead to a population size that would exceed available food supply, resulting in famine, pestilence and war. Darwin realised that the sizes of many animal populations were staying the same.

Questions

You will find two types of question in this book.

In-text questions occur now and again in the text. They are intended to help you to think carefully about what you have read and to aid your understanding. You can self-check using the answers provided at the back of the book.

Q5.42 For the benefit of Mendel and Darwin (who have paid a brief visit back to Earth), give a twenty-first century explanation of the word 'gene'.

If Mendel were publishing his ideas today, he would probably send his paper to a scientific journal such as *Nature* or *Proceedings of the Royal Society*. Before publication, Mendel's paper would be sent to two or three scientists for **peer review**. The reviewers would examine his paper very critically, looking at whether he had included proper controls, used statistics appropriately, considered the work of other scientists, and whether his conclusions were valid. It is likely that the paper would be returned to Mendel for some alterations or clarification, though the reviewers would remain anonymous. The paper would then be published – it might also be presented at a scientific conference – and the world would quickly hear about his findings.

Q5.43 Why is it important that the identities of reviewers of scientific papers are not given to the author(s) of the paper?

Checkpoint

5.6 Write a summary explaining how natural selection can lead to evolution.

Boxes containing '**Checkpoint**' questions are found throughout the book. They give you summary-style tasks that build up some revision notes as you go through the student book.

At present the use of plant-derived alcohol and oil to power vehicles is limited, because it remains cheaper to use petrol or diesel. However, rising oil prices are now making biofuels more cost-effective. There has been a strong political movement to increase the use of biofuels, as these are seen as an important contribution to reducing carbon emissions. The European Union has a target of 10% biofuels in transport fuel by 2020, and the UK has its own target of 5% by 2010. However, some major disadvantages of biofuel production have become apparent.

The destruction of rainforests to plant palm oil trees for biofuel production is releasing vast amounts of stored carbon into the atmosphere, more than cancelling out any gains made by burning biofuels, and also harming wildlife. The use of large quantities of edible corn to make corn oil and ethanol for biofuel affects food availability.

Checkpoint

5.7 Make a table for and against the production and use of biofuels for transport. How might the following differ in their views about biofuels: developing countries, Western countries, farmers, oil companies?

Links to the online resources

'**Activity**' boxes show you which activities are associated with particular sections of the book. Activity sheets and any related animation can be accessed from the activity homepages found via 'topic resources' on the software. Activity sheets include such things as practicals, issues for debate and role plays. They can be printed out. Your teacher or lecturer will guide you on which activity to do and when. There may also be weblinks associated with the activity, giving hotlinks to other useful websites.

Activity

Activity 5.17 allows you to find out about coral bleaching linked to rising sea temperatures. **A5.17S**

A final activity for each topic enables you to '**check your notes**' using the topic summary provided within the activity. The topic summary shows you what you need to have learned for your unit exam.

'**Weblink**' boxes give you useful websites to go and look at. They are provided on a dedicated 'weblink' page on the software under 'SNAB communications'.

Weblink

Visit the British Antarctic Survey website to find out about ice core records for carbon dioxide over the last 740 000 years.

'**Extension**' boxes refer you to extra information or activities available in the electronic resources. The extension sheets can be printed out. The material in them will not be examined.

Extension

In **Extension 5.5** you can find out about what might have caused extinction of the dinosaurs. **X5.05S**

'**Support**' boxes are provided now and again, where it is particularly useful for you to go to the student support provision within the electronic resources, e.g. biochemistry support. You will also be guided to the support in the electronic resources from the activity home pages, or you can go directly via 'student support'.

Support

To find out more about energy transfer in chemical reactions, look at the Biochemistry support on the website.

At the end of each topic, as well as the '**check your notes**' activity for consolidation of each topic, there is an interactive '**Topic test**' box. This test will usually be set by your teacher / lecturer, and will help you to find out how much you have learned from the topic.

On the wild side

Why a topic called On the wild side?

Like the polar bear (Figure 5.2), we are all skating on thin ice. It is not entirely clear when the ice will finally break. We are totally dependent on our environment for food to eat, water to drink, air to breathe, and materials to build a home. As the human population grows and consumption and pollution increase, we are putting more and more pressure on the planet's ability to cope.

What evidence do we have that the planet and its ecosystems are changing? What is causing these changes, and are we partly responsible? In this topic you will address these questions by looking at how ecosystems work and the ways in which humans affect them. It is easy to imagine the fate of polar bears should their icy environment melt away, but are we as aware of the effects closer to home?

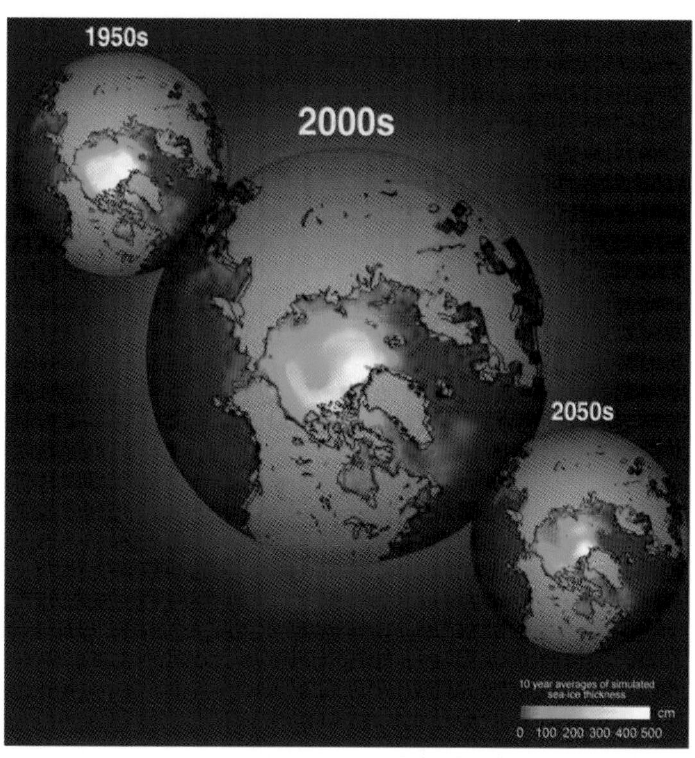

Figure 5.1 Computer model predictions of Arctic sea-ice extent – past, present and future. *Source: NOAA's Geophysical Fluid Dynamics Laboratory.*

Figure 5.2 Our largest living land carnivore: current population 25 000 (2008).

One of the main threats to our future welfare is climate change. The rate of climate change at the poles is greater than nearer the Equator. So polar bears and other Arctic wildlife are likely to be giving us the clearest signals about global warming. The Arctic forms a vast ecosystem where few people live, much of it still in its pristine state – think of the Alaskan and Siberian wildernesses and the vast tracts of Canadian tundra. The area is populated by animals such as Arctic hares, reindeer, moose, musk-ox, wolves, seals, whales, wild geese – and polar bears. But it is not just far off ecosystems that will be affected; the species in your local nature reserve may well face local extinction as species distribution changes.

The study of climate change is vital to an understanding of the changes occurring. We already have computer models capable of explaining past trends and predicting future climate change. For example, models can predict the extent of summer ice melt in the Arctic (Figure 5.1). The amount of ice is critical to polar bears; they use the ice to travel to new feeding grounds, and as a platform from which to catch seals. Since the turn of the millennium, summer Arctic ice melt has been even greater than predicted by the best models, with an especially extreme melt in 2007 – bad news for both climate modellers and polar bears.

Polar bears are large and slow-breeding so are unlikely to be able to adapt very quickly. How will polar bears and other species cope if the present climatic trends continue? Can new species arise in response to environmental change?

Overview of the biological principles covered in this topic

The topic starts by looking at examples of ecosystems, building on the ideas met in Topic 4. You will see how both biotic and abiotic factors affect ecosystems, and understand more about how ecosystems are continuously changing.

Changes to ecosystems are partly a result of natural processes such as succession. You will look at examples of succession and the ways in which human activity, such as agriculture, can interrupt succession.

In this topic you will also be studying the way in which photosynthesis happens at the molecular level and provides the energy that supports nearly all ecosystems. Ecosystems are supported by the energy of sunlight through the process of photosynthesis. Without photosynthesis, life on Earth would be limited to chemosynthetic organisms found in habitats such as thermal vents in the ocean. You will look in detail at the biochemistry of photosynthesis, and see how energy is passed on through the trophic levels of a food chain.

Throughout this topic you will see how scientists attempt to determine the possible impacts of global warming on natural ecosystems and agriculture. Building on the knowledge gained earlier in the course, you will consider the effect of temperature on enzyme activity.

You will critically examine the different types of evidence used to support the theory that the Earth's climate is changing as a result of human activity. This leads on to a consideration of how scientific knowledge is produced, and to the controversies that can arise if the predictions of science clash with politics and business. You will examine how scientists use models of climate change.

You will look at the process of evolution by natural selection, including speciation, and discuss whether populations of living organisms are able to respond to rapid environmental change.

Lastly, recalling the basics of the carbon cycle studied at GCSE, you will gain a more detailed understanding of it, and see how the use of biofuels and reforestation might help reduce atmospheric carbon dioxide levels.

5.1 What is an ecosystem?

The part of the Earth and its atmosphere that is inhabited by living organisms is called the **biosphere**. Within the biosphere there are numerous different **ecosystems**. Ecosystems have distinctive features that affect the organisms living there. In an ecosystem there is the **abiotic** component – physical and chemical factors like the climate and soil type – and the **biotic** component – factors determined by organisms, such as **predation** and **competition**. It can be difficult to define the extent of an ecosystem but, in principle, ecosystems tend to be fairly self-sustaining. So a lake remains as a lake (at least for a few hundred years), a wood remains as a wood, and so on.

The bottle of brine shrimps shown in Figure 5.3 is an ecosystem, although an unusually small and artificial one. It contains a community of organisms that interact with each other and with their physical environment in such a way as to make up a self-sustaining system (more or less!). The small crustaceans (the brine shrimps) in the bottle feed on the microscopic algae; the algae never run out, but multiply as fast as the brine shrimps eat them; the carbon dioxide respired by the shrimps is taken up in algal **photosynthesis**; and so on.

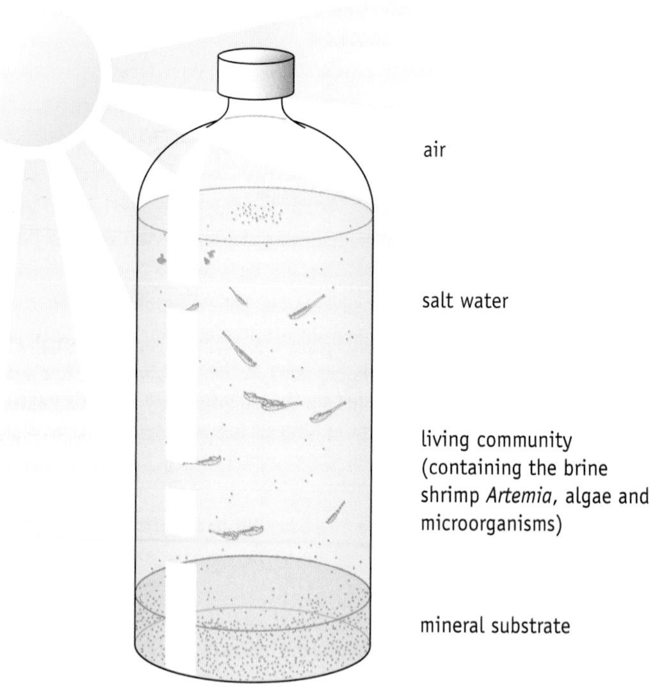

Figure 5.3 A brine shrimp bottle ecosystem.

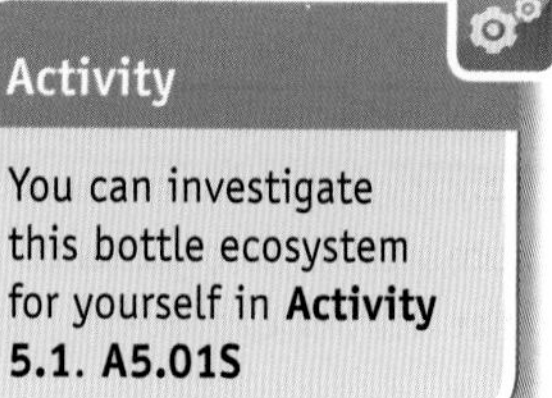

Activity

You can investigate this bottle ecosystem for yourself in **Activity 5.1. A5.01S**

Q5.1 For the brine shrimp bottle:

a list some of the populations making up the biotic component of the ecosystem

b describe what makes up the abiotic component.

Habitats

Within an ecosystem there can be many different **habitats**. As you learned in AS Topic 4, a habitat can be thought of literally as the place with a distinct set of conditions where an organism lives. Within a pond ecosystem some organisms will live on the water surface, some anchored to the banks, and others floating or swimming in the water. Each of these places constitutes the habitat of those organisms. Within a habitat there may be many tiny microhabitats, again each with distinct conditions. For example, the underside of a stone on the bottom of the pond will have a different set of conditions from the upper surface. Even the icy wastes of the Arctic contain many different habitats.

Communities

Within a habitat there will be several to many populations of organisms. Each **population** is a group of individuals of the same species found in an area. The various populations sharing a habitat or an ecosystem make up a **community**. So, for example, a woodland community consists of all the organisms that live in a wood.

You will recall from AS Topic 4 that if two species live in the same habitat and have exactly the same **niche** within the habitat – the same food source, the same time of feeding, the same shelter site, and so on – they will compete with each other. The better adapted of the two species will, sooner or later, out-compete the other and exclude it from the habitat. Two species are able to share the same habitat only if they occupy different niches (Figure 5.4).

Figure 5.4 More than one species can feed on the strand line using the same food source as long as they feed at different times. This ensures that the different species occupy different niches.

The word 'ecosystem' was first used in 1935 by the English botanist Sir Arthur George Tansley, one of the founders of the British Ecological Society. Tansley realised that a community of organisms could not be separated from the habitat in which it lives, and that neither can be fully understood in isolation.

What determines which species occur in a habitat?

A particular species lives in a particular habitat (unless deliberately introduced and maintained by humans) because it is adapted, and is able to survive and reproduce there. Woods, moorland, mud flats, meadows and roadside verges may look peaceful but they are all the scene of intense competition within and between species. Species that come second don't gain a silver medal – they face extinction. The conditions in which species compete for survival are defined by the ecological factors in their habitat. As you have seen, ecological factors can be classified into abiotic and biotic ones.

Abiotic factors

Abiotic factors are non-living or physical factors; they include the following:

- **Solar energy input** is affected by latitude, season, cloud cover, and changes in the Earth's orbit. Light is vital to plants – it is the energy source for photosynthesis. Light also has a role in initiating flowering, and in some species it is required for seed germination. In many animals, light affects behaviour. For example, changes in day length can be a cue for reproduction.
- **Climate** includes rainfall, wind exposure and temperature.
- **Topography** includes altitude (which affects climate), slope, aspect (which direction the land faces) and drainage (see Figure 5.5).
- **Oxygen availability** is particularly important in aquatic systems. For example, fast-flowing streams are often better oxygenated than stagnant pools.
- **Edaphic** factors are connected with the soil, and include soil pH and mineral salt availability – itself affected by geology. The underlying geology of an area can have a significant effect on plant distribution. Edaphic factors also include soil texture. Sandy soils are well-drained; they dry out easily in a drought, but are well-aerated and rarely waterlogged in wet weather. Clay gets easily waterlogged, but it retains water well, which can be an advantage in a drought.
- **Pollution** can be of the air, water or land.
- **Catastrophes** are infrequent events that disturb conditions considerably. Examples are earthquakes, floods, volcanic eruptions and fires.

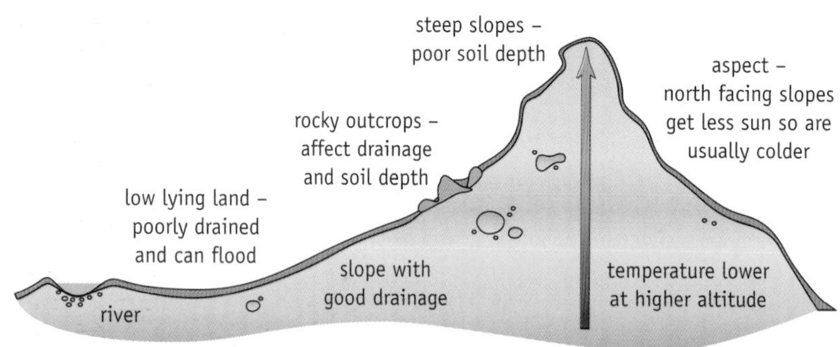

Figure 5.5 Topography affects conditions in a habitat.

Extension

Find out about the effects of fire on a habitat in **Extension 5.1. X5.01S**

Biotic factors

Biotic factors are 'living' factors. These include the following:

- **Competition** for resources like food, light, water and space can be **interspecific** (between species) or **intraspecific** (within species).
- **Grazing**, **predation** and **parasitism** are all relationships between two organisms where one benefits at the other's expense.
- **Mutualism** is a relationship in which both partners benefit.

Biotic factors are usually **density dependent**: the effects are related to the size of the population relative to the area available. The larger the population density, the greater the competition for food, space and so on.

The abundance and distribution of organisms in a habitat are controlled by biotic and abiotic factors. There is normally a complex interaction of biotic and abiotic factors within a habitat. For example, if poor weather reduces the survival rate of one animal species, this has a knock-on effect on all its **predators**. If the pH of the soil changes, this may affect the survival of bacteria in the soil, and the rate of decomposition and recycling of material may alter. Trees affect such abiotic factors as the water content of the soil, the humidity, and light available for smaller plants. In turn, these abiotic factors influence other organisms in the ecosystem.

Sublittoral zone
- rarely uncovered except at extreme tide so conditions relatively constant
- wave action intense
- low light due to depth of water

Lower shore
- relatively stable conditions
- exposed only at low tides so plenty of nutrients available
- less light penetrates water at high tide

Middle shore
- submerged for half the day so less desiccation than upper shore

Upper shore
- submerged only for short periods
- high desiccation
- wide variation in temperature and salinity

Splash zone
- rarely submerged
- extreme temperature variation and desiccation
- high salt content from evaporation
- dead organic matter accumulates on strand line

Laminaria species (brown alga)
- blade (or lamina) often found washed up on beaches because shed each autumn to reduce surface area
- holdfast attaches alga firmly to rocks
- adapted to low light intensities; additional photosynthetic pigments absorb maximum light

Chondrus crispus (red alga)
- grows around base of *Laminaria*
- shade tolerant
- adapted to low light intensities; additional photosynthetic pigments absorb maximum light

Blue rayed limpet
- herbivore
- feeds on *Laminaria*

Toothed wrack (brown alga)
- thin cell walls, intolerant of desiccation
- additional photosynthetic pigments absorb maximum light
- flat fronds in layers prevent the lower ones from drying out

Corallina (red alga)
- grows below brown algae or in rock pools
- shade tolerant
- intolerant of desiccation

Dog-whelk
- carnivore
- on exposed shores develops a larger muscular foot to prevent being dislodged by waves
- shell size and thickness depend on wave action

Bladder wrack (brown alga)
- air bladders float fronds towards the light
- intolerant of desiccation

Flat periwinkle
- herbivore; feeds on wrack
- not tolerant of temperature variation
- several different colours of shell provide good camouflage against bladders of wrack
- eggs laid in a gelatinous mass to prevent drying out

Limpet
- herbivore
- adheres to rock with a muscular foot; clamps down onto rock to prevent drying out
- reduced metabolism when exposed

Channel wrack (brown alga)
- rolled fronds trap water and reduce water loss
- oily layer on surface of fronds slows desiccation
- thick cell walls shrink with drying
- survives low nutrient input
- grows very slowly

Spiral wrack (brown alga)
- spiralling fronds trap water
- thinner cell walls and no oily layer so loses water faster than channel wrack; often occurs lower on the shore
- shades out channel wrack

Rough periwinkle
- herbivore
- tolerant of high temperatures and desiccation by cementing itself to the rock and respiring without oxygen
- gills modified to absorb oxygen from the air

Lichens
- made up of a fungus and an alga or blue-green bacterium which have a mutualistic relationship; the algae or bacteria are protected from desiccation by the fungus and they provide the fungus with photosynthetic products

Black periwinkle
- grazes on lichens
- found in crevices
- tolerant of very variable salt and temperature conditions
- gill cavity modified to form a lung

Greater diversity of organisms as conditions become more stable and more nutrients are available.

Figure 5.6 These examples illustrate how the distribution of organisms on the rocky shore is affected by biotic and abiotic factors. They also show how species are adapted to conditions on the area of shore where they live.

Anthropogenic factors are those arising from human activity. They can be either abiotic or biotic. It is important to recognise that our ecological footprint, the impact of humans on the world environment, is at present far greater than that of any other species.

The landscapes of much of Britain would still be largely wooded were it not for deforestation, moor-burning and grazing. Grazing might be regarded as just a biotic factor since wild animals such as deer and rabbits also graze. But the introduction of sheep and rabbits, and the removal of predators of grazing animals, such as wolves, are all the result of human actions. Grazing by domesticated animals such as sheep is also accompanied by high stocking densities, fencing, the introduction of cultivated types of grass, and the use of fertilisers. The result is that the environment is no longer 'natural'.

Q5.2 Look at the abiotic and biotic factors listed on page 6, and decide which are also anthropogenic.

Q5.3 Look at Figure 5.6 and list the major abiotic factors affecting this habitat.

Activity

Activity 5.2 lets you study the relationship between the distribution of plant species and variations in abiotic factors. It includes case studies and considers some of the adaptations that enable particular species to survive in some places but not others. **A5.02S**

Checkpoint

5.1 Define the term 'niche' and produce a mind map showing how niches affect the abundance and distribution of organisms in a habitat.

Adapted for survival

Species survive in a habitat because they have adaptations that enable them to cope with both the biotic and abiotic conditions in their niche.

For example, the polar bear in Figure 5.7 has adaptations that aid its survival, such as thick fur providing insulation. The outer hairs are long and oily, easily shedding water, but the inner hairs are finer and provide insulation by trapping air.

Q5.4 Look at Figure 5.7. Suggest how each of the adaptations highlighted might aid the survival of the polar bear.

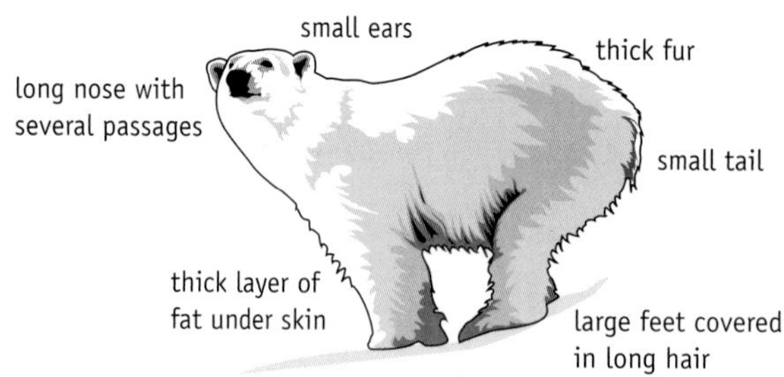

- Hunt seals by standing very still by water's edge
- Mate in summer
- Female builds a den, remaining inside over the winter
- Cubs are born in November or December and emerge from the den in the spring

Figure 5.7 Polar bears are well adapted to the conditions in the Arctic.

Polar bears are the top carnivores in the world's most northerly land ecosystems, collectively known as the tundra. The vegetation of the tundra is only able to grow for about two months of the year when the surface thaws. Plant life includes dwarf willows, cranberry shrubs, sedges, grasses and lichens. These plants are grazed by vast herds of caribou, which migrate over vast areas.

Q5.5 The tundra vegetation cannot withstand heavy grazing. Suggest why **a** the vegetation cannot cope with heavy grazing **b** the vegetation can cope with the grazing by caribou.

The grasses of the tundra – like all grasses – are better able than other tundra plants to withstand grazing, since they grow from the base (like your hair) rather than from the tips. The tough grasses cause the teeth of the caribou to wear down as they age. However, Arctic hares – like our familiar rabbits – have continually growing incisors, so as they wear down, new grinding surfaces emerge. Many hares and rabbits eat their own faeces: grass is very hard to digest and, to gain maximum nutrition from their food, they digest their food twice. They produce soft pellets the first time food passes through their gut; these are immediately eaten to digest the material further. Hard pellets are then produced as waste. These examples illustrate some of the adaptations of **herbivores**.

Q5.6 What other adaptations do some plants have to escape grazing?

Q5.7 The familiar daisy (Figure 5.8) not only survives grazing but relies on it. Explain **a** why the daisy is well adapted to survive close grazing and mowing **b** why it becomes less abundant in an undisturbed habitat.

Q5.8 Rabbits are not only well adapted for a diet of grass, they also have adaptations that help them avoid predators. What features of a wild rabbit help it to escape from predators?

Q5.9 Look at the organisms described in Figure 5.6. For one species in each trophic level, describe how it is well adapted to survive in the area of shore where it is found.

Figure 5.8 The daisy's growing point is at ground level, and its leaves form a basal rosette. Long roots enable it to find water when the soil gets dry (as often happens in grazed grassland on shallow soil).

Ecosystems are dynamic

During the winter months, caribou rely on lichens as a source of food. Lichens grow freely on the bare rock surfaces and the caribou are able to find them by pushing aside the snow. Lichens consist of a mutualistic relationship between a fungus and a green protoctist. The fungus provides protection for the green cells, which in turn provide the fungus with the products of photosynthesis.

It is easy to find lichens much nearer to home, growing on roof tiles, gravestones, pavements and tree trunks, making them look grey or orange. They are often the first organisms to colonise bare ground, and can cope with the harsh conditions on bare rock. They create a community that changes over time, in a process known as **succession**.

Bare ground can sometimes arise through the formation of a completely new island. This happened off the coast of Iceland in 1963 when the island of Surtsey (Figure 5.9A) appeared as a result of volcanic activity on the sea bed. Scientists have catalogued every species to arrive on the island; there are now over 50 species of flowering plants.

Figure 5.9 **A** Surtsey, the 1.5 km^2 island lying off the south coast of Iceland. **B** Dwarf willow is found growing on Surtsey.

Q5.10 Plants were first discovered growing on Surtsey in 1965. The first shrubs, such as the dwarf willow (Figure 5.9B) appeared in 1995.

a Suggest how plants managed to get to the island.

b How do you think the soil formed in which the tree is growing?

Primary succession

A **primary succession** starts in newly formed habitats where there has never been a community before. This may occur, for example, on bare rock, on material on the seashore like sand and shingle, and in open water. Unless prevented, succession then continues until, sometimes thousands of years later, a relatively stable community is established.

The pioneer phase: enter the botanical mavericks

The first organisms to colonise bare rock are lichens and algae – **pioneer species**. These are the only species that can cope with the extremes of temperature and lack of soil, water and nutrients (Figure 5.10).

The pioneers start to break up the rock surface, allowing some organic material to accumulate with this broken up rock as the beginnings of soil. They change the conditions in the habitat just enough to make them suitable for other species. Wind-blown moss spores start growing.

Figure 5.10 Lichens on a rock surface can start the process of breaking it up to form soil.

Succession continues

The mosses build up more organic matter in the soil, which can then hold water. The development of a soil enables seeds of small, shallow-rooted plants to establish.

As the conditions in the habitat improve, seeds from larger, taller plants appear. They compete with the plants already present in the habitat and, winning the competition, they replace the existing community.

Eventually a community usually dominated by trees is reached, and this stable **climax community** often remains unchanged unless conditions in the habitat change (Figure 5.11). The nature of the climax community depends very much on the environmental conditions, such as climate, the soil and which species are available. In much of Britain below 500 m, the natural climax is forest; this is oak-dominated in many areas, but with beech or ash on limestone and birch on more acidic soil.

The dominant species of a community is the one that exerts an overriding influence over the rest of the plant, microbe and animal species. (Sometimes several species will share the role of being dominant and are said to be co-dominant.) The dominant species is usually the largest and/or most abundant plant species in the community; this often gives the community its name, for example, an oak forest.

As the succession progresses, the number of niches increases, as does the number of species present. However, it is not unusual for the climax community to have lower biodiversity than preceding stages in the succession. Figure 5.11 shows the stages in a typical sand dune succession.

Sand is unstable. It lacks organic matter so dries out very quickly. It is low in nutrients and has a high salt content.

Only pioneer plants such as sea rocket or prickly saltwort can establish. Xerophytic characteristics allow them to survive: for example, thick fleshy leaves store water; hairs and a low density of stomata reduce transpiration. The pioneer species are salt tolerant.

The establishment of armed shrubs such as brambles, wild roses and hawthorn which are not grazed allows tree seedlings to become established. These may include willow, alder, oak, ash or pine species.

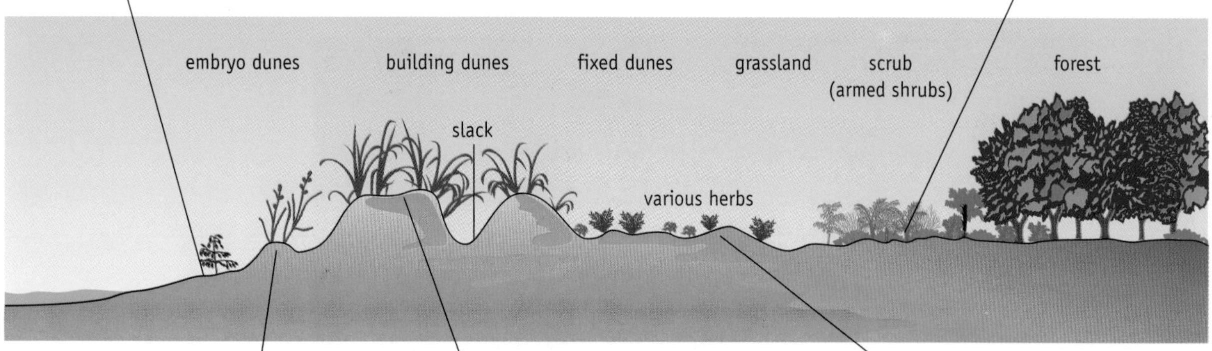

Sea couch grass may then become established. Sand is deposited around the sea couch grass creating an embryo dune. The grass grows faster when buried by sand.

The couch grass roots and horizontal rhizomes (underground stems) bind the sand and add organic matter so improving nutrient content and water-holding capacity.

Marram grass starts to grow. The blade-shaped leaves of marram grass roll inward in dry conditions. All the stomata are on this inward-rolling surface so water loss by transpiration is reduced.

Marram grass grows rapidly when buried by sand. The dunes build upwards, held together by the marram grass root and rhizome system. More organic matter accumulates, further improving nutrient content and water-holding capacity. Salt content declines as rainwater leaches salt from the developing soil.

As conditions become less harsh other species can now become established. Some of these species, e.g. restharrow, are legumes and add nitrates to the soil. The surface of the soil becomes stabilised and marram, which may rely on new roots forming in the freshly deposited sand to supply minerals, dies out.

A wide variety of plant species arrive, creating a grassland community which is grazed by rabbits or domesticated animals.

Figure 5.11 Sand dune succession. The land now covered by forest was originally the sandy foreshore and has gone through all the stages described to become the climax community. The succession literally pushes the sea back and turns a seashore into a forest.

Activity

Use the interactive tutorial in **Activity 5.3** to follow succession in one habitat.
A5.03S

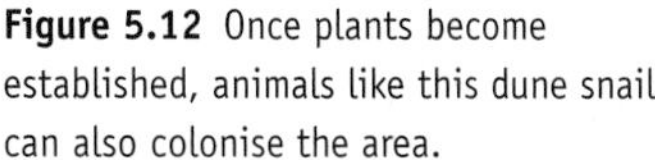

Figure 5.12 Once plants become established, animals like this dune snail can also colonise the area.

A climax community without trees

There are some ecosystems, such as those in the tundra, and sphagnum moss bogs in northern Scotland, where the climax community is not a forest or woodland.

The large area of bare, stony soil on a hill in Shetland called the Keen of Hamar (Figure 5.13) looks like an abandoned motorway construction site that will soon grass over as succession gets under way. But succession does not continue, apparently because of the properties of the soils and underlying rock. One possibility is that the soil is so sandy and freely draining that it suffers drought conditions after only a few days without rain. The vegetation resembles that of a low-competition pioneer phase of a more typical succession. But in this case it is the climax community, and probably has been for over 10 000 years.

The Keen of Hamar serpentine debris habitat has high biodiversity including rare species and is the best example of its kind in north-west Europe. The distinctive species are slow-growing and could not compete successfully with vigorously-growing grasses. Their survival depends on the habitat remaining an open pioneer-type habitat.

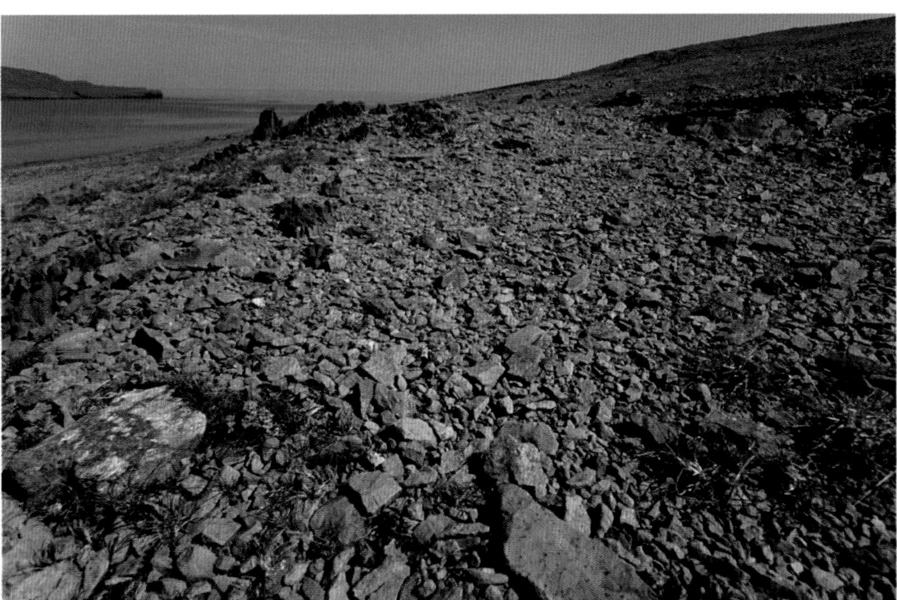

Figure 5.13 Keen of Hamar National Nature Reserve, Unst, the Shetland Islands. Succession is not occurring on this rocky debris.

Secondary succession

On bare soil where an existing community has been cleared, **secondary succession** occurs. When natural plants start to grow in a ploughed field or after a forest fire, they mark an early stage in secondary succession. In the absence of human interference, secondary succession would, in many parts of the UK, lead to the re-establishment of a forest climax community. In the same way, the lost Inca cities were swallowed up by the South American rainforest.

Figure 5.14 Groundsel: a common weed that is a pioneer species.

Bare soil does not stay bare for long in nature. Seeds of many species will already be lying dormant in the soil (as a seed bank), and others will be brought by the wind or animals. Groundsel (*Senecio vulgaris*) is an example of a pioneer species adapted to take advantage of newly bare soil where there is little or no competition (Figure 5.14). These adaptations include:

- seeds widely dispersed by the wind
- rapid growth
- short life cycle
- abundant seed production.

Pioneer species like groundsel cannot compete with slower-growing species such as grasses. Groundsel succeeds by getting in quickly (effective, widespread dispersal), growing rapidly, and flowering within a few weeks.

Deflected succession

A community that remains stable only because human activity prevents succession from running its course is called a **deflected succession** (Figure 5.15). For example, sheep grazing in Britain prevent many grasslands from developing into woodland. When, one day, you come out of your house and notice a wood where once there was a garden, you have really put off mowing the lawn too long!

Many habitats, such as chalk grassland of the South Downs or heather moors, need to be actively managed to prevent succession that would result in a loss or change in biodiversity. Grazing, mowing or burning may be used to deflect succession. There can be similar effects in natural grasslands such as those in the Serengeti in Tanzania, where frequent fires and grazing by vast herds of wildebeest and zebra help to prevent succession.

Figure 5.15 A golf course or grazing on the grassland behind sand dunes deflects succession, preventing the formation of a climax community.

Checkpoint

5.2 Draw up a table of comparison for primary and secondary succession, and explain what is meant by deflected succession.

5.2 Ecosystems rely on energy transfer

Figure 5.16 Reindeer eating lichen. This lichen, known as reindeer moss, is also an important food for caribou.

As with all ecosystems, the Arctic food chain begins with photosynthesising plants. One of the most important is 'reindeer moss', which is a kind of lichen (Figure 5.16). In the Arctic oceans, nutrient-rich cool water rises to support microscopic plants (**phytoplankton**) (Figure 5.17), which then provide food for animal plankton. Animal plankton in turn feeds huge fish populations as well as whales.

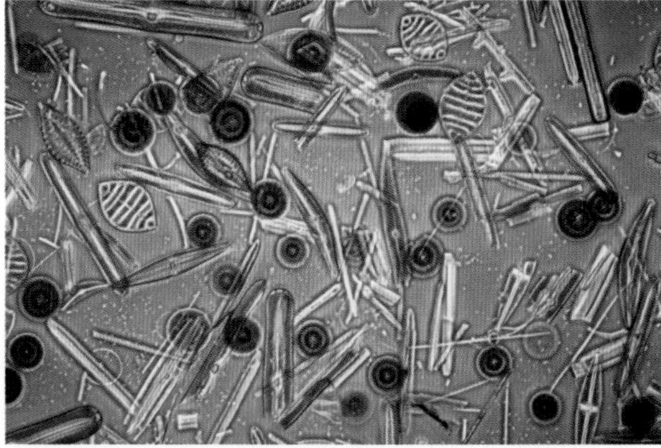

Figure 5.17 Phytoplankton trap the energy that all organisms in the marine ecosystem rely upon.

Producers and productivity

The rate at which energy is incorporated into organic molecules in an ecosystem is called the **primary productivity** of the ecosystem.

Producers, also known as **autotrophs**, are organisms that can make their own organic compounds from inorganic compounds. Green plants, algae and some bacteria are producers. The algae in the bottle ecosystem (Figure 5.3) are producers. When light falls on producers, some energy is transferred to a chemical energy store by producing organic fuels such as glucose.

Not all primary producers are photosynthetic. Some are **chemosynthetic autotrophs**. These make organic molecules using energy released from chemical reactions (Figure 5.18).

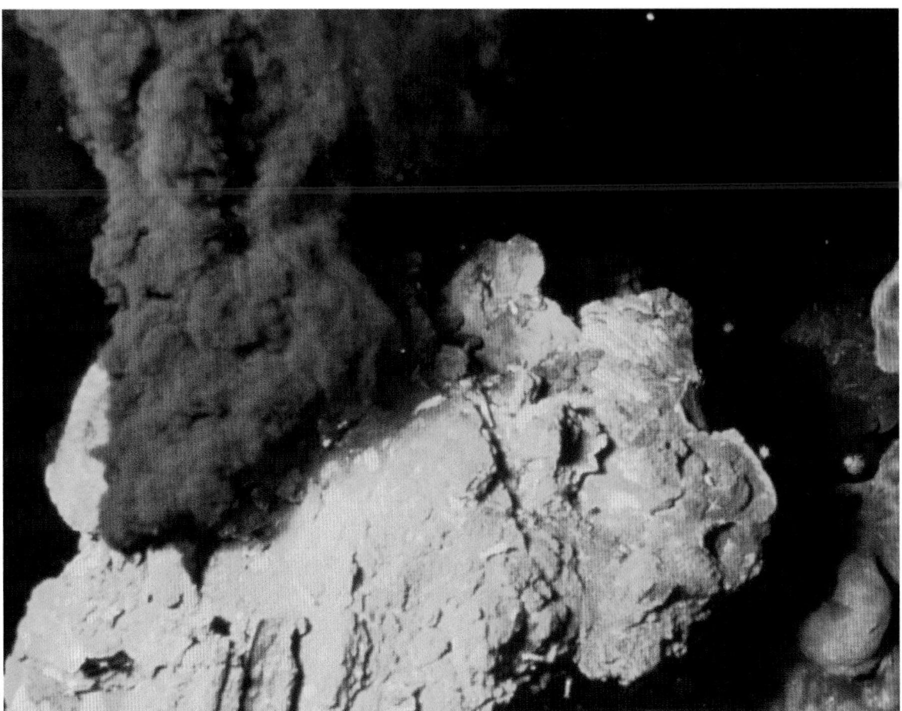

Figure 5.18 Chemosynthetic bacteria near volcanic vents in the deep ocean supply energy to food chains without using light.

Did you know?

Productivity and biodiversity are linked

The primary productivity of terrestrial vegetation is usually positively correlated with plant diversity. Productive ecosystems have a greater diversity of plants than ecosystems with low primary productivity.

Animal diversity is often linked to plant productivity, so that there is higher animal diversity in ecosystems with higher primary productivity. However, there is not always a positive correlation between plant productivity and animal diversity. For example, in lakes with high nutrient input (eutrophic lakes) there is increased algal productivity but a lower diversity of animals. This is due to the changing conditions accompanying **algal blooms**, principally depletion of oxygen as algae die and decompose.

Photosynthesis

An overview

You are probably familiar with the overall equation for photosynthesis:

$$6CO_2 + 6H_2O \xrightarrow[\text{in the presence of chlorophyll}]{\text{energy from light}} C_6H_{12}O_6 + 6O_2$$

Carbon dioxide is reduced as hydrogen and electrons from water are added to it, creating a carbohydrate.

Support

To find out more about energy transfer in chemical reactions, look at the Biochemistry support on the website.

The reactions in photosynthesis require an input of energy from light. The energy needed to break the bonds within carbon dioxide and water is greater than the energy released when the products – glucose and molecular oxygen – are formed. Therefore, the products of the reaction (glucose and oxygen) are at a higher energy level than the reactants (carbon dioxide and water), and act as a store of energy. You can think of this as being similar to stretching (or compressing) a spring to store energy. In photosynthesis, oxygen is a waste product and is released into the atmosphere. Glucose is a fuel, which can later be oxidised during respiration to release energy.

You may remember from your chemistry lessons that, when hydrogen is burned in air, it makes a loud 'pop' as the hydrogen reacts with oxygen, and water is produced. If pure oxygen is used rather than air, the reaction is even more explosive, releasing large amounts of energy (Figure 5.19). In photosynthesis the hydrogen is separated from water and stored within a carbohydrate. When energy is required in the cell, the hydrogen stored in the carbohydrate reacts with oxygen during respiration, releasing a large amount of energy.

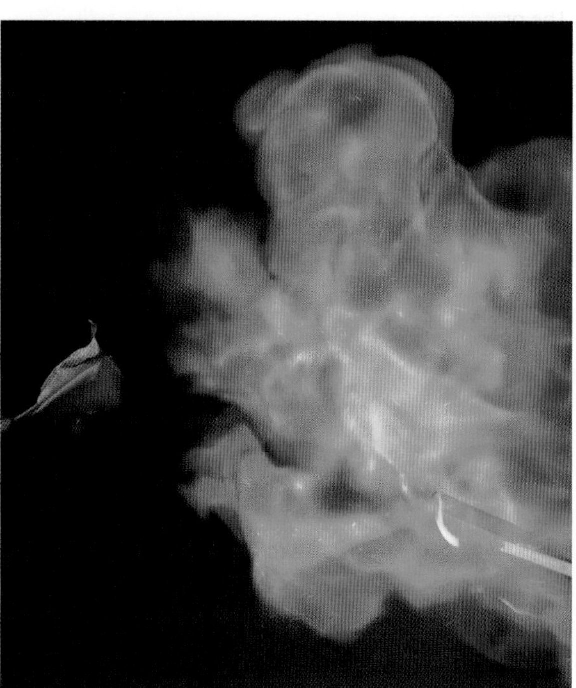

Figure 5.19 The reaction of hydrogen and oxygen releases large amounts of energy. See this reaction happening in the video clip that accompanies Activity 5.4.

Releasing hydrogen from water

The splitting of water into hydrogen and oxygen requires energy. Photosynthesis uses energy from sunlight to split water. This process is known as the **photolysis** of water, because of the involvement of energy from light in the reaction (*photo* means 'light'; *lysis* means 'splitting').

Storing hydrogen in carbohydrates

The hydrogen reacts with carbon dioxide in order to 'store' the hydrogen. Carbon dioxide is reduced to form the carbohydrate fuel glucose, which can be stored or converted to other organic molecules.

Using the glucose

The fuel has the potential to release large amounts of energy when the hydrogen stored in the carbohydrate reacts with oxygen during respiration. In aerobic respiration, glucose is pulled apart; the hydrogen combines with oxygen to make water, and energy and carbon dioxide are released.

Some of the glucose is converted into the variety of chemical substances needed by the plant to grow new cells. For example, glucose can be converted, using enzymes, into starch, cellulose, fats, amino acids/proteins and nucleic acids. The extra elements needed to make some of these compounds, such as nitrogen and phosphorus, are taken up by the roots of the plant from the soil (see AS Topic 4).

Q5.11 **a** Name the two elements not present in glucose needed to make proteins.

b Name the two elements not present in glucose needed to make nucleic acids.

The importance of photosynthesis

The organic molecules made during photosynthesis are passed on through food webs to other organisms. Animals cannot make all the organic compounds they need for themselves; they are entirely dependent on plants for their existence. And of course, plants release oxygen gas into the atmosphere, without which no organism would be able to carry out aerobic respiration.

How photosynthesis works

Photosynthesis is not a single reaction, but a series of reactions controlled by enzymes. These reactions occur in two main stages (Figure 5.20):

- **Light-dependent reactions** use energy from light and hydrogen from photolysis of water to produce **reduced NADP**, **ATP** and the waste product oxygen. The oxygen is either used directly in respiration or released into the atmosphere.
- **Light-independent reactions** use the reduced NADP and ATP from the light-dependent reactions to reduce carbon dioxide to carbohydrates.

What is meant by reduction?

During chemical reactions, existing bonds are broken and new bonds are formed, often involving the transfer of electrons. Electrons are also transferred when atoms turn into ions, or ions into atoms. The loss of electrons from a substance is known as oxidation, and the gain of electrons is known as reduction.

Remember this using OIL RIG: **o**xidation **i**s **l**oss, **r**eduction **i**s **g**ain. When a substance is in an oxidised form it has lost electrons. In a reduced form a substance has gained electrons. The **co-enzyme NADP** is reduced when electrons are added during photosynthesis.

Support

To find out more about reduction and oxidation, have a look at the Biochemistry support on the website. See the Biochemistry support on 'co-enzymes' to learn more about NADP.

Reduction can be carried out by the addition of an electron on its own, or by the addition of a whole hydrogen atom, which consists of a hydrogen ion (H^+) and an electron. Hydrogen ions cannot carry out reduction, since they do not carry an electron – a hydrogen ion is just a proton. However, hydrogen ions do control pH.

CO_2 and H_2O have no direct contact

Although the overall equation for photosynthesis suggests that carbon dioxide and water react with one another, the carbon dioxide and water never come into direct contact with each other. The hydrogen, electrons and energy needed for the reduction of CO_2 are transferred indirectly using reduced NADP and ATP (energy transfer molecule, see page 21).

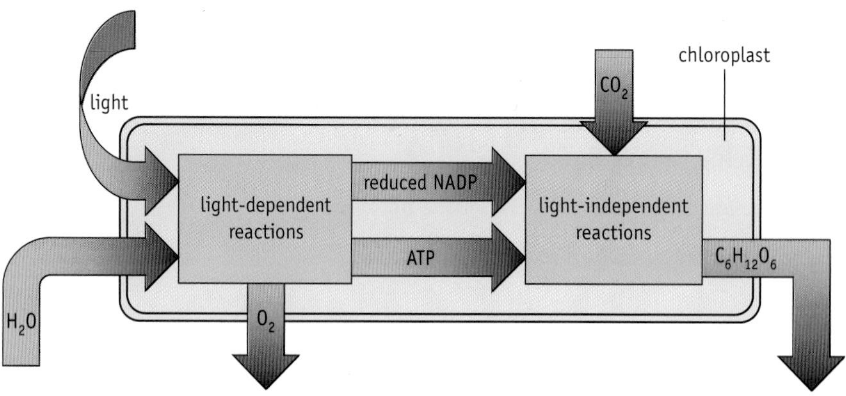

Figure 5.20 The two stages of photosynthesis.

Q5.12 Look at the summary of photosynthesis shown in Figure 5.20 and then answer the following questions.

a Which molecule provides the source of hydrogen for photosynthesis?

b What is the source of energy for photosynthesis?

c Which molecule provides the hydrogen for the light-independent reactions?

d Which molecule provides the energy for the reduction of carbon dioxide to glucose in the light-independent reactions?

> **Activity**
>
> The interactive tutorial in **Activity 5.4** guides you through the process of photosynthesis. **A5.04S**

Where does photosynthesis take place?

The site of photosynthesis is the **chloroplast**. A **palisade mesophyll** cell in a leaf can contain as many as 50 chloroplasts. Each chloroplast is made up of membranes, arranged in a very precise, organised way as shown in Figure 5.21.

The light-dependent reactions

When light is absorbed by chlorophyll in the **thylakoid membranes** of the chloroplast, the following events occur:

1. Energy from the light raises two electrons in each chlorophyll molecule to a higher energy level. The chlorophyll molecules are now in an 'excited' state.
2. The electrons leave the excited chlorophyll molecules and pass along a series of **electron carrier** molecules, all of which are embedded in the thylakoid membranes. These molecules constitute the **electron transport chain**.
3. The electrons pass from one carrier to the next in a series of oxidation and reduction reactions, losing energy in the process. The energy is used in the synthesis of ATP, in a process called **photophosphorylation**. Details of electron transport chains are covered in Topic 7.
4. The electrons lost from the chlorophyll must be replaced if the flow of electrons along the electron transport chain is to continue.
5. Within the thylakoid space, an enzyme catalyses the splitting of water (photolysis) to give oxygen gas, hydrogen ions and electrons. These electrons replace those that were emitted from the chlorophyll molecule, so it is no longer positively charged. The hydrogen ion concentration within the thylakoid space is raised as a result of photolysis.
6. The electrons that have passed along the electron transport chain combine with the co-enzyme NADP and hydrogen ions from the water to form reduced NADP.

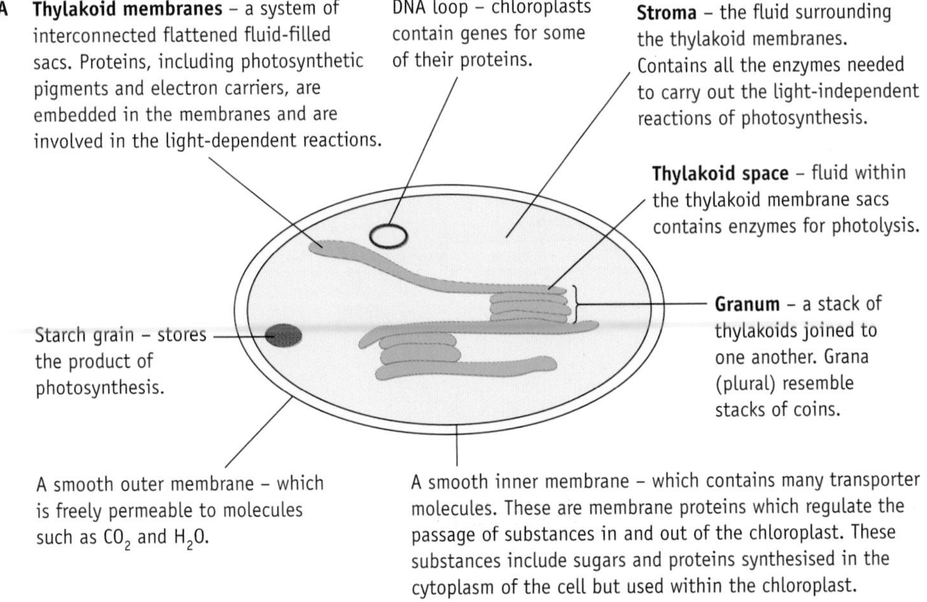

A
Thylakoid membranes – a system of interconnected flattened fluid-filled sacs. Proteins, including photosynthetic pigments and electron carriers, are embedded in the membranes and are involved in the light-dependent reactions.
DNA loop – chloroplasts contain genes for some of their proteins.
Stroma – the fluid surrounding the thylakoid membranes. Contains all the enzymes needed to carry out the light-independent reactions of photosynthesis.
Thylakoid space – fluid within the thylakoid membrane sacs contains enzymes for photolysis.
Granum – a stack of thylakoids joined to one another. Grana (plural) resemble stacks of coins.
Starch grain – stores the product of photosynthesis.
A smooth outer membrane – which is freely permeable to molecules such as CO_2 and H_2O.
A smooth inner membrane – which contains many transporter molecules. These are membrane proteins which regulate the passage of substances in and out of the chloroplast. These substances include sugars and proteins synthesised in the cytoplasm of the cell but used within the chloroplast.

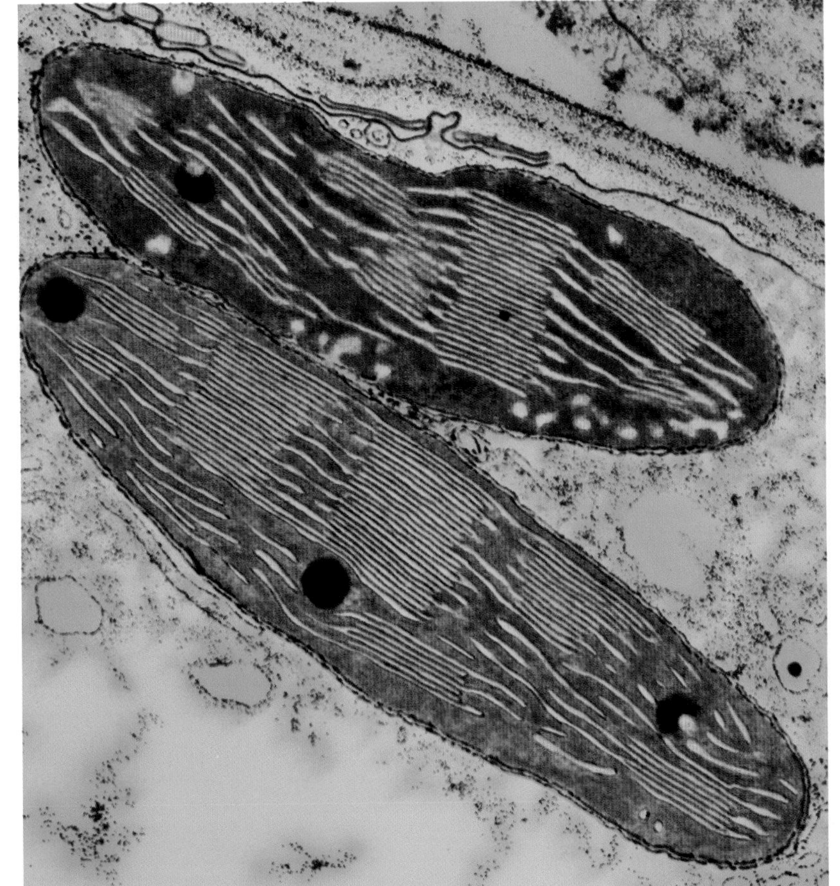

Figure 5.21 **A** The structure and functioning of chloroplasts. **B** Electron micrograph of two chloroplasts. Magnification × 35 000.

The ATP and reduced NADP created in the light-dependent reactions are used in the light-independent reactions.

The light-dependent reactions are summarised in Figure 5.22.

Weblink

Visit the interactive tutorial on cell structure and function in **Activity 3.1** to check out the structure of chloroplasts.

Extension

In **Extension 5.2** you can look in detail at the structure of the leaf. **X5.02S**

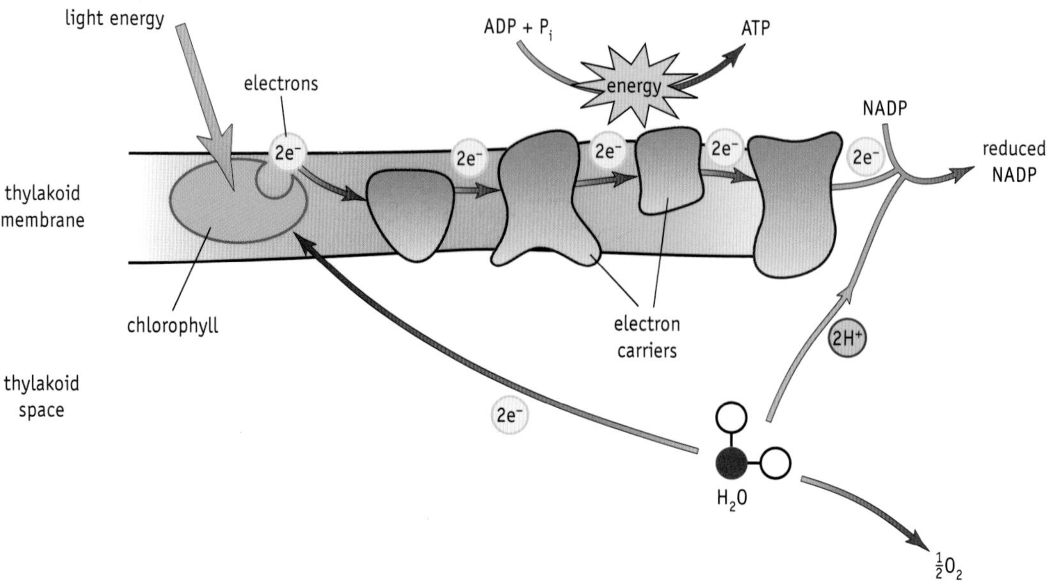

This is often drawn to show the change in energy level of electrons:

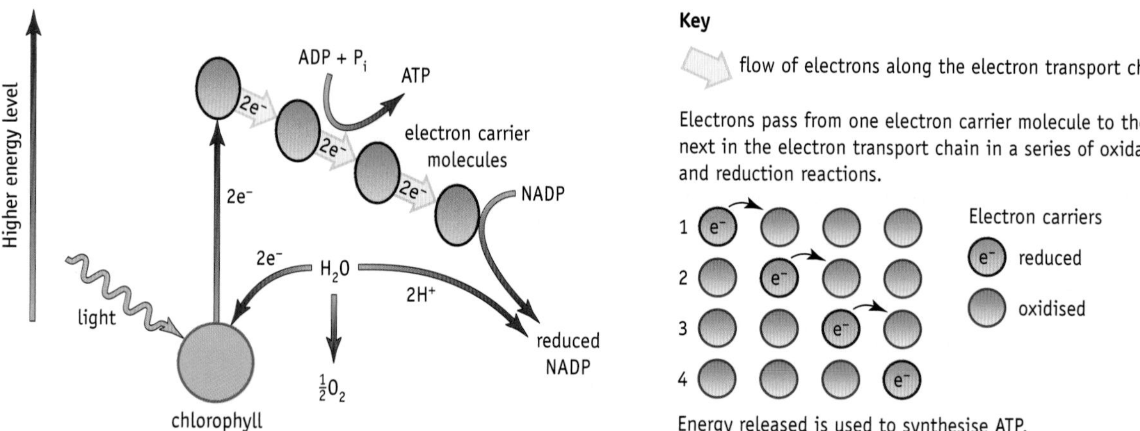

Figure 5.22 In the light-dependent reactions of photosynthesis, the energy from the Sun excites chlorophyll molecules. High energy electrons leave the chlorophyll and pass along a series of carrier proteins in the electron transport chain. The ionised chlorophyll (chlorophyll^{2+}) causes the photolysis of water; electrons from water pass to the chlorophyll^{2+}, and the hydrogen ion and oxygen concentration in the thylakoid space rises. ATP is also formed as a phosphate group is added to ADP.

Q5.13 Write equations summarising:

a the splitting of water

b the reduction of NADP.

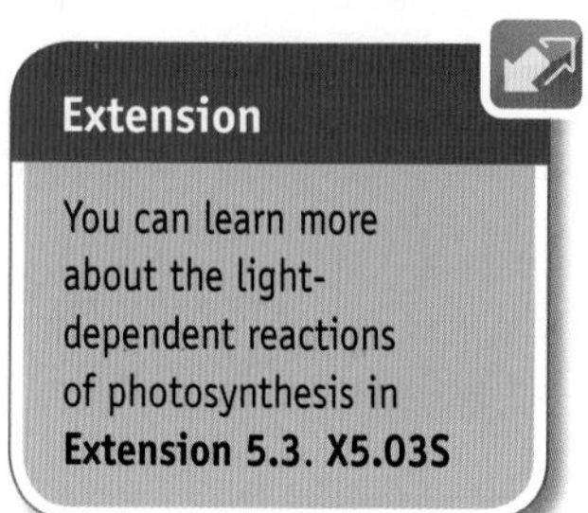

Extension

You can learn more about the light-dependent reactions of photosynthesis in **Extension 5.3. X5.03S**

Key biological principle: The role of ATP

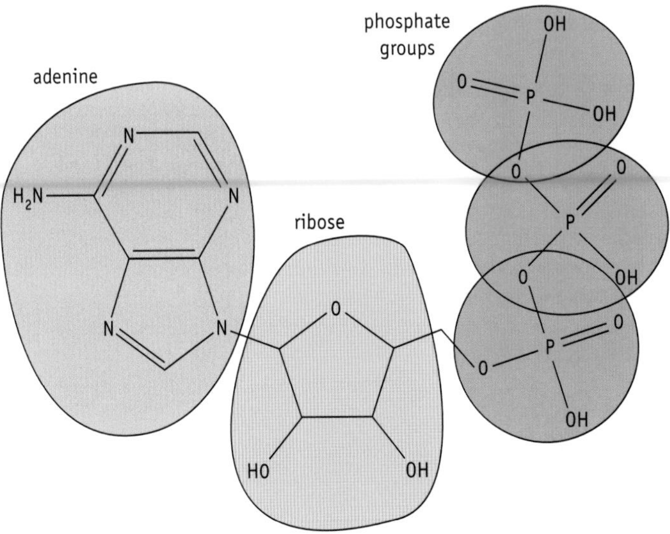

Figure 5.23 The structure of ATP.

ATP (**adenosine triphosphate**) is the most important energy transfer molecule within cells. It moves energy around the cell from energy-yielding reactions to energy-requiring reactions.

ATP consists of adenine (an organic base), ribose (a 5-carbon sugar) and three phosphate groups (Figure 5.23). The third phosphate group on ATP is only loosely bonded to the second phosphate, so is easily removed. When this phosphate group is removed from ATP, **adenosine diphosphate** (**ADP**) forms. Once removed, this phosphate group becomes hydrated, forming bonds with surrounding water molecules. A lot of energy is released as bonds form between water and the phosphate group. This energy can be used to drive energy-requiring reactions in the cell. **ATPase** catalyses the breakdown of ATP to ADP.

ATP in water → ADP + hydrated P_i + energy

ATP is created from ADP by the addition of inorganic phosphate (P_i). This addition of phosphate to ADP is known as phosphorylation. In order to make ATP, phosphate must be separated from water molecules, and this reaction requires energy. ATP in water is higher in energy than ADP and phosphate ions in water, so ATP in water is a way of storing chemical potential energy. Formation of ATP separates the phosphate and water. The phosphate and water can then be brought together in an energy-yielding reaction each time energy is needed for reactions within the cell. In this way ATP transfers energy around the cell.

Support

To find out more about ATP, visit the Biochemistry support on the SNAB website.

Key biological principle: Metabolic pathways

Reactions in cells can involve simple molecules linking to form more complex molecules. You saw this in the creation of polysaccharides in AS Topic 1 and the formation of proteins in AS Topic 2. Reactions can also involve the breakdown of complex molecules into simpler molecules.

Reactions rarely occur in a single step such as A → D. Instead there is generally a series of smaller reactions, each controlled by a specific enzyme, forming a metabolic pathway. For example:

A → B → C → D

This allows the rate of the overall reaction to be controlled, as each step is controlled by a specific enzyme. A range of intermediate products can be produced, which each might be useful as end-products or take part in other reactions. For example:

A → B → C → D
↓
E → F

The light-independent reactions

The light-independent reactions of photosynthesis take place in the **stroma** of the chloroplasts, using the reduced NADP and ATP from the light-dependent reactions. Carbon dioxide is reduced to carbohydrate. NADP acts as a hydrogen carrier, keeping hydrogen loosely bonded so that it can't react with oxygen as it is transferred from water to carbon dioxide.

As with all metabolic pathways, there is a series of reactions. The reactions form a cyclical pathway called the **Calvin cycle** after the scientist who first worked it out. A simplified version of the cycle is shown in Figure 5.24. This diagram shows how carbon dioxide combines with a 5-carbon compound. An unstable 6-carbon molecule forms, and almost immediately breaks down into two 3-carbon compounds. Some

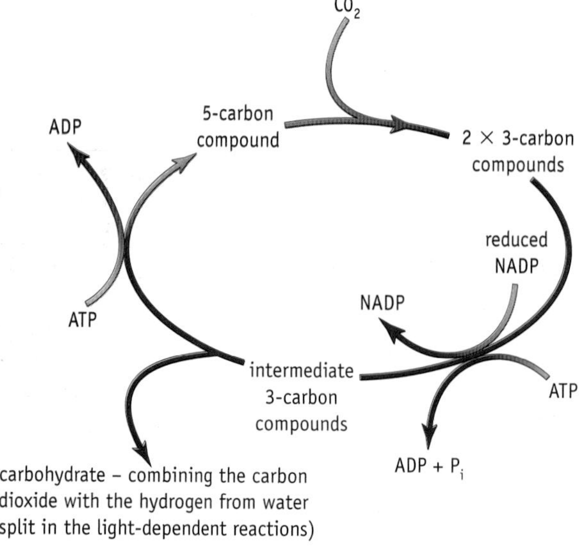

Figure 5.24 The key reactions that occur in the Calvin cycle.

of the 3-carbon compounds created are reduced to form carbohydrate, using the hydrogen from the reduced NADP created in the light-dependent reactions. Some of the 3-carbon compounds are used to regenerate the original 5-carbon compound that accepted the carbon dioxide. This recycling is a very economical use of resources by the cell.

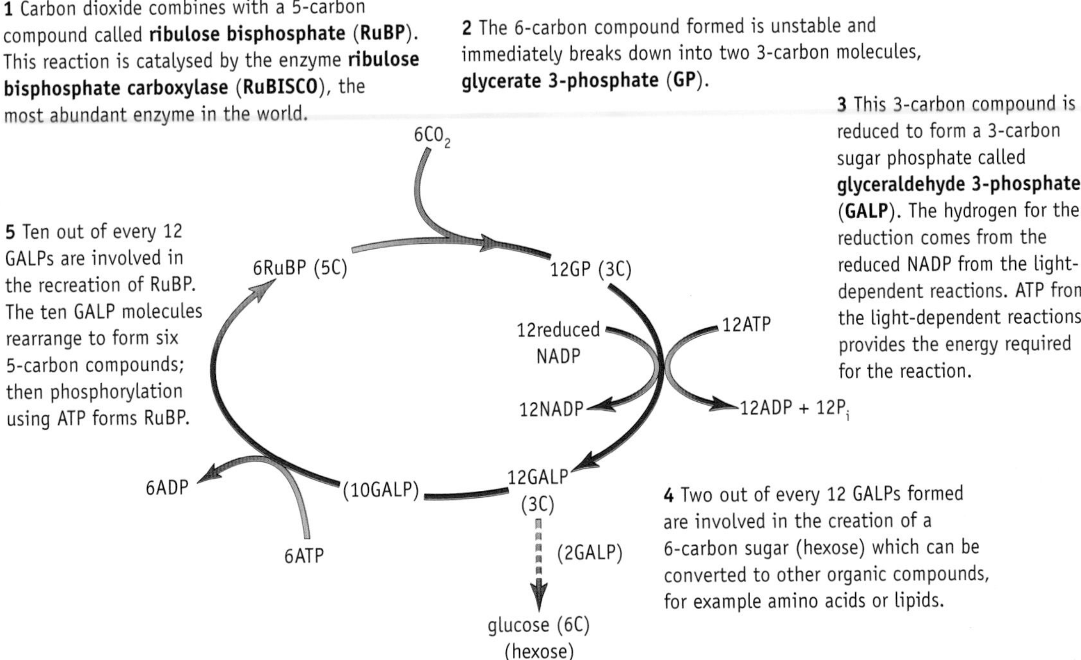

1 Carbon dioxide combines with a 5-carbon compound called **ribulose bisphosphate** (**RuBP**). This reaction is catalysed by the enzyme **ribulose bisphosphate carboxylase** (**RuBISCO**), the most abundant enzyme in the world.
2 The 6-carbon compound formed is unstable and immediately breaks down into two 3-carbon molecules, **glycerate 3-phosphate** (**GP**).
3 This 3-carbon compound is reduced to form a 3-carbon sugar phosphate called **glyceraldehyde 3-phosphate** (**GALP**). The hydrogen for the reduction comes from the reduced NADP from the light-dependent reactions. ATP from the light-dependent reactions provides the energy required for the reaction.
4 Two out of every 12 GALPs formed are involved in the creation of a 6-carbon sugar (hexose) which can be converted to other organic compounds, for example amino acids or lipids.
5 Ten out of every 12 GALPs are involved in the recreation of RuBP. The ten GALP molecules rearrange to form six 5-carbon compounds; then phosphorylation using ATP forms RuBP.
$6CO_2$
6RuBP (5C)
12GP (3C)
12reduced NADP
12ATP
12NADP
$12ADP + 12P_i$
6ADP
(10GALP)
12GALP (3C)
6ATP
(2GALP)
glucose (6C) (hexose)

Figure 5.25 The light-independent reactions of the Calvin cycle (also known as the C3 pathway).

Figure 5.25 shows the reactions of the Calvin cycle in more detail. Read the annotations to discover exactly what is happening at each stage of the cycle. Even this is a simplification. In reality there are large numbers of intermediate reactions.

Q5.14 Look at Figure 5.25, which shows the light-independent reactions.

a Name a substance that has been phosphorylated (has had phosphate added to it).

b Name a substance formed by reduction.

c Explain why the diagram shows six RuBP molecules combining with six carbon dioxide molecules, rather than just one of each.

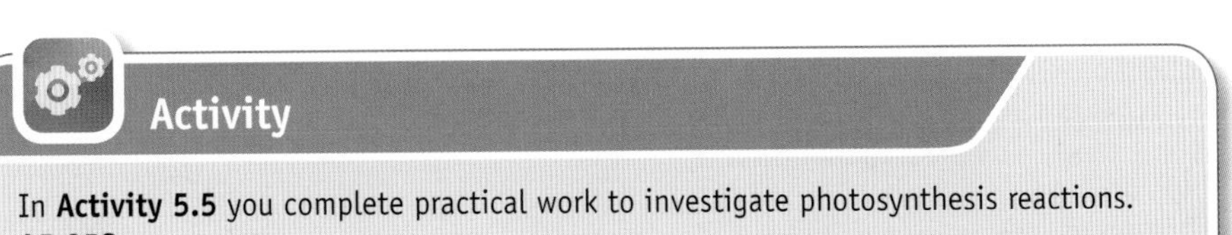

Activity

In **Activity 5.5** you complete practical work to investigate photosynthesis reactions. **A5.05S**

In **Activity 5.6** you examine the results of Calvin's lollipop experiment. **A5.06S**

Checkpoint

5.3 Produce a flow chart or bullet point summary describing the steps in the light-dependent and light-independent reactions of photosynthesis.

Key biological principle: Why does photosynthesis take place inside chloroplasts?

Thylakoids and the light-dependent reaction

Photosynthesis consists of a series of enzyme-controlled reactions, in which some of the energy falling on the plant surface is initially stored as chemical potential energy within ATP in water. The formation of ATP occurs in the light-dependent reactions, as a result of a series of oxidation and reduction reactions. These reactions involve the transfer of electrons between electron carrier molecules. The electron carriers are located within the thylakoid membrane in the chloroplast. Their positioning within the membrane creates an electron transport chain, allowing electrons to pass efficiently from each electron carrier to its neighbour.

The stroma and the light-independent reaction

The ATP molecules that are formed act as energy carriers within the cell, allowing small amounts of energy to be transferred and used where needed. In photosynthesis, the chemical potential energy within ATP in water is used in the reactions of the light-independent stage, in the fixing of carbon dioxide to form organic molecules. These reactions are dependent on the collision of substrate(s) and the appropriate enzymes to catalyse the reactions.

Maintaining a high concentration of each enzyme throughout the cell would be very 'costly' in terms of synthesis of enzyme. On the other hand, low concentrations would reduce the rate of reaction and efficiency of photosynthesis. The compartmentalisation of these reactions within the chloroplast stroma means that the substrates and enzymes can be at concentrations that allow the reactions to be catalysed quickly.

Transfer of energy through the ecosystem

Energy transfer and feeding relationships

Some of the energy fixed within organic molecules by **autotrophs** (also known as producers) is transferred to other organisms in the ecosystem. Organisms that obtain energy as 'ready-made' organic matter by ingesting material from other organisms are known as **heterotrophs**. Heterotrophs include all animals, all fungi, most bacteria and some protoctists.

Heterotrophs cannot make their own food; instead they must consume it. All heterotrophs are **consumers** and depend on producers (autotrophs) for their food.

- **Primary consumers**, also called **herbivores**, are heterotrophs that eat plant material.
- **Secondary consumers**, also called **carnivores**, feed on primary consumers.
- **Tertiary consumers** (also carnivores) eat other consumers. The carnivores at the top of the food chain are sometimes called top carnivores (Figure 5.26).

Figure 5.26 A polar bear eating a seal. The polar bear is the top carnivore in the Arctic tundra.

Animals that kill and eat other animals are known as predators and carnivores ('flesh eaters'). Animals that eat plants and other animals are known as **omnivores.**

Energy is transferred from producers to primary consumers, then to secondary consumers and then, sometimes, to tertiary consumers. Such feeding relationships can be shown in a **food chain** (Figure 5.27). The position a species occupies in a food chain is called its **trophic level**. Energy is transferred from one trophic level to the next trophic level by consumers. In reality an ecosystem has a complex **food web** in which each organism eats or is eaten by several other organisms.

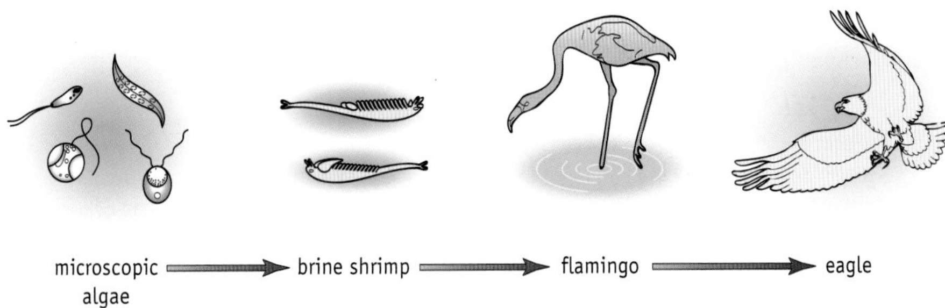

microscopic algae → brine shrimp → flamingo → eagle

Figure 5.27 The brine shrimp food chain. How many trophic levels are there?

Q5.15 In the brine shrimp food chain in Figure 5.27, which organism occupies the position of:

a producer

b primary consumer

c secondary consumer

d tertiary consumer?

Q5.16 Using the information in this topic, draw a food chain with five trophic levels, with the polar bear being the fifth.

Detritivores are primary consumers that feed on dead organic material called detritus. Woodlice, earthworms and freshwater shrimps are examples of detritivores.

Decomposers are species of bacteria and fungi that feed on the dead remains of organisms and on animal faeces. Like animals, they are heterotrophs. They secrete enzymes and digest their food externally, before absorption takes place.

Decomposers and detritivores play an important role in the recycling of organic matter from dead remains and waste. In the bottle ecosystem (Figure 5.3), bacteria living in the bottle cause the decomposition of dead brine shrimps and brine shrimp droppings. The organic compounds are broken down into inorganic substances, which are taken up by the algae. The algae can use them to grow and reproduce, replacing those eaten by brine shrimps.

Activity

You can construct a food web and investigate trophic levels in **Activity 5.7. A5.07S**

Checkpoint

5.4 Define the terms:

- habitat
- population
- community
- ecosystem
- abiotic
- biotic
- autotroph
- heterotroph
- producer
- primary consumer
- secondary consumer
- predator
- trophic level
- decomposer.

How efficient is the transfer of energy through the ecosystem?

The productivity (and to some extent the species biodiversity) of an ecosystem will depend on how much energy is captured by the producers, and how much is transferred to the higher trophic levels.

Plants are not actually as good as you might think at absorbing light, and only a small fraction of the light energy reaching a plant is used in photosynthesis. It is estimated that 1×10^6 kJ m^{-2} year^{-1} of energy is intercepted by plants in the UK, but that less than 5% of this available energy is captured in photosynthetic products.

For a start, if you look at Figure 5.28 you can see that most energy reaching the plant is not even absorbed. This is largely because chlorophyll can only absorb certain wavelengths, as shown by the absorption spectrum in Figure 5.29.

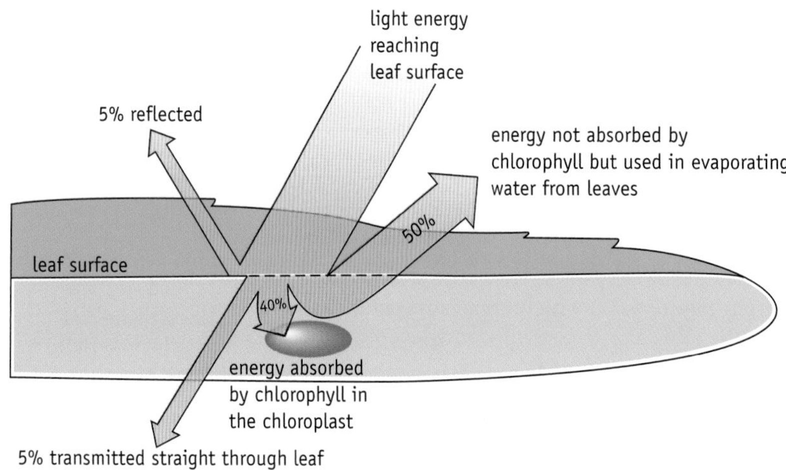

Figure 5.28 Not all the light energy falling on the leaf gets absorbed by the chlorophyll.

Only about 40% of the energy reaching the leaf is absorbed by the chlorophyll. Much of this energy is used to make organic molecules, but some is lost during photosynthesis and transferred to the environment.

Limiting factors will also influence the rate of photosynthesis. The law of limiting factors states that when a process is affected by more than one factor, its rate is limited by the factor furthest from its optimum value. So photosynthesis can be limited by temperature in cool conditions, and by light in overcast conditions. In the UK during the summer, the main limiting factor is carbon dioxide concentration. This means that not all the light energy falling on the plant surface can be used to make organic molecules, even if it is absorbed by the photosynthetic pigments.

Q5.17 Look at Figure 5.29. Which parts of the visible spectrum are absorbed by chlorophyll?

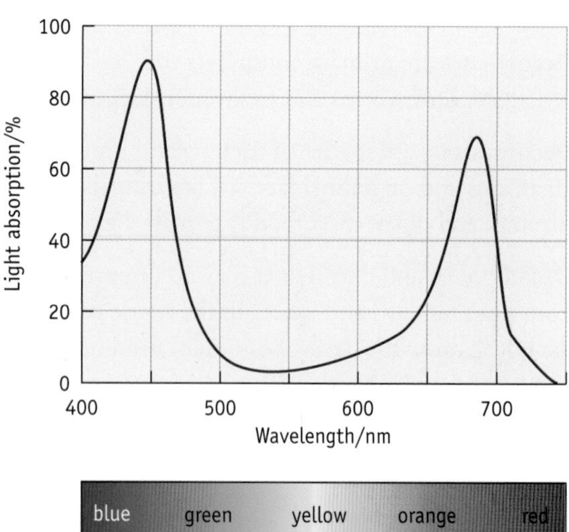

Figure 5.29 Absorption spectrum for chlorophyll.

Q5.18 Suggest which factor is the limiting factor in photosynthesis in each of the following situations.

a Plants growing on the Canadian tundra in summer.

b The trees of a tropical rainforest at noon.

c A field of wheat on a still summer's day at noon.

d Plants growing on the Keen of Hamar (page 12) in summer.

Q5.19 Outline how increased carbon dioxide might affect the growth rate of plants in each of the above situations.

The rate at which energy is incorporated into organic molecules by an ecosystem is the **gross primary productivity** (**GPP**). GPP is usually expressed as units of energy per unit area per year (e.g. kJ m^{-2} y^{-1} or MJ ha^{-1} y^{-1}).

The percentage efficiency of photosynthesis can be calculated as GPP divided by amount of light energy striking the plant multiplied by 100. Very little of the light energy reaching a plant is fixed in carbohydrates. For example, in a grassland community where 2×10^9 kJ m^{-2} y^{-1} of sunlight energy reaches the plant and GPP is 25×10^6 kJ m^{-2} y^{-1}, the efficiency of photosynthesis is 1.25%.

Some of the carbohydrates produced are quickly broken down in respiration. This provides the energy for the plant's life processes, such as cell division and active transport. The rest of the carbohydrates are incorporated into the proteins, chromosomes, membranes and other components of new cells, becoming new plant **biomass**. The rate at which energy is transferred into the organic molecules that make up the new plant biomass is called **net primary productivity** (**NPP**). The fate of energy absorbed by chlorophyll is summarised in Figure 5.30.

Net primary productivity, gross primary productivity and respiration are related to each other by the equation:

$$NPP = GPP - R$$

where R is plant respiration.

The net primary productivity is available to the rest of the ecosystem.

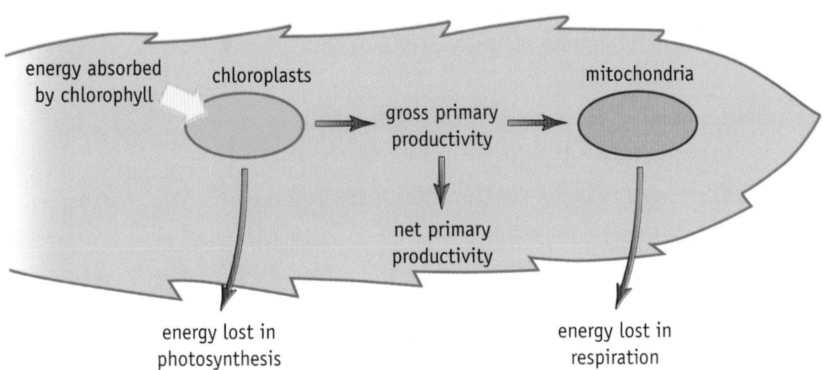

Figure 5.30 Transfer of energy absorbed by chlorophyll.

Q5.20 **a** Using the figures in Table 5.1, calculate the missing values **i** to **iii** for each of the ecosystems. Comment on the data.

b 2×10^6 kJ of sunlight energy falls on each square metre of the maize field described in Table 5.1; calculate the efficiency of photosynthesis.

c Explain why so little of the sunlight energy is used in photosynthesis.

d A desert was found to have a NPP of 836 kJ m^{-2} y^{-1}. Suggest reasons why this value is so low.

Figure 5.31 Will this pine forest ecosystem or a tropical rainforest have the higher NPP? Complete Question 5.20 to find out if you were right.

Ecosystem	GPP/ kJ m^{-2} y^{-1}	Respiration/ kJ m^{-2} y^{-1}	NPP/ kJ m^{-2} y^{-1}
Tropical rainforest	23 140	17 820	**i**
Young pine forest	5100	1960	**ii**
Maize field	**iii**	8000	26 000

Table 5.1 Productivity of different ecosystems.

Disappearing energy

Producers to primary consumers

Transfer of energy from producers to primary consumers is also not very efficient. Only about 2–10% of the energy in the producers goes to make new herbivore biomass. What happens to the rest?

- *Not all the available food gets eaten.* This may be due to limitations of the animals' feeding methods. When a bull eats grass, for example, it can only eat long, untrampled grass because it needs to wrap its tongue around the grass and rip it up. Some parts of the plants, such as the roots, twigs and parts protected by spines or thorns, will not get eaten by herbivores.
- *Some energy is lost in faeces and urine.* The main component of plant material, apart from water, is the cellulose of the cell walls. Cellulose is tough stuff to deal with, and mammals have no enzymes of their own to help break it down. Even in ruminants, animals whose guts contain microbes producing cellulase enzymes, much of the cellulose still passes through the gut intact and comes out in the faeces.
- *Much of the energy absorbed by the consumers is used in respiration* for movement and chemical reactions in the body, and is lost to the environment as heat.

Primary consumers to secondary consumers

The transfer of energy from primary to secondary consumers – from herbivores to carnivores – is often more efficient. Often, over 10% of the energy in herbivores ends up in carnivores. This is because most of a herbivore may be eaten by a carnivore, and the protein-rich diet is easily digested so there is less lost in faeces.

The fate of energy within a trophic level

The energy entering a trophic level must equal the amount used or lost by that trophic level. We can summarise this in a simple equation:

energy entering the trophic level (i.e. consumed) = energy lost in respiration + energy lost in faeces + energy lost in urine + energy in new biomass

Q5.21 Look at the cow in Figure 5.32 and work out:

a how much energy is transferred to new biomass

b what percentage this is of the energy consumed.

Activity

Activity 5.8 lets you work out net primary productivity for an ecosystem, and explore the transfer of energy through the ecosystem. **A5.08S**

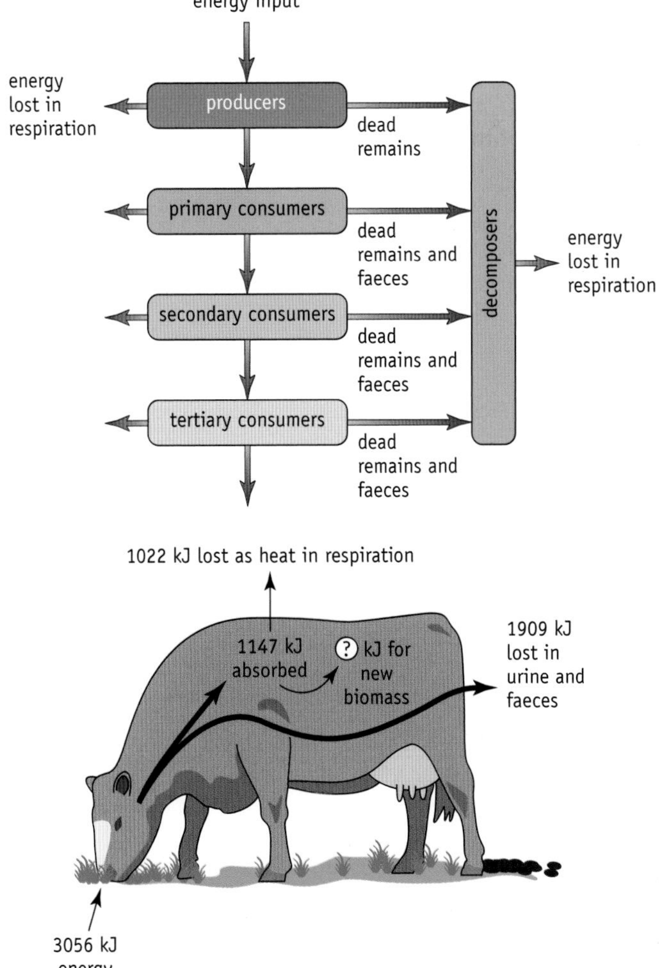

Figure 5.32 The flow of energy through the ecosystem.

Energy flow and energy 'loss' explain why food chains and food webs rarely have more than four or five trophic levels. What always gets less and less as you go up a food chain is *the transfer of energy to the next trophic level.* If a diagram is drawn with a bar representing the energy transferred from one trophic level to the next, in kJ m^{-2} y^{-1}, a pyramid is formed. The bars representing the higher trophic levels are always smaller (Figure 5.33). Such pyramids allow comparisons of the efficiency of energy transfer between trophic levels.

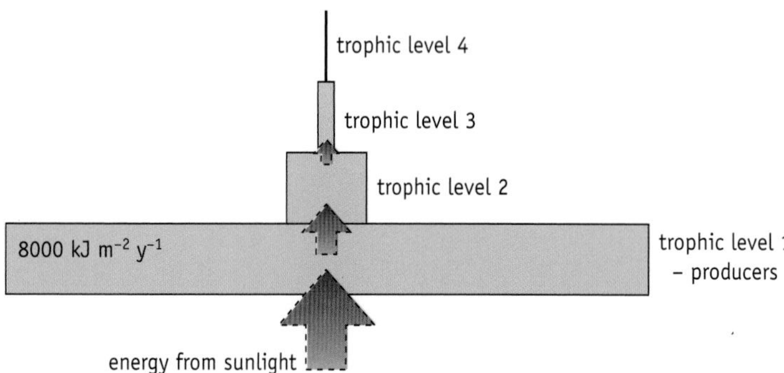

Figure 5.33 A pyramid of energy for a grazed pasture (drawn to scale). Each bar represents the energy transferred from one trophic level to the next, in kJ m^{-2} y^{-1}. The bottom bar represents the energy from sunlight captured by producers in trophic level 1; the second bar is the energy transferred from level 1 (producers) to the primary consumers in trophic level 2; and so on.

Q5.22 Look at Figure 5.33 and work out the efficiency of transfer between:

a trophic level 1 and trophic level 2

b trophic level 2 and trophic level 3.

5.3 Is the climate changing?

Evidence for climate change

One of the biggest challenges facing ecosystems is climate change. Older people often claim that the weather in their childhood was different – usually much better than today! Is the climate really changing, and has it changed in the past? How can we tell? Evidence from personal memory is termed **anecdotal**; it is often unreliable and can only go back less than 100 years. We cannot rely on anecdotal reports; we need reliable scientific evidence.

We seem at present to be in a period of global warming. This view is supported by evidence from a range of sources including:

- temperature records
- pollen in peat bogs
- dendrochronology (tree-ring studies).

Temperature records over long periods

Long sequences of temperature records exist for a number of places, for example central England from 1659 to the present and Toronto, Canada from 1780 to the present. Old data sets are very important in the study of climate change, even though they may have been collected with equipment that was not as accurate as that used today.

> **Activity**
>
> In **Activity 5.9** you analyse some long data sets for yourself.
> **A5.09S**

Q5.23 Look at Figure 5.34. The graph shows a long data set of temperature measurements for central England. Do the data show an increase in temperature? To help you decide, you could place a piece of tracing paper over the graph and draw a straight best fit line and a separate curved best fit line. These will help you to pick out any trends in the data.

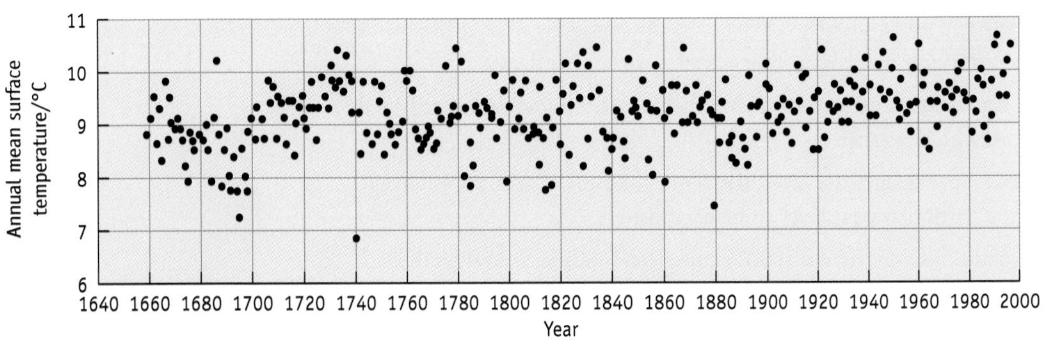

Figure 5.34 Is the temperature rising?
Source: Open University.

Studying peat bogs

Records of temperatures measured with a thermometer only take us back two or three centuries at the most. One way of finding out about the climate at least back to the last Ice Age (which ended in Britain around 12 000 years ago) is to study plant and insect remains preserved in peat (Figure 5.35). Peat is an accumulation of partially decayed organic matter, mainly the remains of dead plants. There are still extensive peat bogs in Ireland and some upland areas of Britain.

When plant material dies it normally decays. However, in the **anaerobic** and often acidic conditions of a peat bog, the decay rate is slowed or stopped altogether.

Q5.24 Why do anaerobic and acidic soil conditions slow the decay rate?

Figure 5.35 Scientists take a core from a peat bog. The type of pollen found in each layer of the core provides information about the conditions when the peat was deposited.

Pollen from the past

Pollen grains (Figure 5.36) are particularly well preserved in peat, and can be used to determine climate conditions in the past.

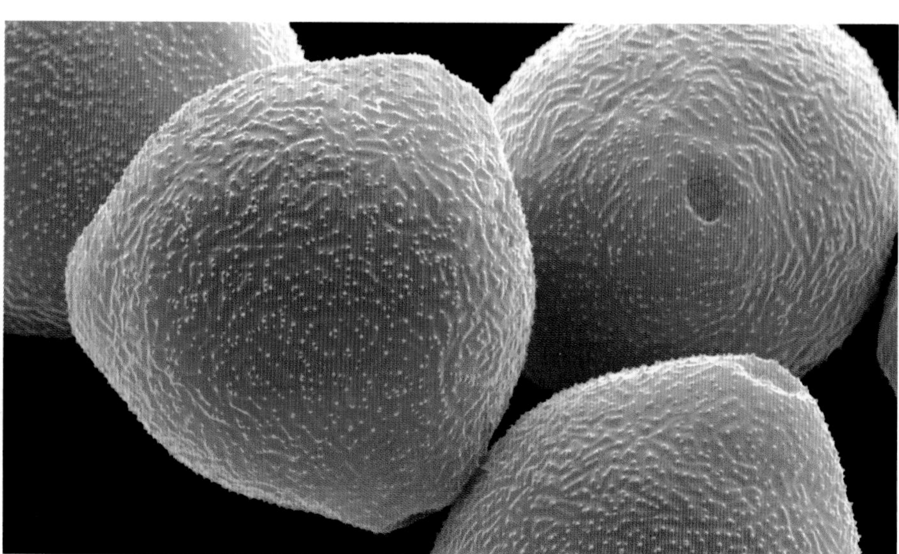

Figure 5.36 Hazel pollen grains from peat. Magnification × 2900.

Pollen from peat is useful for reconstructing past climates because:

- Plants produce pollen in vast amounts. Throughout the spring and summer, countless millions of pollen grains fall from the air onto the ground, including the surface of peat bogs, as 'pollen rain'.
- Pollen grains have a tough outer layer that is very resistant to decay.
- Each species of plant has a distinctive type of pollen, allowing us to identify the plant species from which it came.
- Peat forms in layers: the deeper the layer, the older the peat. Carbon-14 dating allows the age of a particular peat layer to be established.
- Each species of plant has a particular set of ecological conditions in which it flourishes best. If we find pollen from a species favouring warmer conditions, we can infer that the peat was laid down when the climate was warmer.

Q5.25 Alder trees grow in damp soil, often near rivers, lakes and marshes. What would the appearance of a lot of alder pollen in a particular peat layer suggest about the climate at the time it was deposited?

Activity

Collect your own pollen core, without getting your feet wet, and reconstruct climate change over the past 10 000 years, using virtual pollen analysis in **Activity 5.10.**
A5.10S

From the abundance of pollen grains in a sample, a pollen diagram can be constructed (Figure 5.37).

Q5.26 Figure 5.37 shows a pollen diagram taken from Hockham Mere in Norfolk. Describe the changes that have occurred in the species abundance as you move forward in time towards the present.

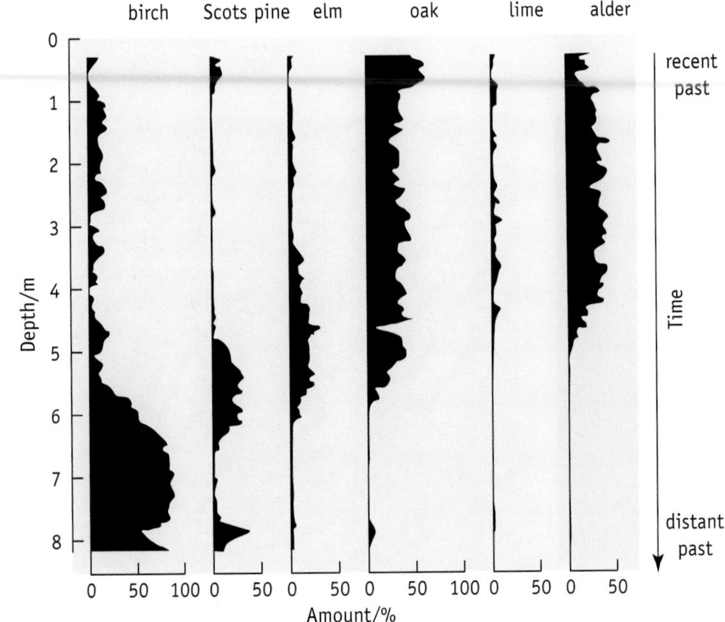

Figure 5.37 A tree pollen diagram from Hockham Mere in East Anglia.
Source: Open University.

Did you know?

Bog beetles

Plants respond rather slowly to changes in climate, but insect populations respond much faster. Studies of the exoskeletons of beetles in peat can therefore give a more precise measure of climate change.

Q5.27 Different species of beetle thrive in specific temperature conditions, so their remains in peat and lake sediments can be used as indicators of past climate conditions. Table 5.2 shows data obtained in this way.

a What is the advantage of giving temperatures separately for February and July rather than as an average for the year?

b What conclusions can you draw about climate change over this period of time? Sketching a graph may help.

Table 5.2 Temperatures estimated from beetle remains preserved in layers of peat from different periods.

Period/years before present	Average July temperature/°C	Average February temperature/°C
18 000–13 000	10	−20
13 000–12 000	17	0
12 000–11 000	15	−5
11 000–10 000	10	−20
10 000–7500	16	3
7500–5000	18	5
5000–2500	17	4
2500–present	15	5

Tree-ring analysis – dendrochronology

Every year, trees produce a new layer of xylem vessels by the division of cells underneath the bark. The diameter of the new xylem vessels varies according to the season when they are produced: wide vessels in spring when the tree grows quickly, followed by narrow vessels in summer. Little if any growth takes place in autumn and winter. The different widths of the vessels create a pattern of rings across the trunk, which can easily be seen when a tree is cut down, with a ring for each year of tree growth (Figure 5.38). Instead of cutting down a tree to see the rings, a core sample can be taken and examined.

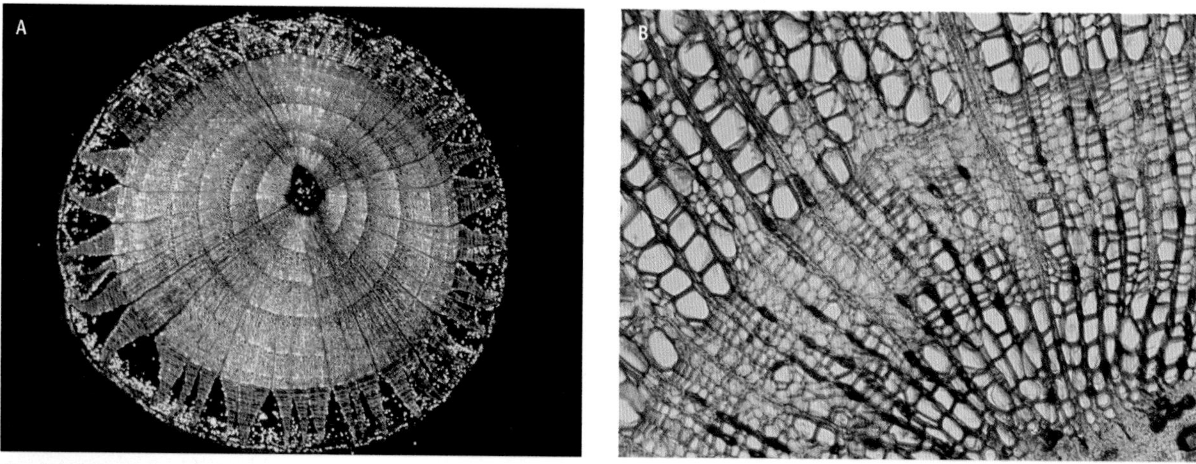

Figure 5.38 **A** Cross-section of the trunk of a lime tree showing tree rings. **B** Close-up of the xylem vessels. The larger vessels are produced in spring; the narrower vessels in the summer.

Q5.28 Where will the youngest rings be located, on the outside or deeper in the tree trunk?

If you cut a tree down or take a core sample in autumn 2006, the outermost ring will have come from growth in 2006. By counting inwards you can date the year each ring was formed. To date preserved trees or wood samples accurately, experts find common patterns of tree-ring growth that allow cross-dating (Figure 5.39).

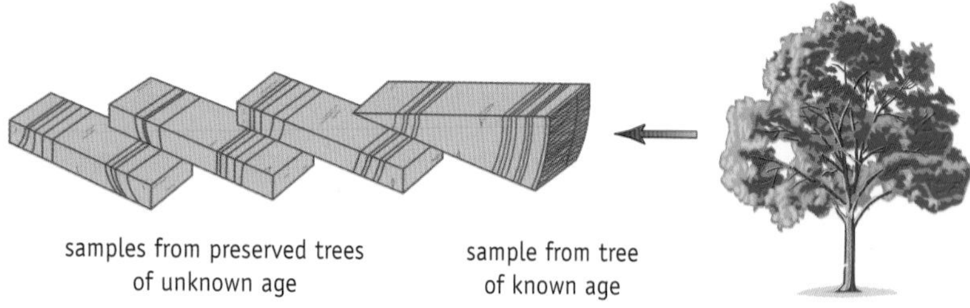

Figure 5.39 Matching the pattern of tree rings allows preserved wood samples to be dated.

If you find that the ring made in the year 1326 was wider than that made in 1327, it means that the tree grew more in 1326 than in 1327; this probably indicates that the conditions in 1326 were better for tree growth. It was probably warmer or wetter (Figure 5.40). So tree rings can give not only precise dates but also strong clues about past climates. The study of tree rings is known as **dendrochronology**.

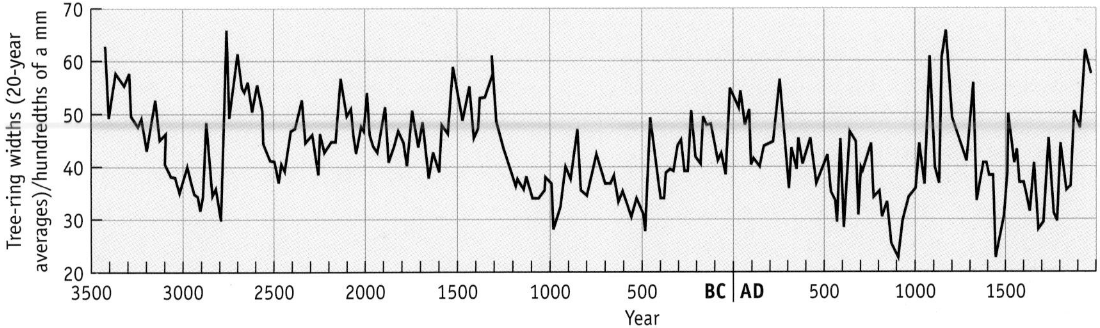

Figure 5.40 Graph showing the change in average tree-ring width with time.

Q5.29 Look at the graph of tree rings in Figure 5.40. The width of tree rings is related to their conditions of growth.

a Which two separate years were particularly cold or dry?

b Which was probably the warmest or wettest?

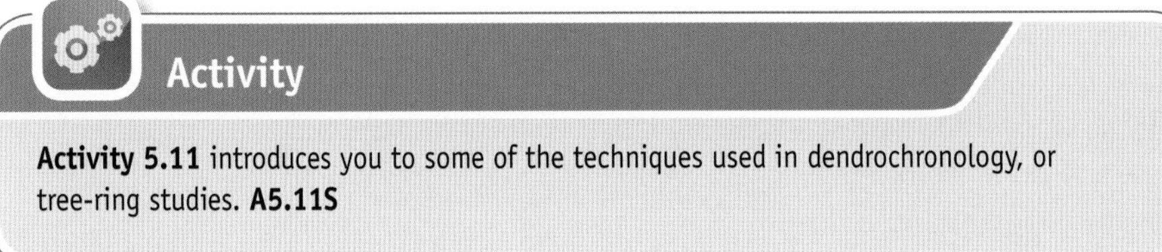

Activity

Activity 5.11 introduces you to some of the techniques used in dendrochronology, or tree-ring studies. **A5.11S**

Weblink

Ice cores and mud deposits can also be used to investigate climate conditions in the distant past. See the NOAA paleoclimatology website for further information.

Putting the data together

Historical temperature records made by people provide climate information from the present back to about 1650. Tree-ring studies extend this record back hundreds of years, 3000 years in some cases. Pollen gives us information going back some 20 000 years. Ice cores are used to find out what happened before this. As water freezes, bubbles of air become trapped within the ice. The ratio of different oxygen isotopes in the trapped air is measured, and this gives an estimate of the average air temperature when the ice was formed. The carbon dioxide concentration of the air can also be determined from these bubbles.

Combining the evidence obtained using all these techniques allows us to build up a picture of the conditions over the last 160 000 years. Figure 5.41A shows the changes in average global surface temperature over this period. It is clear that major fluctuations in climate have occurred regularly. Changes during the past century indicate that we are in a period of global warming (Figure 5.41B).

A

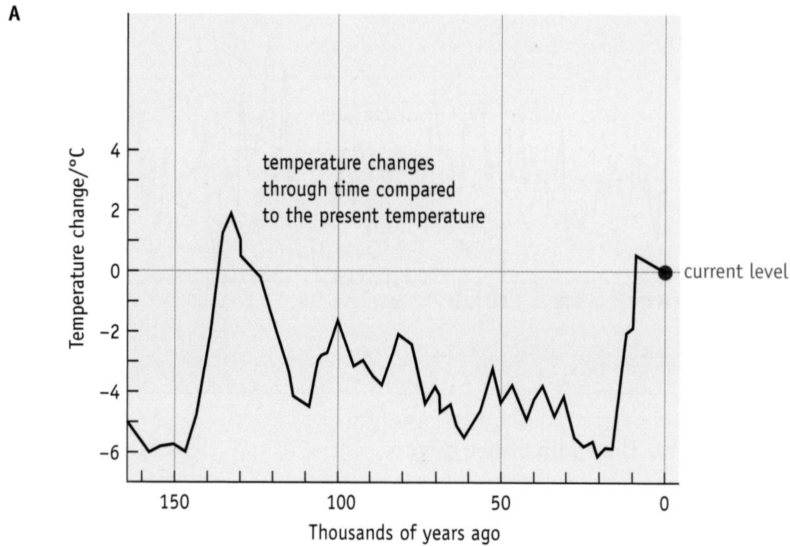

B

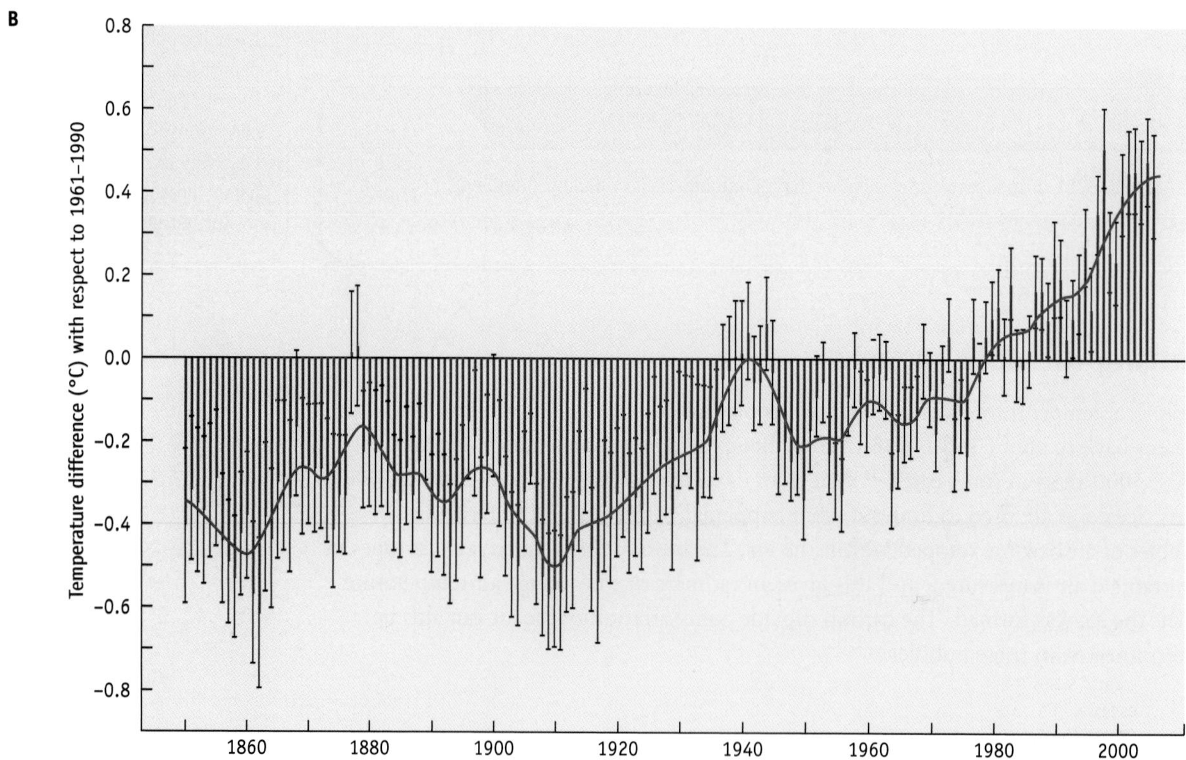

Figure 5.41 A Major temperature variations have occurred over the last 160 000 years. **B** Changes in average global surface temperature over the last century relative to the global average for 1961–1990.

5.4 Why are global temperatures changing?

Climate patterns have been fluctuating for many thousands of years, with temperature and rainfall both subject to great variations over time. There seems to be a body of reliable data that suggests that changes in the atmosphere are linked to climate change.

The atmosphere is a thin layer of gases extending 100 km above the Earth's surface and held in place by gravity. Table 5.3 shows the average composition of the bottom 15 km of the atmosphere. It has been predicted that without an atmosphere, the temperature of the Earth's surface would fluctuate between very hot days and very chilly nights, the nights being –40 °C or lower. The atmosphere has an important role in helping keep the Earth's average temperature stable and suitable for living organisms.

Keeping the Earth warm – the greenhouse effect

The Sun radiates energy, largely as visible light, and the Earth absorbs some of this energy. The Earth warms up and in turn radiates energy back into space as infrared radiation (Figure 5.42). Some of the energy that is radiated from the Earth's surface is absorbed by gases in the atmosphere, warming it. In Figure 5.43 you can see how this is similar to the way that glass traps energy in a greenhouse. The gases in the atmosphere that stop the infrared radiation from escaping are called **greenhouse gases**. They create the **greenhouse effect**, which keeps the Earth warm.

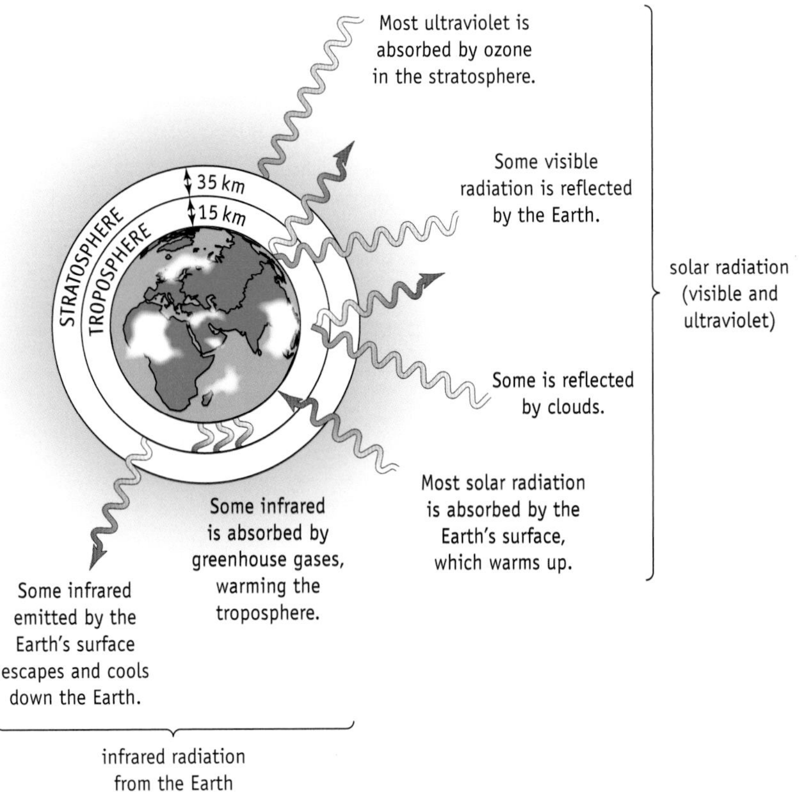

Figure 5.42 Inputs and outputs of energy to the Earth's atmosphere.

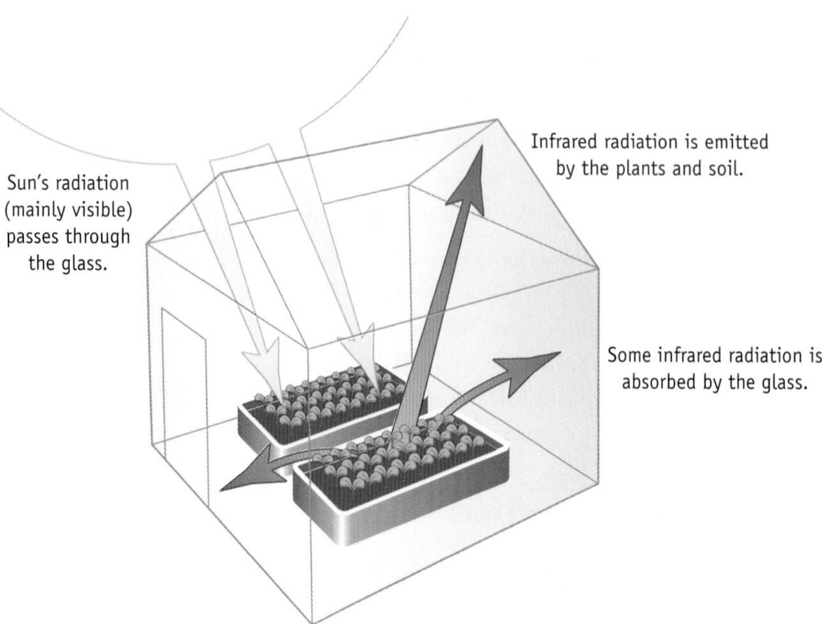
Sun's radiation
(mainly visible)
passes through
the glass.
Infrared radiation is emitted
by the plants and soil.
Some infrared radiation is
absorbed by the glass.

Figure 5.43 The greenhouse effect.

Which are the greenhouse gases?

Not all gases are greenhouse gases. In Table 5.3 you can see the relative contribution of the different gases in the atmosphere to the greenhouse effect. The 'global warming potential' is a measure of the greenhouse effect caused by that gas relative to the same amount of carbon dioxide over a given time, which is given a value of 1. Although carbon dioxide does not have the largest global warming potential, it is so much more abundant than the more potent greenhouse gases that it has the largest effect.

Table 5.3 The contribution of different gases in the bottom 15 km of the atmosphere to the greenhouse effect. The global warming potential (GWP) gives an indication of the effect, over a 100 year period, of releasing 1 kg of the gas into the atmosphere compared with 1 kg of carbon dioxide. The GWP for water is not calculated, because it is considered that human activity has little influence over the concentration of water vapour in the atmosphere.

Gas	Abundance in atmosphere/ % by volume	Global warming potential	Average lifetime in the troposphere/ years
nitrogen	78	negligible	
oxygen	20	negligible	
argon	1	negligible	
carbon dioxide	3.8×10^{-2}	1	150
methane	1.8×10^{-4}	25	12
nitrous oxide	3.2×10^{-5}	298	114
chlorofluorocarbon CCl_3F	2.5×10^{-8}	4750	45

Q5.30 Look at Table 5.3. The relative contribution of a gas to the greenhouse effect can be calculated by multiplying the percentage abundance by the global warming potential.

a Work out the relative contribution for each gas.

b Which gas would it be best to control to help reduce the greenhouse effect?

Methane (CH_4) is produced by anaerobic decay of organic matter in waterlogged conditions, for example in bogs and rice fields. Decay of domestic waste in landfill sites and the decomposition of animal waste are other sources of methane. It is also produced in the digestive systems of animals such as cattle and released when they belch and fart. Incomplete combustion of fossil fuels also releases methane.

Table 5.4 shows that, since pre-industrial times, the atmospheric concentration of methane has more than doubled. A molecule of methane absorbs more infrared radiation than a molecule of carbon dioxide does. Unlike carbon dioxide, however, it does not stay in the atmosphere very long. In less than 12 years a methane molecule will have reacted with oxygen in the air to form carbon dioxide and water, but of course more methane keeps being made.

Methane emissions could be reduced through better waste recycling and by using it as a biofuel (see Section 5.8). When methane burns and produces carbon dioxide and water vapour, one greenhouse gas changes into two less serious ones. It would eventually have turned into these gases anyway, and the energy that comes from burning methane from waste can replace energy from fossil fuel.

Significant increases in carbon dioxide levels in the atmosphere (Table 5.4), through processes such as the combustion of fossil fuels, have been linked to changing temperatures.

Q5.31 Look at Table 5.4.

a Give the percentage rises in the global concentrations of carbon dioxide, methane, nitrous oxide and CFCs between 1750 and 2005.

b Which one of these four gases has declined in concentration between 1998 and 2005?

Q5.32 **a** What does the footnote to Table 5.4 show about the rate of increase in global carbon dioxide concentration?

b Suggest why the average change in concentration over ten or more years is given, rather than quoting yearly changes.

Q5.33 What can you deduce from the data in Table 5.5 about the UK's success in controlling greenhouse gas emissions?

Q5.34 Suggest three pieces of information you would need to help you to decide whether current levels of UK carbon dioxide emissions are sustainable (not causing a rise in global concentration).

Table 5.4 Changes in the concentrations of key greenhouse gases.
Source: Intergovernmental Panel on Climate Change.

Date	Carbon dioxide CO_2/ppm*	Methane CH_4/ppb	Nitrous oxide N_2O/ppb	Chlorofluorocarbons CFC/ppt
Pre-industrial (1750) (approximate)	280	700	270	zero
1998	365	1745	314	268
2005	379	1774	319	252

*Rate of increase in concentration 1960–2005 1.4 ppm yr^{-1}; 1995–2005 1.9 ppm yr^{-1}

Table 5.5 Estimated UK emissions of greenhouse gases/million tonnes per year.
Source: Department for Environment, Food and Rural Affairs.

Date	Carbon dioxide	Methane	Nitrous oxide
1990	590	4.9	2
1995	549	4.3	0.17
2000	549	3.3	0.14
2005	557	2.4	0.13

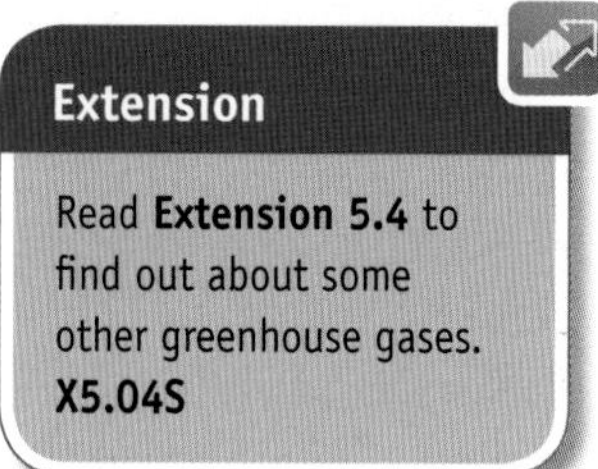

Extension

Read **Extension 5.4** to find out about some other greenhouse gases. **X5.04S**

Does increased carbon dioxide cause global warming?

Compare the changes in temperature over the last 160 000 years with the changes in carbon dioxide concentration over the same period (Figure 5.44). There appears to be a correlation, but is there a causal relationship? Is one (carbon dioxide) causing the other (warming)?

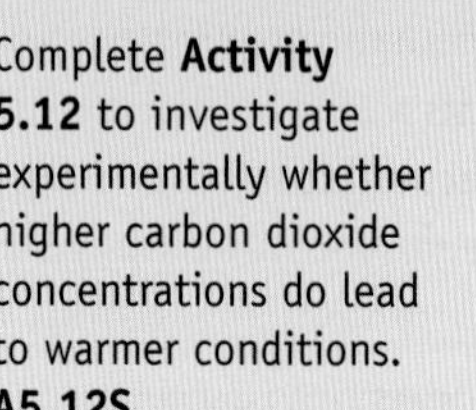

Activity

Complete **Activity 5.12** to investigate experimentally whether higher carbon dioxide concentrations do lead to warmer conditions. **A5.12S**

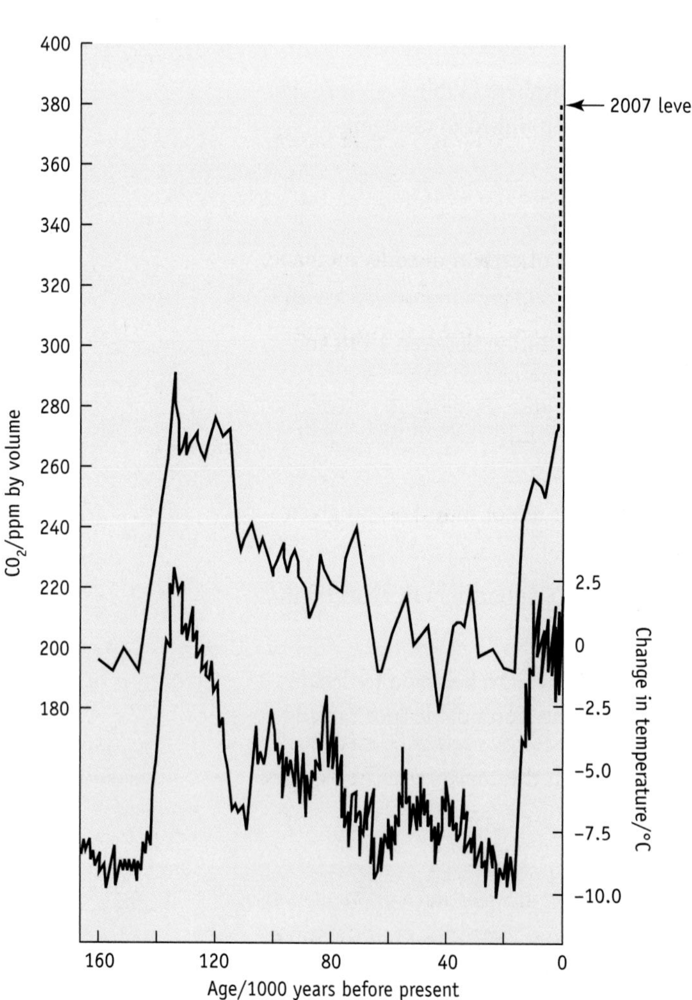

Figure 5.44 Changes in temperature and carbon dioxide concentration over the last 160 000 years.

There is high correlation between temperature and carbon dioxide levels. Before the industrial revolution (1750), changes in carbon dioxide level are the result of changes in the Earth's orbit around the Sun, changes in solar radiation, and volcanic eruptions.

A rise in temperature is followed by a rise in carbon dioxide released from the oceans, which causes further warming.

The rise in carbon dioxide since the industrial revolution did not follow a rise in temperature. The concentration in 2007 (383 ppm) is 37% higher than in 1750. Current levels of carbon dioxide are estimated to be 30% higher than at any time over the last 400 000 years. The levels will continue to increase rapidly if emissions are not reduced. The correlation between carbon dioxide levels and temperature does not *prove* that one causes the other. However, there is so much scientific evidence supporting the link between the current rise in global temperatures and rising levels of greenhouse gases that this is now a widely accepted theory.

Scientists use the word 'theory' in a different way from its use in everyday speech. 'Theory' to some people implies a speculative idea, a weak hypothesis, or an idea that lacks much evidence or reliability. In science, 'theory' implies a well-tested and widely-accepted idea or principle supported by a great deal of evidence. In science, theories are the most reliable types of ideas we have.

Activity

In **Activity 5.13** you compare carbon dioxide levels and global temperatures. **A5.13S**

Weblink

Visit the British Antarctic Survey website to find out about ice core records for carbon dioxide over the last 740 000 years.

A controversial issue

Global warming has been described by many people as a controversial issue. An issue is controversial when alternative points of view about it can reasonably be held. Global warming remains controversial because:

- Science cannot prove theories – scientific methods can only disprove them. Using scientific methods, an idea (hypothesis) is proposed to explain an observation, and it is then tested. If the results disprove the idea it is rejected; if the results support the idea it does not actually prove it – there could be alternative explanations.
- There is incomplete knowledge of how the climate systems of our planet work, and the data sets used in making predictions about climate change have their limitations. For example, there is no way to measure precisely how much carbon dioxide is added to the atmosphere by fossil fuel combustion. This doesn't mean that the numbers reported are mere guesses. They are estimates based on scientifically defensible procedures. But some people use this uncertainty to dismiss the link between rising carbon dioxide levels and global warming.

Despite much controversy, there is widespread scientific consensus that temperatures are rising, and that rising levels of greenhouse gases are at least partly responsible. This consensus results from a gradual build-up of a large body of scientific evidence supporting the theory. But some people do not agree with the consensus. They suggest instead that the changes in temperature currently observed are part of a natural cycle of climate variations or are due to changes in the Sun's activity.

When presenting and interpreting scientific evidence, almost everyone is influenced by his or her own particular values and viewpoint. Political and economic considerations may affect how individuals, organisations and countries interpret the evidence. Where cuts in carbon dioxide emissions to reduce global warming might harm business, a company or country may not accept the causal link between the two factors. In the US the oil and power industries support organisations that advocate such scepticism. Some of these organisations go further, and suggest that rising carbon dioxide levels will be a benefit through increased plant productivity.

Ethical arguments are often quoted when considering the issue of global warming. These arguments include:

- We all have the right to choose for ourselves whether we use fossil fuels to achieve a good standard of living.
- We have a duty to allow others to improve their standard of living (which is often equated with industrialisation).
- We have a duty to preserve the environment for the next generation.

Q5.35 Suggest a further ethical argument based on the framework of utilitarianism (see AS Topic 2).

Activity

Activity 5.14 allows you to critically evaluate views expressed in articles about global warming. **A5.14S**

In **Activity 5.15** you debate the different views expressed on global warming. **A5.15S**

The fact that climate change cannot be conclusively proved leaves us with a dilemma. If we therefore do nothing, the consequences could be very serious. Scientists use mathematical models to try and predict what these consequences might be.

Did you know?

Kyoto Protocol

The Kyoto Protocol was an international agreement to cut back on greenhouse gas emissions across the globe. It came about as a result of discussions at the Earth Summit in Rio, Brazil, in 1992, and was agreed as a formal strategy in Kyoto, Japan, in 1997. It has been very difficult to get countries to agree on just how much emissions need to be cut in order to prevent serious damage to our climate. However, in 2001 agreements were reached and 178 countries signed the treaty.

There were two major exceptions: the USA and Russia, who failed to sign up. This was partly because American politicians accepted the arguments of big businesses, particularly the oil industry, that the Kyoto Protocol would be bad for the American economy. As these are two of the world's largest producers of carbon dioxide, this does cause some concern. For the Kyoto treaty to be effective we need to include the USA in the long term.

In December 2007, the United Nations Climate Change Conference was held in Bali, Indonesia, with 180 nations in attendance (Figure 5.45). The 'Bali Roadmap' was agreed, whereby new negotiations for reductions in greenhouse gas emissions will be completed by 2012, when the Kyoto agreement ends. It is generally accepted that future targets will need to be much more ambitious than the Kyoto agreement, which requires developed nations to reduce greenhouse gas emissions to 5% below 1990 levels by 2012. Targets of 50–60% reductions have been proposed by climate modellers in order to contain global warming within acceptable limits.

CONTINUED ▶

Weblink

Find out about international developments since Kyoto by visiting the UN Framework Convention on Climate Change website and the World Resources Institute's website.

Figure 5.45 The 'Bali Roadmap' was agreed at the United Nations Climate Change Conference in Indonesia in December 2007. Many people do not hold out much hope that the agreed reductions will be achieved.

5.5 Predicting future climates

Enhanced global warming due to rising carbon dioxide levels is only one of several factors likely to affect the future climate. Scientists use computer models to study the interaction of many factors in an attempt to make predictions about likely climate change.

Making mathematical models

A really reliable model

If a square has one side of 10 cm, can we 'predict' the length of the other three sides? That one's easy. By definition all the sides of a square are equal in length, and so the sides are each 10 cm long. This model gets the right answer every time!

Extrapolation

Figure 5.46 shows the actual data for atmospheric carbon dioxide covering the period 1958 to 1988, recorded in the field. Since we are interested in the long-term view, we can ignore the annual fluctuations (due to different levels of photosynthesis in winter and summer), and apply a smoothed best-fit line. This produces a steep upward trend as in Figure 5.46.

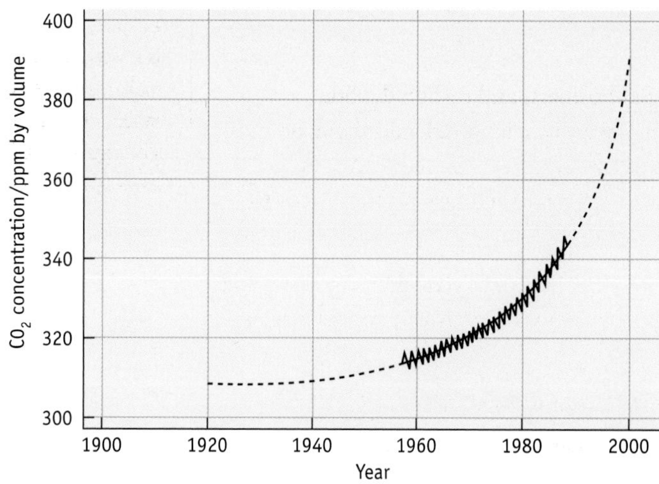

Figure 5.46 Changes in atmospheric carbon dioxide 1958–1988 measured at Mauna Loa observatory in Hawaii, with extrapolation back to 1920 and forward to 2000.

Q5.36 Use Figure 5.46 to estimate the carbon dioxide concentration in:

a 1940

b 1995.

In Figure 5.46, the smoothed best-fit line is extended (dotted line) back to 1920 and forwards to 2000. Extending a line is called **extrapolation**. In extending the line we make the assumption that:

- we have enough data to establish the trend accurately
- present trends continue.

The easiest line to extrapolate is a straight one – you just use a ruler. To extrapolate a curve you need to identify its shape. Computers can extrapolate curves mathematically more exactly than humans can.

An extrapolation is often the basis for predictions; the graph in Figure 5.46 acts as a mathematical model of carbon dioxide change. However, 2000 has come and gone without the mean carbon dioxide level reaching the level predicted by Figure 5.46. We can therefore conclude that predictions based on this model are not very accurate. This Mauna Loa data is the longest set of direct measurements of atmospheric carbon dioxide concentrations in existence, but estimations based on this data tend to turn out too high (see 'The mystery of the disappearing carbon dioxide', Section 5.8).

The present trend for increasing carbon dioxide may not continue. For example, if steps were taken to reduce CO_2 emissions, if the efficiency of petrol and diesel engines were improved, and if the rising cost of energy encouraged people to improve the insulation of their homes, then 'present trends' might not continue. On the other hand, increases in living standards in countries such as China and India (which between them have over two billion people) might cause atmospheric CO_2 levels to rise even faster than present trends suggest.

Taking into account many factors at once

If we could understand all the key factors that affect climate and how they interact, we could produce a computer model that is much more mathematically sophisticated than the graph in Figure 5.46. This would give us the opportunity to make more reliable predictions.

Activity

Activity 5.16 provides the opportunity to make use of a computer model of climate change. **A5.16S**

Modelling climate change is a very complicated business, and carbon dioxide concentration is only a part of the story. Many factors are involved, and if one or more is missed out of a model then its predictive accuracy will be reduced. It is also not enough to include all the factors; the model must take into account interactions between them.

Carbon dioxide is an important factor, because it is one that is changing at present and one that humans can influence. But it isn't by any means the only one. Other factors that may affect climate change include:

- other greenhouse gases such as methane, CFCs and nitrous oxide (N_2O)
- aerosols – extremely small particles or liquid droplets found in the atmosphere
- the degree of reflection from those parts of the Earth's surface that are free of ice and snow
- the fraction of the Earth covered with ice and snow
- the extent of cloud cover
- changes in the Sun's radiation.

Weblink

You can view more information about the Hadley Centre and their climate models by visiting their website via the Met Office site. This also provides lots of links to other useful sites.

The Hadley Centre for Climate Prediction and Research, part of the UK Met Office, has several climate models that they use to predict the surface temperatures, precipitation, soil moisture content, sea level change, sea-ice area, and sea-ice volume. The maps in Figure 5.47 assume that no measures are taken to reduce greenhouse gas emissions, and that there is mid-range economic growth resulting in a doubling of emissions of greenhouse gases over the course of the twenty-first century.

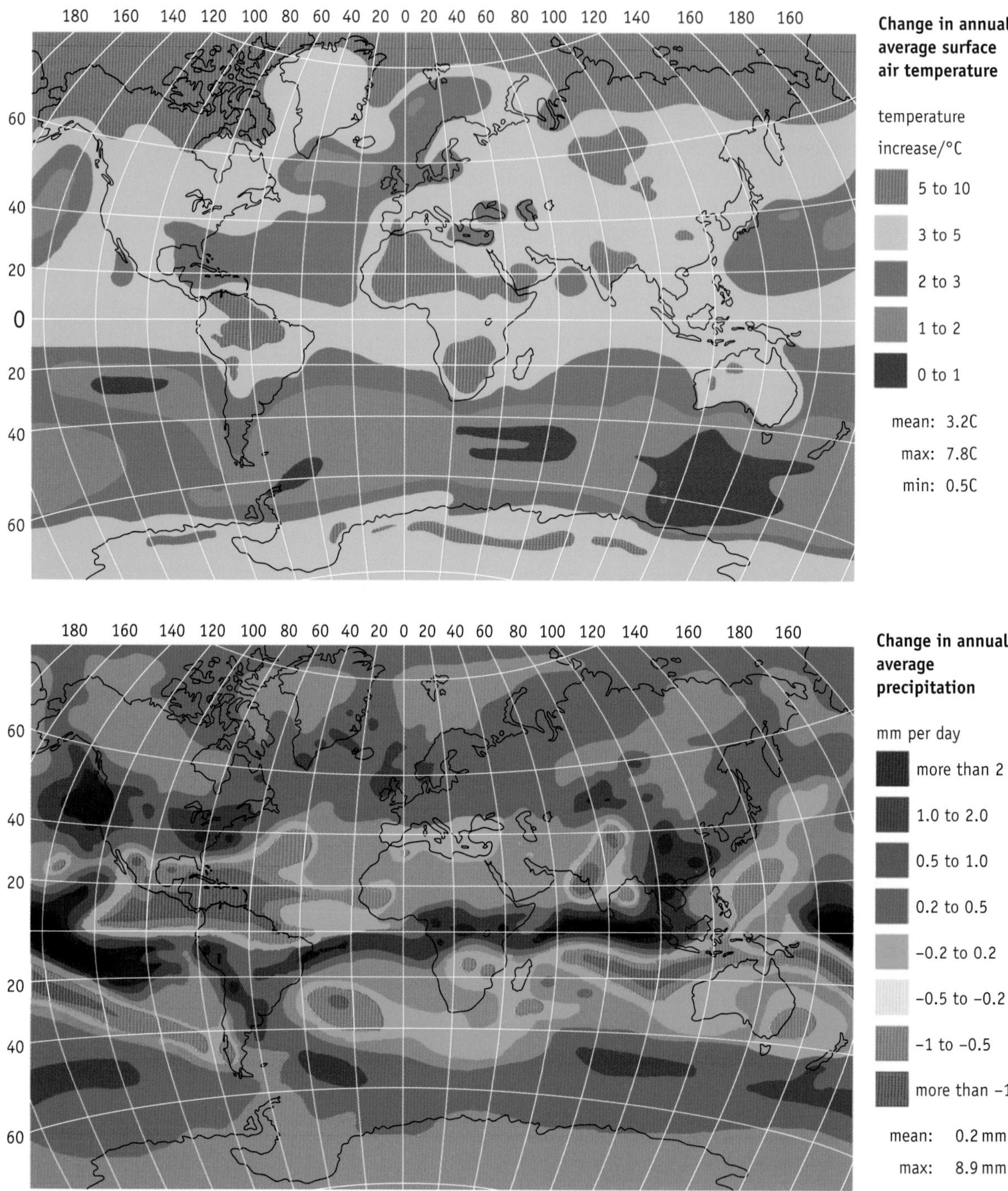

Figure 5.47 The Hadley Centre computer model predictions for climate at the end of the twenty-first century relative to current climate, defined as 1960–1990. Source: Met Office.

Don't expect models to be perfect

Several major climate models are currently in use around the world. They do not always give the same answers to the same question, and sometimes all the models 'get it wrong'. The predictions may be incorrect because of:

- limited data
- limited knowledge of how the climate system works
- limitations in computing resources
- failure to include all factors affecting the climate
- changing trends in factors included, e.g. faster than expected loss of snow and ice cover or greater carbon dioxide emissions.

The models are, however, continually improving, using bigger data sets and incorporating more factors and more sophisticated interactions.

Taking data from years past, feeding the data into the models and seeing whether the predictions match what actually happened can test climate models. Of course, even if a model works when tested with old data, we can't be sure that it will be reliable in future. For one thing, other factors may change in ways we never expected. Models are not expected to predict the future precisely, but to make the best prediction based on all the evidence available.

Climate model predicts a colder UK

Winters in Britain are much milder than in other places on a similar latitude. For example, Goose Bay in Canada is the same latitude as Manchester, UK, but the winter temperature in Goose Bay can drop to –50 °C. Apart from Alaska, the whole of the US lies south of Britain, yet the UK has milder winters than much of the US. This is because of the Gulf Stream and its extension, the North Atlantic Drift, which brings warm water from the Gulf of Mexico to north-west Europe. In the North Atlantic, surface water cools, contracts and sinks. This cooled water flows deep below the ocean surface and eventually goes back to the Gulf of Mexico, where it is warmed again, expands and rises to the surface. This sinking of water in the North Atlantic acts as a 'pump', causing warm surface water to flow northwards.

As we have seen, the extent of the Arctic ice sheet is reducing. The melting ice means an influx of fresh water into the North Atlantic. Fresh water does not behave like salt water; at 0 °C, fresh water starts to freeze and expands. This water rises to the surface rather than sinking, as salt water of the same temperature does. At some critical point, the North Atlantic Drift could break down and no longer bring warm water from the Gulf of Mexico. If this were to happen, the average temperature across Britain would fall. This has not happened yet and may never do so, but there is evidence that the salt concentration of the North Atlantic has been falling over the past two decades. We can't be sure it's due to global warming, but this is certainly a possible cause that needs to be taken very seriously.

Stefan Rahmstorf of the Potsdam Institute for Climate Impact Research has predicted (using a computer model) that global warming could result in a reduction of surface temperature in north-west Europe. A 5 °C fall in the global mean surface temperature is predicted, which is quite substantial – not quite an Ice Age but getting that way.

So for Britain there are two scenarios that we could ask our climate models about. What would happen if surface temperature fell by 5 °C, and what would happen if it rose by 5 °C? Each could form the basis for a disaster movie.

5.6 Coping with climate change

The evidence shows that the Earth's climate has changed considerably over the past few hundred years. We are in a period of global warming that influences not only temperature but also rainfall and wind patterns. How will organisms cope with changing conditions?

Some animals can migrate to avoid the changing conditions. Plants will change their distribution over time, even though as individuals they are rooted in one spot. The changes already being attributed to climate change can be divided into two categories:

- changing distribution of species
- altered development and life cycles.

Changing distribution of species

Changing communities and alien invaders

Within any community, some species will cope with change while others will fare less well. Climate change will therefore cause the balance between species in the community to shift. Some species may benefit from the new conditions and become **dominant**, while others may be lost from the community altogether due to competition with existing or invading species. If they are mobile or have good seed dispersal they may migrate to more favourable conditions. In other words, the distribution of some species will probably change. For example, plants that currently reach their northern limit of distribution in England or southern Scotland, such as the ground thistle, will begin to expand northwards. Others, such as heath bedstraw (Figure 5.48), may retreat north from the southern limits of their ranges in Britain. Future communities may well be quite different from those present today.

Activity

Activity 5.17 allows you to find out about coral bleaching linked to rising sea temperatures. **A5.17S**

Q5.37 Suggest why heath bedstraw distribution will move further north with climate change.

A study of 35 non-migratory European butterfly species found that the ranges of 63% of them have shifted northwards by between 35 and 240 km during the twentieth century. Figure 5.49 shows how the speckled wood butterfly (Figure 5.50) has moved north in Britain. This is a result of more successful colonisation at the northern edge of their range. Only 3% of the species studied had shifted further south. The changing distribution may be a direct response to rising temperature, or the result of a shift in distribution of the plants they feed on.

With global warming, higher temperatures early in the year could extend the growing season, and crops such as olives and citrus plants that are sensitive to frosts could be grown much further north.

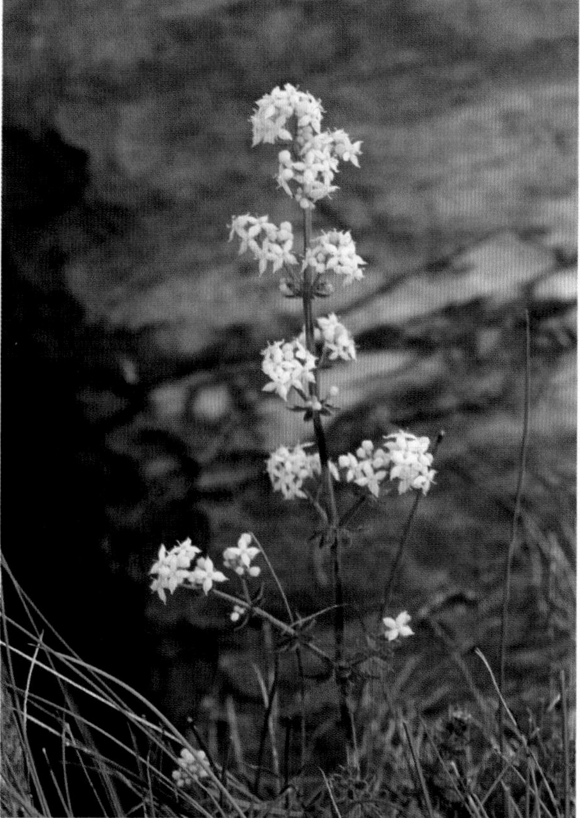

Figure 5.48 Heath bedstraw distribution will move north if conditions in the south become warmer.

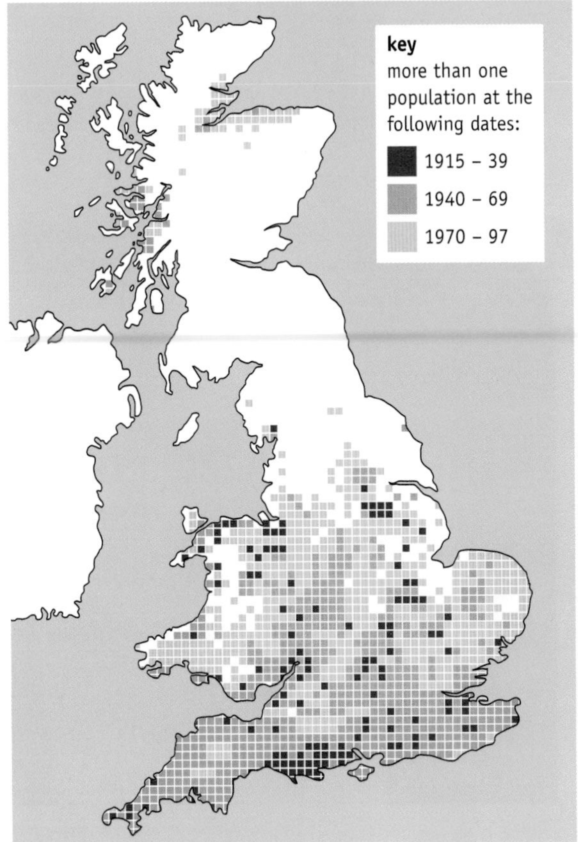

Figure 5.49 The speckled wood butterfly has extended its range north. *Source*: Nature.

Figure 5.50 The speckled wood butterfly.

Q5.38 In mountainous regions, how might the ranges of plants and animals shift with warmer conditions?

Butterflies and plants like the ground thistle – whose seeds are carried by the wind – are easily able to disperse long distances to find suitable habitats. However, many animals and some plants are poor at dispersal. As so much of the environment is now used for intensive agriculture or is built over, organisms under pressure from climate change may find it impossible to reach new areas and their populations will simply die out.

Q5.39 Suggest what particular problems the community that inhabits the northern tundra will face with global warming.

A particular problem for some communities may be the invasion of exotic or alien animal or plant species from other regions of the world. Species from Mediterranean regions might take advantage of the warmer conditions and invade southern England, pushing out the current inhabitants. Some exotic species such as scorpions already survive in local areas in Britain and could spread rapidly given a suitable climate.

Pests and diseases may also spread to new areas and act to reduce crop yields. Witchweed is a **parasitic** weed that infects cereal crops including maize and sorghum in Africa (Figure 5.51). It already causes huge amounts of damage to subsistence farms, but requires an air temperature of at least 28 °C to grow effectively. Unlike most weeds, which merely compete with crops, a parasitic plant like witchweed taps directly into its host plant and absorbs nutrients and moisture. As a result the host plant is less productive, destroying 30–100% of the crop. If global temperatures increase, it may spread to cereal crops in new areas and cause even more damage.

Figure 5.51 The parasitic witchweed already causes serious damage in Africa and may become more widespread if conditions become warmer.

In Britain, fruit crops could be threatened by the easier spread of fungal diseases in more humid conditions.

Some invertebrate pests are also likely to change their distribution. For example, the nematode *Longidorus caespiticola*, which feeds on grass, is currently restricted to England, Wales and south-east Scotland. A 1 °C rise in temperature could extend its range to cover the whole of Scotland. The brassica pod midge *Dasineura brassicae*, a major pest of oil seed rape, is also expected to move further north. Not only is it likely that the distribution of such pests will change, but the warmer temperatures may allow them to produce more new generations in a single year, increasing their effect on crops. However, it is hard to predict what the impact will really be, due to interactions with other animals and any change in the life cycle of the host plant.

It is not only changing temperature that affects species distribution. Changing rainfall patterns, soil moisture, winds and rising sea levels are all likely to influence species and communities.

Extension

In **Extension 5.5** you can find out about what might have caused extinction of the dinosaurs. **X5.05S**

Altered development

Faster photosynthesis – faster growth

Scientists have established that a few degrees of warming will lead to an increase in temperate crop yields, but greater temperature rises could reduce crop yields. As we have seen, the rate of photosynthesis is determined by a number of limiting factors including both carbon dioxide concentration and temperature. In cooler climates where photosynthesis is temperature-limited, a rise in temperature will result in faster photosynthesis. But the situation is a little more complicated because, above an optimum temperature (which varies for different species), plant enzymes work more slowly. Figure 5.52 shows how the rate of an enzyme-controlled reaction is affected by temperature.

Key biological principle: The effect of temperature on enzyme activity

Temperature affects the rate of an enzyme-catalysed reaction, as Figure 5.52 shows. At low temperatures the reaction is very slow. This is because the enzyme and substrate molecules move slowly and don't collide very often. As the temperature increases there are more collisions. Therefore the substrate binds with the enzyme's active site more frequently, thus increasing the rate of reaction. The rate of collisions, and so the rate of reaction, approximately doubles for each 10 °C rise in temperature; this means that the shape of the graph is a curve. The temperature at which the rate of reaction is highest is called the **optimum temperature** (Figure 5.52). If the temperature continues to rise, the enzyme molecule vibrates more, and bonds holding it in its precise three-dimensional shape break. The substrate no longer fits easily into the active site; this slows the reaction. Eventually the shape of the active site is lost and the enzyme-substrate complex no longer forms; no reaction can occur. The enzyme is said to be **denatured**.

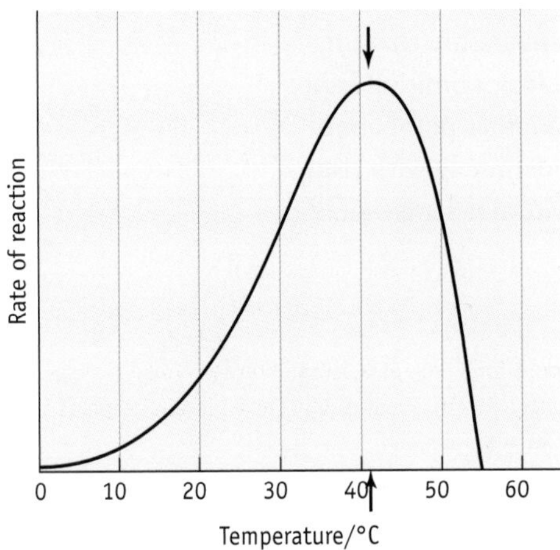

Figure 5.52 The effect of temperature on the rate of an enzyme-catalysed reaction. The optimum temperature is indicated by the arrows.

Activity

Activity 5.18 lets you investigate the effect of temperature on enzyme activity. **A5.18S**

Q5.40 **a** Look at Figure 5.53. Decide on the optimum temperature for wheat photosynthesis under high light conditions.

b Most animal species are much more mobile than plants, and some migrate between regions with very different temperatures. Animals such as migratory fish already show the ability to cope with a range of temperatures. They often synthesise slightly different forms of the same enzyme, termed isoenzymes, as they move from one region to another. How might having different forms of the same enzyme help them cope with changing conditions?

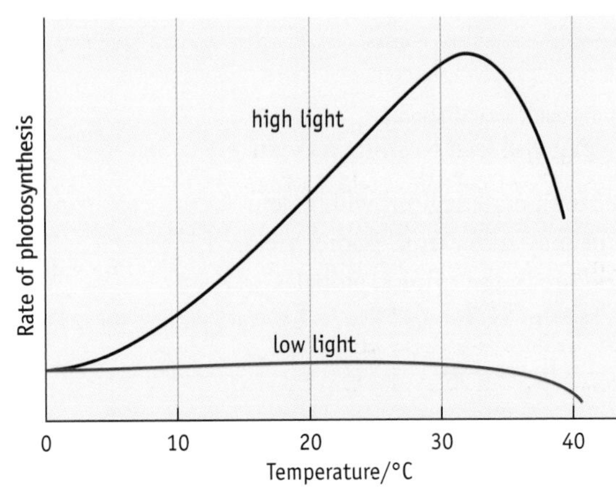

Increases in temperature result in higher rates of photosynthesis when no other factors are limiting. Above an optimum temperature, where the rate of photosynthesis is at a maximum, rate declines. Under natural conditions the optimum is rarely achieved because CO_2 and light are limiting.

Figure 5.53 The response of photosynthesis in wheat to increasing temperature at two light intensities.

A complex picture

In today's tropics, many crops are at the limit of their temperature tolerance. Even a small rise in temperature would be likely to decrease yield, owing to the direct effect of the higher temperature along with the resulting fall in soil moisture.

Activity

Activity 5.19 allows you to analyse research data and design an experiment to investigate the effect of changing environmental conditions on plant growth. **A5.19S**

Predictions about the ecological and agricultural consequences of global warming are difficult to make with great confidence, as there are so many interacting factors. If the concentration of carbon dioxide in the atmosphere rises, the rate of photosynthesis in many plant species may also increase, irrespective of any rise in temperature resulting from the rising carbon dioxide levels. This is because carbon dioxide is a limiting factor for photosynthesis. However, the graph in Figure 5.54 illustrates another point about limiting factors – the rate of photosynthesis would not increase indefinitely with rising carbon dioxide, because other factors such as light intensity, water or nutrients would start to limit photosynthesis.

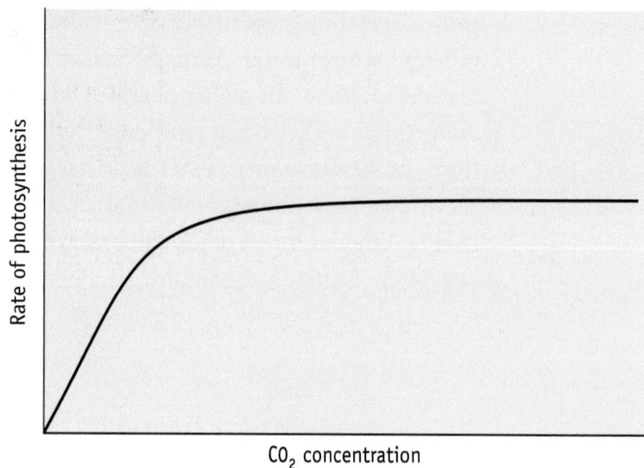

Figure 5.54 The response of potato leaves to increased carbon dioxide concentration.

Overall, it is probable that crop production in cooler temperate regions will benefit from climate change, whereas warmer tropical regions may suffer from poorer yields. Given that some of the poorest countries in the world are in the tropics, global warming may increase global inequality.

Disrupted development and life cycles

Animals are likely to be affected if temperature acts as an environmental cue or trigger for their development or behaviour. In the case of salmonid fish (salmon, trout and sea trout), it is their spawning, hatching and growth rates that are most likely to be disrupted. Recent research has shown that brown trout stop growing in late summer when the river temperatures have warmed up. At present, this is not a problem because they have already completed most of their growth by this stage in the year. However, if river temperatures were to warm up earlier, they might stop growing earlier, resulting in underweight fish with a reduced chance of surviving the winter.

The egg incubation temperature of some reptiles determines the sex of the offspring. In leatherback turtles, higher temperatures in the nest result in females (Figure 5.55). Usually, there are sufficient nests at or below the critical temperature for the sex ratio to remain reasonably close to 50:50.

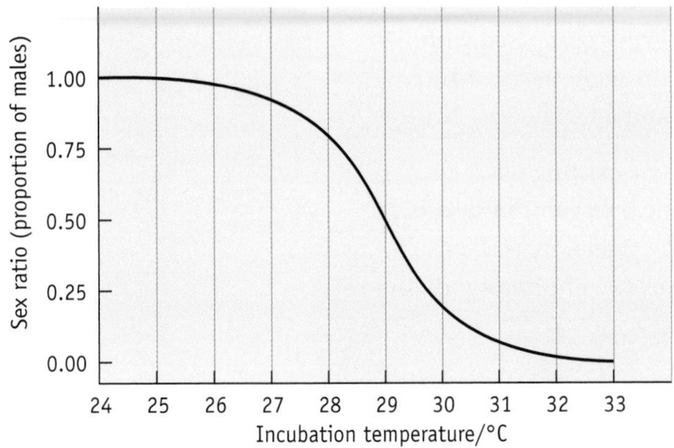

Figure 5.55 In leatherback turtles, temperature determines the sex of the hatchlings.

Q5.41 What would be the implications of warmer temperatures for leatherback turtles?

> **Activity**
>
> **Activity 5.20** allows you to investigate the effect of temperature on the hatching success of brine shrimps.
> **A5.20S**

Phenology: changing with the seasons

Phenology is the study of seasonal events in the lives of animals and plants, such as the time of flowering or fruiting, the time of egg laying or hatching, the first appearance of migrants, and so on. It has been recognised that the timing of such events is a useful biological indicator of global climate change. Many long-term records exist that allow us to see how much change has been experienced in natural systems, particularly those associated with the onset of spring (Figure 5.56).

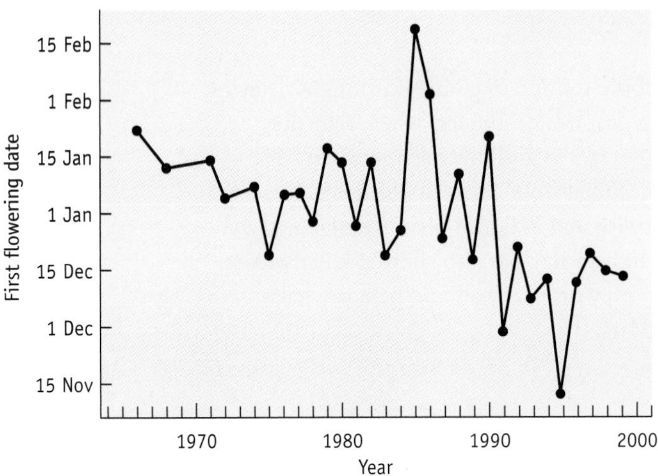

Figure 5.56 First flowering of the aconite in Norwich, UK, from 1966 to 1999. Aconites are members of the *Ranunculus* (buttercup) family. The yellow flowers normally appear alongside snowdrops in the spring.

Of course, evidence from a single location about a single species is not sufficient to substantiate claims that global warming is taking place. However, data sets exist for a range of events from many locations. Trends within them can be identified and tested statistically, and these do indicate that spring is getting earlier for many species.

For example, Richard Fitter recorded the first date on which over 500 plant species flowered in each year from 1954 to 2000 near his home in Oxfordshire. Professor Alistair Fitter, his son, analysed these data. He found that nearly one in six of the species are now flowering significantly earlier than they did before the 1990s. For example, white dead-nettle flowered 55 days earlier in the 1990s compared with the period 1954–90, and can now often be seen flowering throughout the winter. However, not all species responded as dramatically as this, and a few flowered later.

Weblink

Visit the Nature's Calendar website where you can get the latest records for seasonal events.

The UK Phenology Network in Britain aims to build on these existing data sets. Leafing, flowering, summer bird arrival, bird activity and the behaviour of insects, butterflies and amphibians are all being monitored.

Activity

Activity 5.21 involves you analysing an online dataset for the emergence of some butterflies and moths, and visiting the UK Phenology Network website to investigate some seasonal changes that are affected by rising temperatures. **A5.21S**

Making the most of the food supply

For many species, the hatching of eggs or the emergence of adults is synchronised with periods of maximum food availability. For example, the eggs of many marine worms are laid so that hatching coincides with high levels of microscopic plants (phytoplankton) on which the worm larvae feed. The problem is that the worms lay their eggs in response to day length (**photoperiod**), whereas phytoplankton grow in response to temperature. If global temperatures rise, the peak in phytoplankton will occur earlier in the spring, but the worms may still lay their eggs at the usual time because the photoperiod is unchanged. The resulting mismatch between hatching time and peak food availability could drastically reduce the survival rates of the worm larvae. This lack of synchrony is also observed with UK birds such as great tits and their food source, winter moth caterpillars.

Checkpoint

5.5 Write a short summary explaining how rising temperatures, changing rainfall patterns, and changes in seasonal cycles can affect plants and animals.

In Canada's Hudson Bay, the later freezing of the sea due to global warming is delaying the polar bears' departure for the rich feeding grounds on the ice, where they prey on seals. The average weight of the bears has decreased and juvenile mortality has increased. The Hudson Bay population has declined from 1140 to 950 bears in the ten years to 2007. In 2008, the US Government Fish and Wildlife Service designated the polar bear a threatened species. Ice melt in Hudson Bay has also allowed killer whales to penetrate further into the area, where they prey on narwhals and beluga whales. One species' loss is another species' gain.

5.7 Adapt or die

Polar bears are very well suited to the cold conditions in their Arctic habitat. Will they be able to cope with increasing temperatures, or will they be confined to a shrinking habitat? Many organisms unable to move to new habitats will only survive if they can adapt to the new conditions. Otherwise they may face local extinction. Climate change provides a selection pressure for natural selection and evolution.

Changing over time

As you saw in AS Topic 4, species change over time as they adapt to their changing environment. This evolution by natural selection produces organisms that are better adapted to their environment. These ideas were published in 1859 by Charles Darwin in his book *On the Origin of Species by Means of Natural Selection, or the Preservation of Favoured Races in the Struggle for Life*. This set out his theory that organisms might have evolved, gradually changing from one form to another.

One of the things that greatly influenced Darwin in the development of his ideas was the great diversity of life he saw on the Galapagos Islands, where he observed closely-related species. These were clearly similar to, but not the same as, species on the South American mainland. One example that influenced his thinking was the giant land tortoise (Figure 5.57).

Figure 5.57 A giant tortoise on the Galapagos Islands. At dinner one evening Darwin was told that these tortoises have a different-shaped shell on each island.

A summary of Darwin's theory is as follows:

- *Darwin observed*: Organisms produce more offspring than can survive and reproduce. Numbers in natural populations stay much the same over time.
- *He concluded*: There is a **struggle for existence** – competition for survival between members of the same species. As a population increases in size, environmental factors halt the increase (see Figure 5.58). Many individuals die due to predation, competition for food and other resources, or due to the rapid spread of disease resulting from overcrowding.

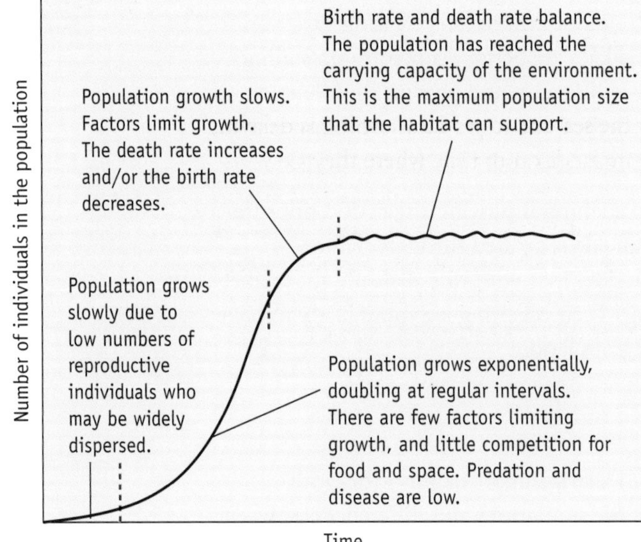

Figure 5.58 Population growth.

- *Darwin also observed*: There is a huge amount of variation within species.
- *He concluded*: Those individuals that are best adapted to conditions in their environment are more likely to survive and breed. They have a selective advantage: natural selection is acting, and there is **survival of the fittest**. Individuals with these adaptive features will be more common in the next generation. Those organisms that are not well adapted are more likely to die before maturity, and so do not produce many offspring. Over a period of time, the character of the species will change to the more adapted form.

Did you know?

Those who influenced the development of Darwin's theory

Darwin's own observations of organisms were key to the development of his ideas on evolution but other scientists influenced his thinking, e.g. Thomas Malthus's ideas about the factors influencing the size of populations. In 1798, Malthus wrote an essay about the potential of the human population to grow exponentially. He argued that this would lead to a population size that would exceed available food supply, resulting in famine, pestilence and war. Darwin realised that the sizes of many animal populations were staying the same.

Darwin had also read the work of the geologist Charles Lyell. He suggested that the world was many millions of years old, rather than mere thousands of years as was widely believed at the time.

Darwin put these ideas and observations together and came up with the theory of evolution by natural selection. For a long time Darwin did not publish his findings. But then another naturalist, Alfred Russel Wallace, who had also read Malthus and had observed the variety of life in South America and Malaysia, sent Darwin an essay outlining a theory that was identical to Darwin's own. They published a short paper together outlining the theory. This stimulated Darwin to put together the mass of material he had gathered in support of the theory and he finally published *The Origin of Species* in 1859.

We now know

Today, we know that new alleles arise through random mutations creating variation within a population. If individuals with these alleles are better adapted to prevailing conditions, natural selection will act and they will survive and pass on the alleles or groups of alleles that are responsible for the inherited variation. Thus evolution is a change in allele frequency in a population of organisms over time (generations).

In AS Topic 4 you saw how the insecticides in shampoos provide a selection pressure on head lice. Some individual head lice have an allele, created through chance mutation, making them resistant to the insecticide. Lice with this allele are more likely to survive, reproduce and so produce offspring. Their offspring are more likely to have the allele, so it becomes more common in the population.

Evidence for evolution

How do we know that evolution is occurring? Darwin supported his ideas with scientific evidence from careful observation of the similarities and differences between organisms and their geographical distribution. Evidence from the fossil record of changes in species also provided evidence.

What Darwin struggled with was how the variation could be inherited – he had no knowledge of genetics. He noticed that human offspring tend to look a bit like each of their parents, so he imagined that the inherited factors must themselves blend in the offspring. But this confused him, as he thought it would eventually make the members of a population all look the same. He spent much of his life puzzling over the problem, studying and cataloguing variation in wild and domestic organisms.

The answer had in fact come in Darwin's own lifetime, yet Darwin apparently did not hear about it. Gregor Mendel, an Austrian monk, discovered that characteristics (in pea plants) are not blended but are passed on unchanged to the offspring. But his ideas about genes were published in an obscure journal and were not rediscovered for many years.

Q5.42 For the benefit of Mendel and Darwin (who have paid a brief visit back to Earth), give a twenty-first century explanation of the word 'gene'.

If Mendel were publishing his ideas today, he would probably send his paper to a scientific journal such as *Nature* or *Proceedings of the Royal Society*. Before publication, Mendel's paper would be sent to two or three scientists for **peer review**. The reviewers would examine his paper very critically, looking at whether he had included proper controls, used statistics appropriately, considered the work of other scientists, and whether his conclusions were valid. It is likely that the paper would be returned to Mendel for some alterations or clarification, though the reviewers would remain anonymous. The paper would then be published – it might also be presented at a scientific conference – and the world would quickly hear about his findings.

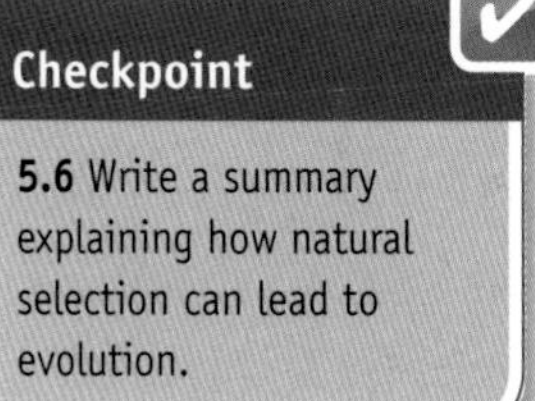

Checkpoint

5.6 Write a summary explaining how natural selection can lead to evolution.

Q5.43 Why is it important that the identities of reviewers of scientific papers are not given to the author(s) of the paper?

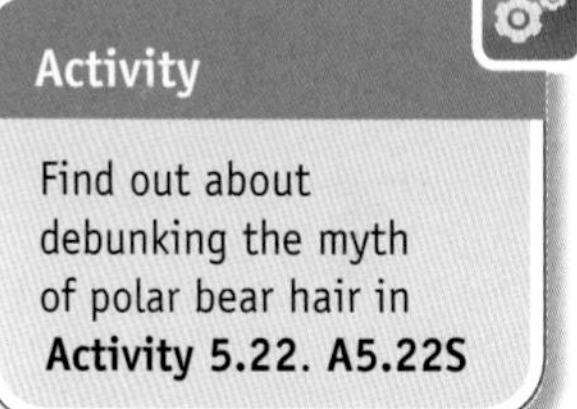

Activity

Find out about debunking the myth of polar bear hair in **Activity 5.22. A5.22S**

Q5.44 Many scientific ideas are now published online without peer review. Outline the benefits and disadvantages of publishing directly online rather than in a peer-reviewed scientific journal. (Many peer-reviewed scientific journals are also available online.)

Q5.45 Suggest three reasons why it is important to present scientific findings in conferences.

Molecular evidence

Darwin would probably be amazed and delighted to see the vast array of research evidence supporting and extending his ideas. The structure of DNA, revealed in 1953 by the work of Rosalind Franklyn, Maurice Wilkins, James Watson and Francis Crick, is the same in all organisms. The genetic code and the mechanism of protein synthesis are essentially the same in all organisms, as is the working of cells at a biochemical level. This uniformity provides important evidence for evolution, revealing genetic continuity and supporting Darwin's idea that all living organisms stem from a common ancestor.

The modern techniques of molecular biology have provided a detailed source of evidence for evolution, revealing information about the evolutionary history of organisms. Proteins and nucleic acids contain a record of genetic changes over time. The study of DNA (**genomics**) and proteins (**proteomics**) are revealing these genetic changes. As you saw in AS Topic 2, the genetic changes are due to random mutations.

DNA hybridisation

When DNA is gently heated, it separates into its component strands as the hydrogen bonds between the base pairs break. If we do this to human DNA and chimpanzee DNA we can then mix them to make hybrid DNA, made up of a strand of the human DNA and a strand of chimp DNA. Since chimp and human DNA will have some different base sequences, not all the bases will be able to pair up. Therefore, when heated, the hybrid DNA will separate (denature) at a lower temperature than pure human or pure chimp DNA. The more similar the base sequences of the two species, the better they will stick together. This technique can be used as a measure of the similarity between the genomes of the two species, without needing to know their base sequences. It has been used to look at the evolutionary split between chimps, gorillas and humans. The results indicated that chimps are more closely related to humans than gorillas are. Darwin had always maintained that humans evolved in Africa, because our closest primate relatives (based on morphology, anatomy and behaviour) are found there.

Q5.46 In DNA hybridisation experiments, a 1 °C temperature difference of denaturing indicates a 1% difference in DNA. Chimps turn out to be about 1.6% different from humans, and gorillas 2.3%. Comment on the reliability of the conclusion that we are more closely related to chimps than to gorillas.

DNA profiling

Restriction enzymes cut DNA at specific sequences. This produces a series of different sized DNA fragments, which can be visualised as a series of bands or peaks on a DNA profile. If mutations have occurred in some of these DNA sequences, the enzymes will not cut the DNA, and the size of the fragments and hence the position of the bands produced will change. The differences in the fragment lengths produced provides information about the genetic differences between individuals and between species. In Topic 6 you will look in detail at how restriction enzymes and gel electrophoresis are used to produce DNA profiles.

Activity

In **Activity 5.23**, use the amino acid sequences in cytochrome C to determine how closely related different animals are in evolutionary terms. **A5.23S**

DNA and protein sequencing

By comparing the sequence of bases in DNA or the amino acid sequence in proteins of different species, it is possible to determine how closely related organisms are in evolutionary terms. If two species have very few differences in the DNA or amino acid sequence, they evolved from a common ancestor more recently than organisms with more differences. For example, in human cytochrome C, an enzyme involved in respiration, there are 104 amino acids. Chimpanzee cytochrome C has the same 104 amino acids in the same order. Rhesus monkeys have one difference in the sequence. Horses have 11 additional amino acids in the sequence.

Extension

Find out in **Extension 5.6** how the sequence of bases in DNA can be determined. **X5.06S**

DNA molecular clocks

The polymerase chain reaction (see Topic 6) can produce millions of copies of DNA. This reaction and the use of automated DNA-sequencing machines allow rapid determination of the DNA base sequence. As species evolve, it is assumed that they accumulate random mutations at a regular rate, becoming genetically more different. By comparing the number of differences between species, it is possible to calculate how long ago they shared a common ancestor.

Evolutionary trees based on molecular clocks support Darwin's theory that organisms have evolved slowly over vast periods of time – measured in thousands or millions of years.

Evolution observed

Figure 5.59 The Asian shore crab.

When a susceptible HIV population is exposed to a new drug, within a few days the virus population becomes resistant to the drug, providing an observable demonstration of evolution by natural selection.

Rapid evolution has been identified in other organisms. For example, the blue mussel growing on the eastern US coast has evolved due to the selection pressure of predation by the Asian shore crab (Figure 5.59). The crab invaded this coast in 1988 and began cracking open and feeding on the blue mussels. After just 15 years, the blue mussels had become adapted to protect themselves, thickening their shells in response to chemicals released by the crab.

Results of experiments by scientists at the University of New Hampshire in the US supported the hypothesis that evolution had occurred in blue mussels. Mussels from northern shores where the crab had not yet reached were tested, as well as mussels from southern shores where the crab had invaded. They were both exposed to the chemicals released by Asian crabs, and to those released by native green crabs that live on both northern and southern shores. In response to the Asian crab chemicals, the southern mussels built up thicker shells. The northern mussels, which had not been exposed to any previous selection pressure, did not significantly change (Figure 5.60). But both sets of mussels thickened their shells in response to chemicals from green crabs.

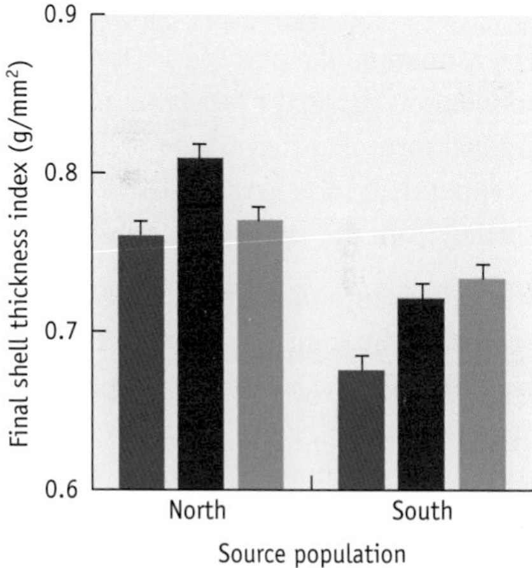

Figure 5.60 The thickness of mussel shells from northern and southern populations grown in cages with the chemical cues from green crabs (*Carcinus maena*) and Asian shore crabs (*Hemigrapsus sanguineus*).

Q5.47 Describe how the southern blue mussels have become adapted in response to the predation by Asian crabs.

Q5.48 The green crab was introduced to the United States in 1817. What conclusions can you draw about mussels and green crabs from the results?

Natural selection and climate change

Organisms experience selection pressure as a result of climate change. The major effect of global warming is considered to be milder winter temperatures, earlier springs and later autumns. The effect is greater nearer the poles, reducing the length and severity of the winters. Scientists consider a major selection pressure associated with climate change to be the increased length of the growing season.

As you saw earlier in this topic, hatching of bird eggs may be synchronised with periods of maximum food availability. Warmer temperatures in the spring have led to earlier development of many insects, providing a selection pressure favouring birds that nest early. In a Europe-wide study it was found that some great and blue tit populations now exhibit earlier onset of egg laying, but others do not. The reproductive maturity of great tits is linked to day length; this has not changed, which may explain why some populations do not exhibit earlier egg laying. The populations that do have earlier egg laying may still be threatened if any shift in egg laying does not keep pace with the change in maximum insect numbers.

Q5.49 Suggest why some populations of great tits have been able to advance their egg laying date but other populations have not.

Rapidly reproducing species like head lice can evolve quickly. What are the chances that long-lived animals like polar bears can adapt to rapidly changing conditions? Polar bears are capable of feeding on a wide range of foods. They quickly learn to take food from human rubbish tips around towns, and even eat seaweed and berries. However, the females need a protein-rich diet to breed successfully, so their future remains very uncertain.

Speciation

The changes in environmental conditions resulting from climate change will provide selection pressure, which can lead to evolution by natural selection. Can new species arise in response to environmental change?

How are new species formed?

The formation of a new species is called **speciation**. Darwin thought that, over time, populations exposed to selection pressure would become so different from their original form that they would eventually become different species. It is now generally accepted that, for a new species to arise, a group of individuals has to be reproductively isolated from the rest of the population.

The most common method of speciation is thought to be isolation of part of a population by some geographical feature that prevents a group of individuals from breeding with the rest of the population, such as a high mountain range, a river, or a stretch of ocean (Figure 5.61). Over time, the two groups will become less like each other as they respond to different selection pressures within their local habitats, and

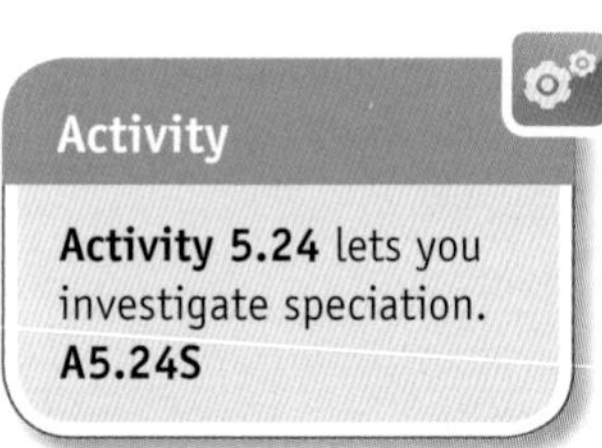

Activity
Activity 5.24 lets you investigate speciation.
A5.24S

1

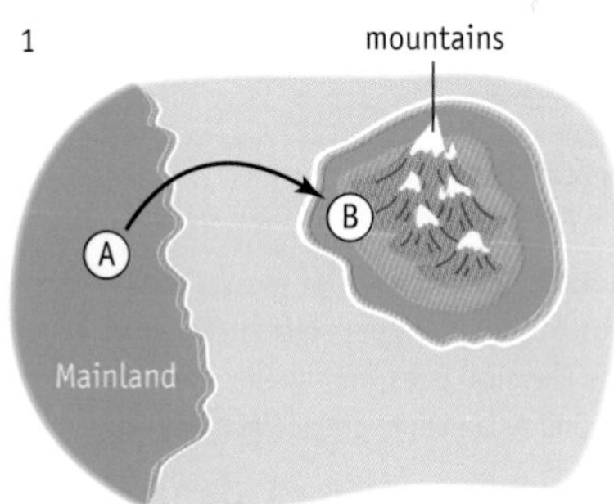

Some of population A moves from the mainland to the island.

2

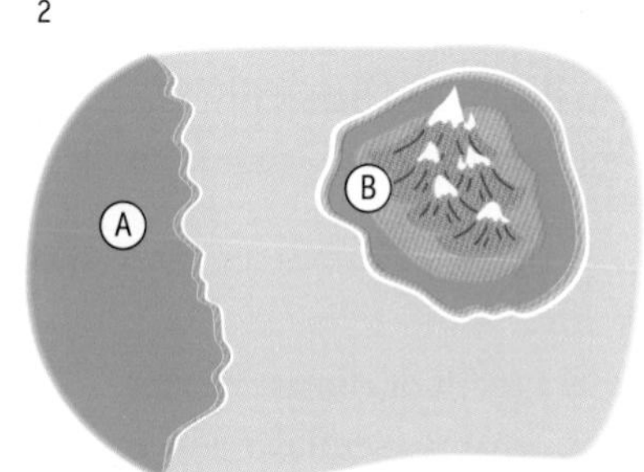

Isolated from population A, the island population eventually evolves to form a new species, B.

3

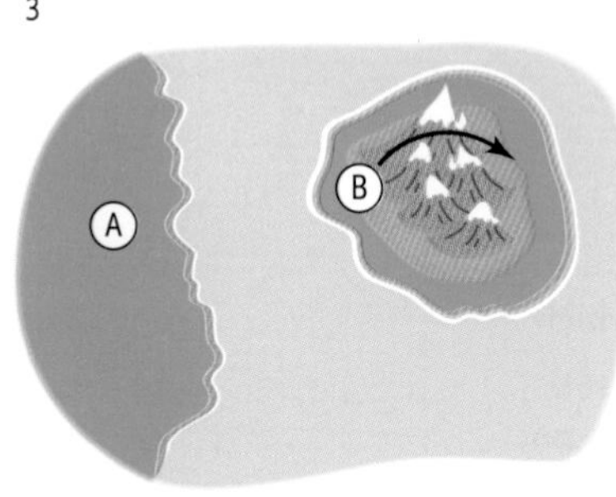

Some of population B moves over the mountains to the opposite side of the island.

4

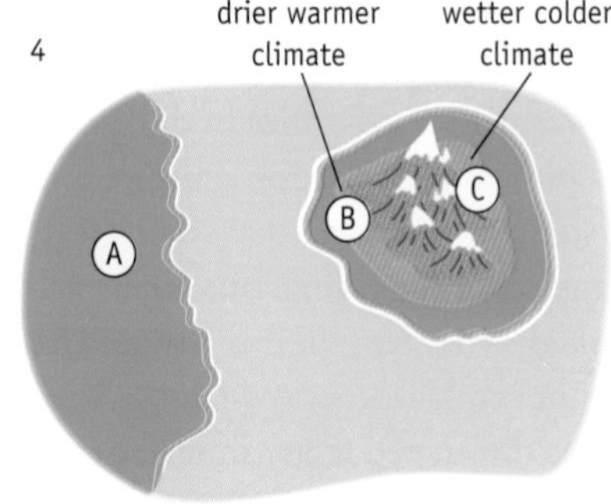

Isolated from population B, a new species, C, evolves due to the different selection pressures on that side of the mountain range.

Figure 5.61 Illustration of how geographical isolation can lead to the evolution of new species.

as random mutations accumulate. When the members of the two groups meet again, they may not be able to interbreed if these differences are great enough. Once the two populations are unable to breed and produce fertile offspring, they are considered to belong to two different species, and there is **reproductive isolation** between them. The longer the two groups are geographically isolated, the more likely it is that speciation will occur.

There are a number of other reasons, in addition to geographical isolation, why two species may not be able to breed and produce fertile offspring. These are shown in Table 5.6. If behavioural, ecological or genetic barriers arose within a population and reproductively isolated a group of individuals it is possible (although rare) for speciation to occur.

Table 5.6 Reasons (in addition to geographical isolation) why two species may not be able to interbreed successfully.

Method of isolation	Description
ecological isolation	The species occupy different parts of the habitat. For example, the violet species *Viola arvensis* grows on alkaline soils, whereas *V. tricolor* only grows on more acid soils (Figure 5.62).
temporal isolation	The species exist in the same area but reproduce at different times.
behavioural isolation	The species exist in the same area, but do not respond to each others' courtship behaviour (e.g. different species of fireflies).
physical incompatibility	Species co-exist, but there are physical reasons that prevent them from copulating (e.g. size or shape of genitals in some insects).
hybrid inviability	In some species, hybrids are produced but they do not survive long enough to breed.
hybrid sterility	Hybrids survive to reproductive age but cannot reproduce (e.g. mules are a cross between donkeys and horses; they are very hardy but cannot reproduce).

Figure 5.62 The violet species *Viola arvensis* (**A**) and *Viola tricolor* (**B**) are reproductively isolated because they always grow on different soil types.

It is not always clear cut!

Ruddy ducks (Figure 5.63A) and white-headed ducks (Figure 5.63B) are accepted as two different species. But when ruddy ducks were introduced from the US to Europe, they interbred to produce fertile hybrids. The white-headed duck is now an endangered species. Under the Convention on Biological Diversity, the UK is obliged to conserve the white-headed duck. The UK Government is introducing measures to protect the white-headed duck, including attempting to eradicate the ruddy duck.

 **a** Should ruddy ducks and white-headed ducks be considered to be one species or two?

b Should the number of ruddy ducks be controlled?

Figure 5.63 These ducks look very different, but are they actually the same species?

Did you know?

For some people the theory of evolution is controversial

Darwin had considered training to be a clergyman and knew full well that his ideas about evolution meant that he seemed to be challenging the Bible. The first two chapters of Genesis tell how all of the Earth's organisms were created by God about 6000 to 10 000 years ago, in much the same form that they exist today.

It is hard for many people to appreciate just how controversial Darwin's ideas were at the time. To this day there are many people in the world who cannot reconcile their religious beliefs and the theory of evolution. Few Muslims accept the theory of evolution and many Christians either reject it or aren't comfortable with it. In the UK around 10% of people are *creationists*. Creationists accept the literal teaching of the Bible, Qur'an or other scriptures and reject the theory of evolution. Creationists believe instead that God specially created Adam and Eve out of the dust of the earth so that humans are quite distinct from all other species. In the USA around 40% of people are creationists. Indeed, worldwide there is no doubt that the idea that all organisms are descended from a common ancestor that lived some three thousand million years ago – which is what evolutionary biologists believe – is a minority position.

The great majority of biologists, geologists and other scientists do accept a modernised version of evolution called *neo-Darwinism*. Neo-Darwinism combines Darwin's and Wallace's theory of natural selection with what we now know about inheritance. Many scientists who accept the theory of evolution in this modernised version also have firm religious beliefs. Such scientists have various ways of reconciling their religious beliefs with the theory of evolution. For example, they may believe that God created the original conditions that enabled the universe to come into existence, and then allowed evolution to take its course. In this understanding, God gives the whole of creation, including us, a certain freedom. Life is not predetermined but open-ended.

Weblink

You can find out more about control of ruddy ducks at the Department for Environment, Food and Rural Affairs website.

Extension

In **Extension 5.7** you can discuss why the theory of evolution is controversial using a role play. **X5.07S**

5.8 Getting the balance right

Key biological principle: The carbon cycle

The amount of carbon dioxide in the atmosphere, approximately 0.03–0.04%, is maintained by a balance between the processes that remove carbon dioxide from the air and those that add carbon dioxide to it. This circulation of carbon is known as the **carbon cycle** (Figure 5.64).

Activity

In **Activity 5.25** you can construct your own carbon cycle. **A5.25S**

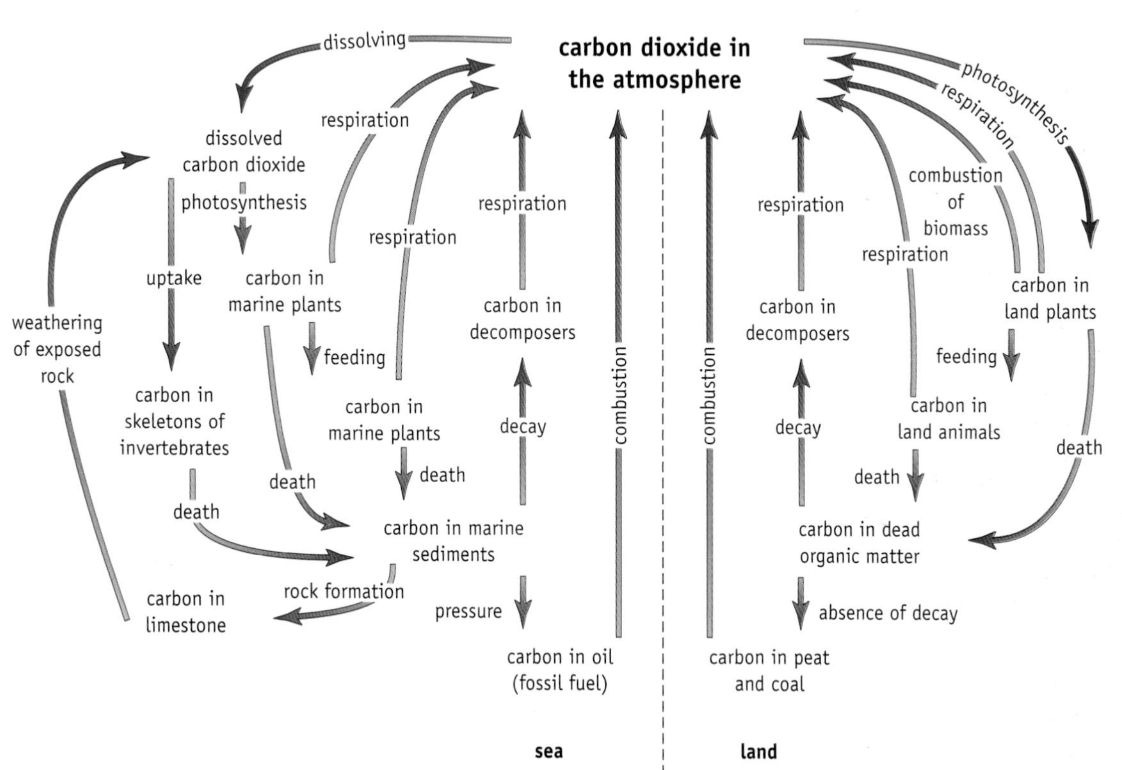

Figure 5.64 The carbon cycle.

Q5.51 For each definition below, decide which term in the carbon cycle shown in Figure 5.64 is being described.

a The reduction of carbon dioxide to organic substances such as sugars using light energy, which is carried out in the chloroplasts of plants and in the cells of some microbes. Some of these sugars are used by the organism directly to provide energy through respiration, some are converted to starch for storage, and the rest are used to make proteins, nucleic acids, lignin and cellulose. Oxygen is given off as a waste product.

b The total amount of living material per unit area. It includes animal and microbial (bacterial and fungal) mass, but by far the biggest proportion of it in an ecosystem is plant material. This includes the wood and foliage of trees and grass, and the countless millions of microscopic phytoplankton (plant plankton) that account for most of the photosynthesis of the oceans.

CONTINUED ▶

c The oxidation of organic substances such as sugars to simpler inorganic compounds such as carbon dioxide and water, with the release of biologically available energy. Carried out by all living organisms.

d Sooner or later plants and animals die and, along with animal faeces, are broken down. Carbon dioxide is released during the breakdown by microbes (types of fungi and bacteria) when they use the dead organic matter as a food to obtain energy by respiration.

e Oxidation of organic molecules outside living cells with the formation of carbon dioxide and water.

The role of microbes in the carbon cycle is vital. If there were no microbes to recycle plant biomass by giving out carbon dioxide during respiration, most of the carbon dioxide in the air would be used up in photosynthesis in less than ten years. This would mean the greenhouse effect would be greatly reduced, and the environment would soon become too cold to support life (not enough global warming). Photosynthesis would slow down greatly as temperatures fell, and so there would be slower plant growth and less food for animals.

Extension

Read **Extension 5.8** for more detail about decay. **X5.08S**

Out of balance

The fact that carbon dioxide concentrations have been rising for over 100 years indicates that the carbon cycle is not in balance. Two factors likely to be responsible for this are:

- combustion of fossil fuels
- deforestation.

Combustion of fossil fuels

We are currently burning fossil fuels much faster than they are forming. Table 5.7 shows the large amounts used in the UK. This represents a significant addition to the carbon dioxide already in the atmosphere.

Coal is a fossil fuel formed from the wood of trees that lived millions of years ago. Therefore it is a product of photosynthesis. The wood did not decay, releasing the carbon dioxide. This is because when the trees died they fell into conditions where fungi and bacteria could not grow, such as the anaerobic conditions of a swamp (Figure 5.65). Carbon remains locked in the coal instead of being returned to the air. Coal thus represents a **carbon sink**, and its accumulation involves the net removal of carbon dioxide from the air over millions of years.

Table 5.7 Energy use in the UK. *Source*: Department for Business Enterprise and Regulatory Reform.

Year	Total energy use/ million tonnes of oil equivalent*
1980	206.2
1990	221.6
2000	237.9
2001	238.7
2002	236.2
2003	236.8
2004	237.7
2005	236.0
2006	234.1

*Energy consumption corrected to allow for very cold or mild winters.

Figure 5.65 A swamp amongst prehistoric forest where coal might have formed 250 000 000 years ago.

When we extract and burn fossil fuel, the carbon released as carbon dioxide has been out of circulation for millions of years. Since the late eighteenth century people have been burning more and more fossil fuel, setting free carbon that has taken millions of years to accumulate.

Deforestation

Most mature forests are very stable ecosystems. They do not accumulate additional biomass or become fossil fuel, so they are not net absorbers of CO_2. Carbon dioxide uptake by photosynthesis is expected to be equal to the release of CO_2 due to respiration (including decay). If a forest were cut down, photosynthesis would drop, and although in the long term respiration would also drop, in the short term more carbon dioxide would be released than absorbed.

Figure 5.66 A recently felled rainforest. If the discarded wood were not burned, it would still decay and release just as much carbon dioxide.

Often only the large pieces of timber from big trees of the species being harvested are removed. Large amounts of branches, small trees, shrubs, and other plants that are of no use to the loggers are discarded. These are either left to rot away or burned (Figure 5.66), releasing carbon dioxide.

What else could upset the carbon dioxide balance?

There are other factors affecting carbon dioxide levels in the air:

- Volcanoes may release CO_2 – an increase in volcanic activity in the future could make a bigger difference.
- Carbon dioxide is continually being lost to sediments in the ocean by various processes such as the incorporation of carbon into the calcium carbonate skeletons and shells of marine organisms, but this is balanced by various erosion processes.
- An increase in acid rain might increase the rate at which CO_2 is released by erosion of limestone, but this is not thought to be a major factor upsetting the balance at present.

Extension

In **Extension 5.9** you can examine the carbon dioxide balance in greater depth. **X5.09S**

Rising global temperatures may themselves affect the carbon dioxide balance. For example, microbial decomposition of peat in soil releases carbon dioxide – an increase in global temperatures prolonging the summer thaw of Arctic tundra may increase decomposition of peat, increasing CO_2 release into the atmosphere. Warm water holds less carbon dioxide – an increase in sea temperatures would release carbon dioxide into the atmosphere.

The mystery of the disappearing carbon dioxide

It is estimated that burning fossil fuels adds 5.4×10^{12} kg of carbon per year to the atmosphere, and that deforestation adds approximately 1.6×10^{12} kg of carbon per year. However, the actual increase in carbon dioxide is only 3.0×10^{12} kg of carbon per year. Some carbon dioxide seems to be disappearing. There is still quite a lot of debate about why the carbon dioxide level doesn't rise as fast as we would expect. One explanation, suggested by the workings of the carbon cycle, is simply that more carbon dioxide means more photosynthesis.

In order to investigate the movement of carbon through an ecosystem, carefully controlled plots can be set up in artificial chambers called ecotrons. Results from these suggest that when CO_2 levels are raised, more carbon is stored in other components of the carbon cycle, for example within the soil as dissolved organic components. There is also evidence of a rise in the rate at which carbon dioxide is dissolving in the ocean. One reason for this could be increased photosynthesis by massive blooms of algae.

Maintaining the balance

Can we help maintain the carbon dioxide balance by:

- using biofuels
- reforestation?

Using biofuels

A **biofuel** is any source of energy produced, directly in plants or indirectly in animals, by recent photosynthesis. This provides a *renewable* energy source and is *carbon dioxide neutral*. For example, when wood burns, the carbon dioxide released to the atmosphere replaces the carbon dioxide previously absorbed through photosynthesis, which would have eventually been released through the processes of decay. The release of carbon dioxide in combustion does not therefore cause a net increase. The exception is when carbon dioxide is released in the process of transporting the biofuel from where it is made to where it is used.

Figure 5.67 Sunflower plants can be grown to supply oil as fuel.

Q5.52 Why is the use of biofuel carbon dioxide neutral?

Examples of biofuels include wood, straw, dried chicken litter, vegetable oil and methane. In 2004, Drax, Europe's largest coal-fired power station, started testing the use of fast-growing willow biomass mixed with coal. Another example of biofuel is ethanol from the fermentation of any kind of cheap and locally available sugar. In Brazil, waste from the refining of sugar cane is used, and the resulting alcohol added to petrol to make gasohol. In the UK we could grow sugar beet to provide the sugar to make alcohol as biofuel. Vegetable oil from, for example, sunflower seeds (Figure 5.67) is a very energy-rich biofuel that can be used as a substitute for diesel in motor vehicles.

At present the use of plant-derived alcohol and oil to power vehicles is limited, because it remains cheaper to use petrol or diesel. However, rising oil prices are now making biofuels more cost-effective. There has been a strong political movement to increase the use of biofuels, as these are seen as an important contribution to reducing carbon emissions. The European Union has a target of 10% biofuels in transport fuel by 2020, and the UK has its own target of 5% by 2010. However, some major disadvantages of biofuel production have become apparent.

The destruction of rainforests to plant palm oil trees for biofuel production is releasing vast amounts of stored carbon into the atmosphere, more than cancelling out any gains made by burning biofuels, and also harming wildlife. The use of large quantities of edible corn to make corn oil and ethanol for biofuel affects food availability.

Checkpoint

5.7 Make a table for and against the production and use of biofuels for transport. How might the following differ in their views about biofuels: developing countries, Western countries, farmers, oil companies?

A committee meeting of UK Members of Parliament in 2008 called for a moratorium on biofuel targets while the merits of biofuels were reassessed. They were seeking a ban on the use of forests, wetlands and permanent grasslands for biofuel production, as these habitats are rich in biodiversity and are not easily reinstated once lost. Fossil fuel reserves will not last forever, but are biofuels part of the answer? It seems the jury is still out.

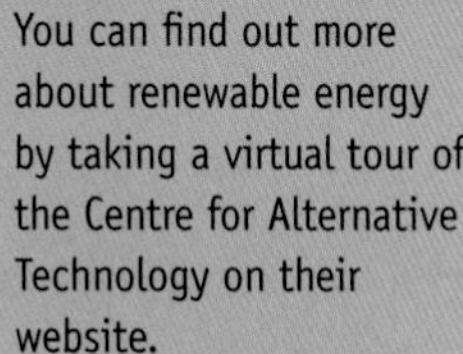

Weblink

You can find out more about renewable energy by taking a virtual tour of the Centre for Alternative Technology on their website.

Methane produced from anaerobic fermentation of human sewage can be used to generate enough electricity to make a modern sewage treatment plant energy self-sufficient. Methane can also be produced from domestic waste and from animal slurry. Methane produced in this way may be referred to as **biogas**, and when burned it is CO_2 neutral. In this it is like solid biofuel, and unlike fossil fuel methane from the North Sea.

Reforestation

In a newly planted forest all the trees are young; all are growing rapidly, turning carbon dioxide into wood. There is very little old wood and relatively little decay, so respiration will be less than photosynthesis. This means that the system is a net absorber of carbon dioxide. As the plantation gets older, the system will move towards a balance between photosynthesis and respiration, and will no longer be a net absorber. It becomes a carbon store with carbon locked up in the biomass.

The Earth's forests could soak up extra carbon dioxide, as increased CO_2 concentrations and higher temperatures stimulate photosynthesis, leading to extra growth. However, there will be a limit to how much or for how long the world's forests can soak up extra carbon dioxide. In the long term more vegetation means more food, therefore more animals and more decay-causing microbes, hence more respiration producing more carbon dioxide. There is an upper limit to tree growth, and the land available for forests is also limited. It has been suggested that an increase in mean global surface temperatures above a certain level (perhaps an increase of 5 or 6 °C) could reduce water availability in rainforests. This would cause them to go into decline and accelerate further increases in atmospheric carbon dioxide, resulting in even more global warming.

Activity

Use **Activity 5.26** to check your notes at the end of this topic. **A5.26S**

Both the microscopic algae in the ocean and the Earth's forests could act as carbon sinks, soaking up extra carbon dioxide and incorporating it into new biomass. Planting trees, along with a range of other actions such as less use of fossil fuels, improving energy efficiency, and capturing carbon dioxide and putting it in the oceans, may all help slow down further increases in atmospheric carbon dioxide concentrations. There are solutions to this complex problem, but they will affect nations' economies and how we live our lives.

Review

Now that you have finished Topic 5, complete the end-of-topic test before starting Topic 6.

Infection, immunity and forensics

Why a topic called Infection, immunity and forensics?

Two deaths, two bodies discovered. There is a mystery: who were the dead man and woman? Forensic biologists and pathologists use a wide range of techniques to help answer such questions, and to determine when people died, and what caused their untimely deaths.

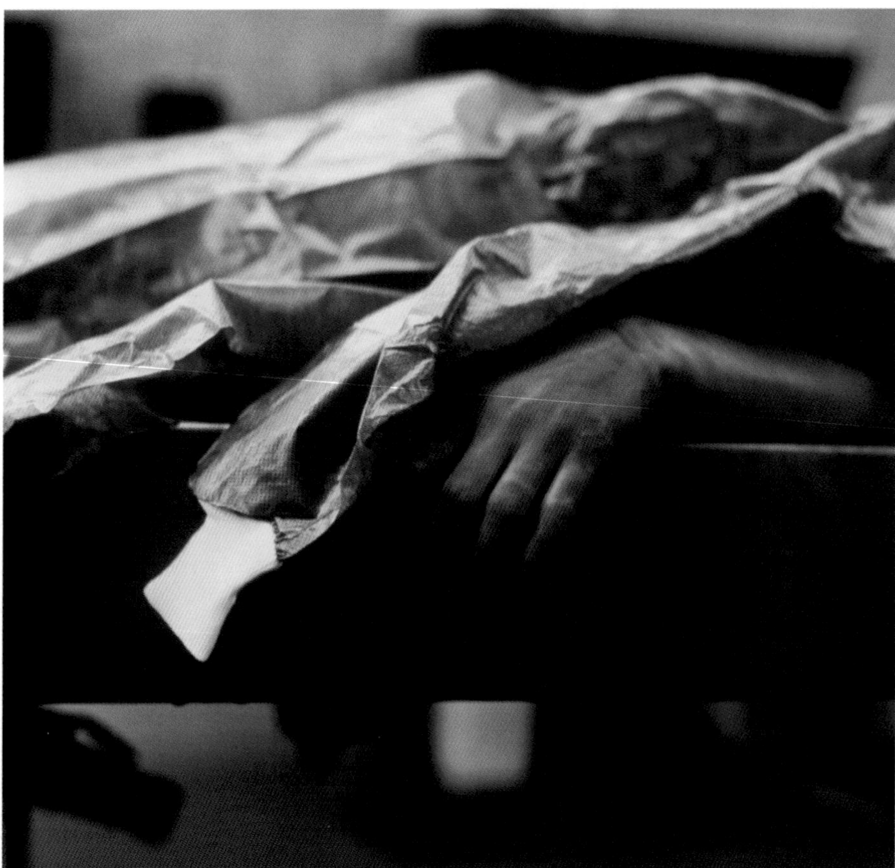

Figure 6.1 Whose body? When and how did they die?

The villains in this particular case were not violent criminals or gun-carrying muggers who made an obvious assault. The victims may have died because of the genes they inherited, the food they ate, or other aspects of their lifestyle. Or microbes may have stolen into their bodies undetected. If a disease-causing pathogen enters your body, it may soon make itself at home, in which case you will probably know all about it.

Usually an invading microbe will be nothing more than a nuisance, causing a cold or sore throat. The immune system swings into action and deals with the uninvited guests. Specialised cells and chemicals, functioning like a well-organised army, destroy the invaders. But sometimes the microbes are so numerous that, without help, the immune system is overwhelmed and the body faces disaster. Did this happen to either of our victims?

Antibiotics, discovered in a lucky accident by Alexander Fleming in 1928, revolutionised the treatment of bacterial diseases. In the 1940s it was envisaged that toothpaste and even lipstick would be impregnated with antibiotics, protecting us from infection. Antibiotics were thought to have the potential to eradicate most of the major infectious diseases. Why has this proved not to be the case? How is the evolution of microbes causing drug developers and hospitals a real problem? Despite this problem, antibiotics do remain an important weapon in the fight against infection. So how do they work?

It is possible to avoid infection by taking preventative measures. Being vaccinated as a child or when travelling abroad as an adult may be the answer. Could this solution have helped save our victims?

Overview of the biological principles covered in this topic

In this topic, you will look at how a dead body can provide evidence for forensic biologists as they try to identify a person, and decide when and how the person died. You will use your knowledge of DNA gained in the AS course to study DNA profiling. In Topic 5 you learned about the principles of succession; here they underpin some of the techniques used by forensic entomologists to determine the time of death. In Topic 5 you also learned about decomposition and recycling by microorganisms; this also provides evidence for how long a person has been dead.

You will revisit the structure of prokaryotic cells in the context of pathogens. You will look at the symptoms people have when they are infected by TB and HIV. You will gain a detailed understanding of the immune system's role in the body's response to infection.

In this topic you will also consider the principles involved in the evolutionary race between pathogens and their hosts (those they infect!). You will examine the evolution of antibiotic resistance in bacteria, and see how drug developers battle to find new treatments. You will also consider the actions being taken by hospitals to overcome the problem of antibiotic resistant bacteria.

6.1 Forensic biology

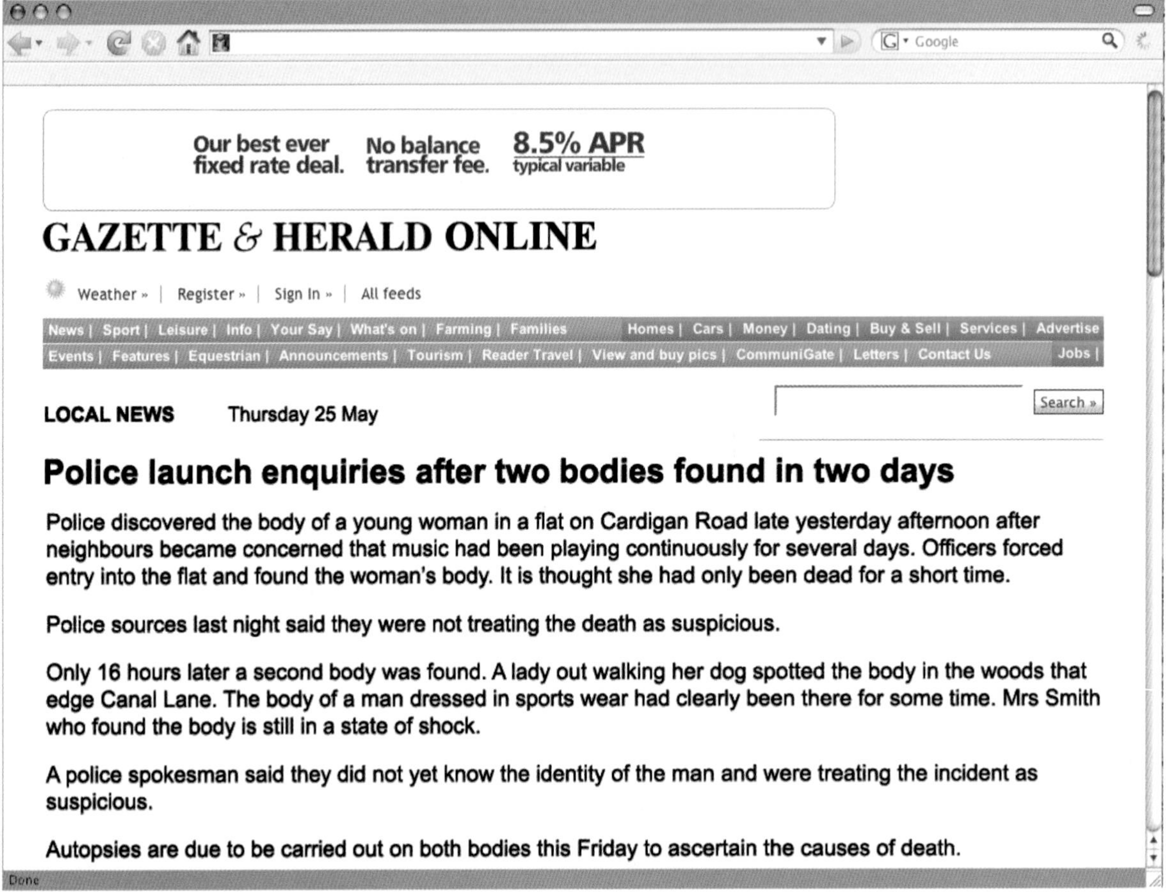

Figure 6.2 Bodies found.

On finding any body (Figure 6.2) the police must first answer three basic questions:

1 Who is the dead person?

2 When did the person die?

3 How did the person die?

Forensic biologists use a variety of techniques to help the police answer these questions.

Identifying the body

Q6.1 What do you think the police would do first to identify a body?

The identity of a person is normally fairly easy to establish, using the papers that most of us carry such as a diary, bank card, bus pass, or receipts. The woman found in Cardigan Road was quickly identified from documents including the photo on her driving licence. She was Nicki Overton, 23 years old, just back from a trip to the US and staying in her friend's flat for a few weeks while the friend visited the Far East. However, if there are no documents, as with the Canal Lane body, forensic techniques may have to be used. Conventional fingerprinting, dental records and, increasingly, DNA profiling are used.

Did you know?

Conventional methods for identification

Fingerprints

Fingerprints are found on all humans and some other animals. They are unique to the individual and don't change over a lifetime. Even identical twins have different fingerprints. Fingerprints are small ridges caused by folds in the epidermis of the skin (see Figure 6.3). The ridges vary in width and length; they can branch or join together forming distinctive patterns.

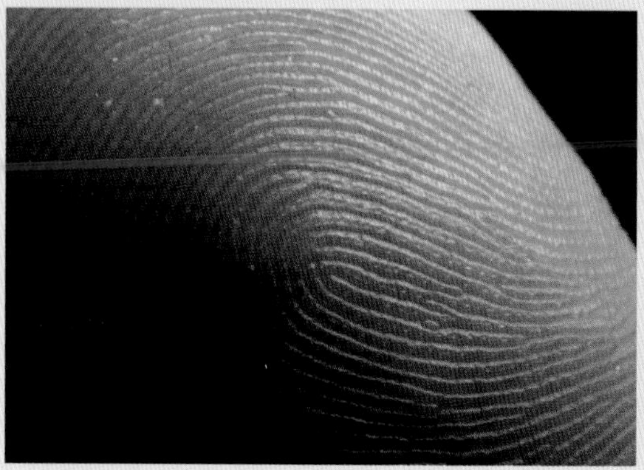

Figure 6.3 Fingerprints form early on in human foetal development, usually by the fifth month.

Sweat and oil secretions leave impressions of our fingerprints on any surface we touch. The oils are secreted from sebaceous glands. There are none of these on our palms or fingers, but the oils are easily transferred to our hands when we touch other parts of our body. There are four main types of fingerprint pattern (Figure 6.4). This is known as the Henry Classification. Arch patterns are rare – the loop is the most common.

A Arch – ridges rise in the centre in a general arching formation.

B Tented arch – one upthrusting ridge tends to cut through the ridges above at right angles, more or less. The arch rises to an angle of 45° or more.

C Whorl – a circular pattern of ridges with at least one ridge making a 360° circle in the centre of the print.

D Loop – ridges make a looping pattern, exiting the same side as they entered.

Figure 6.4 The four main types of fingerprint pattern.

CONTINUED ▶

Weblink

Look in detail at slides through the skin on the University of Illinois College of Medicine interactive histology website.

Fresh prints are made visible using various techniques, depending on the surface and the clarity of the print. Methods include: using fine aluminium, iron or carbon powders that stick to the print; using superglue, which reacts with the water and other substances left by the print; using ninhydrin, which reacts with the amino acids in sweat to produce a purple-coloured fingerprint impression on absorbent surfaces; and using vacuum metal deposition to cover the print with a thin layer of a non-reactive metal. Magnets and iron flakes are sometimes used, as shown in Figure 6.5. Once fingerprints have been obtained, they will be examined by an expert who looks for minute details in the ridge patterns and compares them with other recorded prints. At least 16 points must be identical for fingerprints to match.

Figure 6.5 Iron flakes with an organic coating are applied using a magnetic 'brush'. The iron flakes stick to the grease in the fingerprint. The magnet removes excess powder and, because a brush never physically touches the print, a sharper result can be obtained. This is particularly useful for difficult surfaces.

Computerisation has automated the process and speeds up searching for a match. The UK police use a national computerised database of biometric information called IDENT1. This holds over seven million 'ten print' records and nearly three million palm prints. The same database will also hold other biometric information including iris and facial recognition data. Everyone who is arrested has their prints taken and these are stored on IDENT1, even if the person is later found to be innocent. Some people have concerns about the storage of such a large amount of personal data.

Dental records

Dental records are often the best way to identify individuals who have no fingerprints on file, or whose bodies have been damaged making other means of identification difficult. Teeth and fillings decay very slowly, and are much more resistant to burning than skin, muscle or bone. Dental records can be as reliable as fingerprints for identifying remains, although records may sometimes include dental work that has not been completed.

Q6.2 Look at the dental X-ray in Figure 6.6 and match it to the correct person's records. In reality, a person's actual teeth would be compared with their records.

CONTINUED ▶

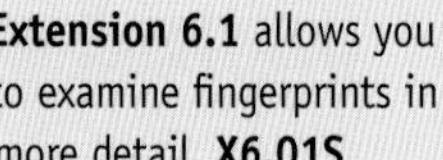

Extension

Extension 6.1 allows you to examine fingerprints in more detail. **X6.01S**

Weblink

Listen to the Radio 4 series on fingerprints by visiting their website.

Find out what the UK Home Office Science and Research team say about biometrics and fingerprinting by visiting their website.

A forensic dentist makes an accurate chart of the teeth, including fillings, other dental work and any missing teeth. This can then be compared with dental records of missing persons. Both paper dental records and dental X-rays are useful in the identification process. The forensic dentist may also look at the development of the teeth and roots, to help determine the age of the unidentified person and narrow down the search.

Dental records were important in the identification of victims of the World Trade Centre attacks in New York on 11 September 2001.

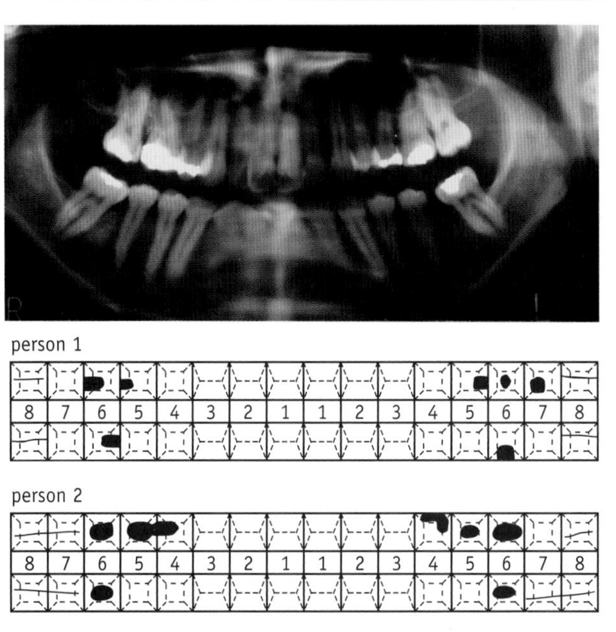

Figure 6.6 Dental X-ray and records. The light marks show fillings, which are highly opaque to X-rays.

Extension

Extension 6.2 uses dental records and other procedures to identify skeletons of the Russian royal family. **X6.02S**

DNA profiling

DNA profiling (also known as genetic fingerprinting and DNA fingerprinting) relies on the fact that, apart from identical twins, every person's DNA is unique. As you saw in Topic 3, the human genome contains about 23 000 genes, with an average size of 3000 base pairs per gene. This should give about 7×10^7 base pairs in total. However, there are over three billion base pairs (3×10^9) in the genome, so it is clear that a large amount of the DNA does not code for proteins. The non-coding blocks are called **introns** and are inherited in the same way as any genes within the coding regions (**exons**).

Within introns, short DNA sequences are repeated many times. The sequences of repeated bases are known as **short tandem repeats** (**STRs**) or satellites. An STR can contain from 2 to 50 base pairs, and can be repeated from five to several hundred times.

The same STRs occur at the same place (locus, plural loci) on both chromosomes of a homologous pair. However, the number of times they are repeated on each of the homologous chromosomes can be different, as shown in Figure 6.7. The number of repeats at a locus also varies between individuals.

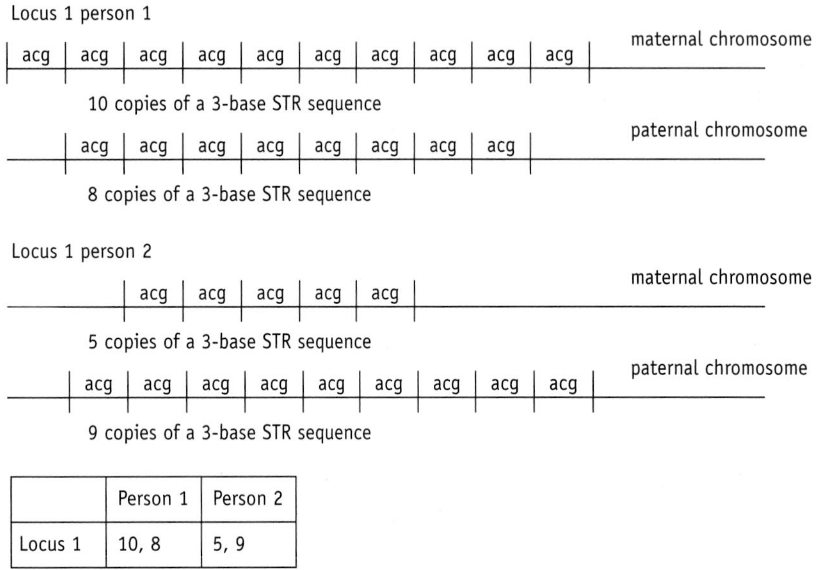

	Person 1	Person 2
Locus 1	10, 8	5, 9

Figure 6.7 Introns at the same locus on each chromosome of a homologous pair may contain different numbers of STRs.

Q6.3 Look at the DNA sequence shown below. Identify the short tandem repeat sequence. Count how many times it is repeated.

aaccatccttagtaagtaagtaagtaagtaagtaagtattgcccat

Table 6.1 shows how the number of repeats might differ at five STR loci for two individuals. Each person has a large number of introns with lots of STR loci. There is a large amount of variation in the number of repeats at each locus. These facts combined mean that two individuals are highly unlikely to have the same combination of STRs. It is this important feature that enables scientists to create a virtually unique DNA profile.

Q6.4 Look at the STRs from two loci shown below. Compare them with the information in Table 6.1 and decide if they are from person A or B. Explain your answer.

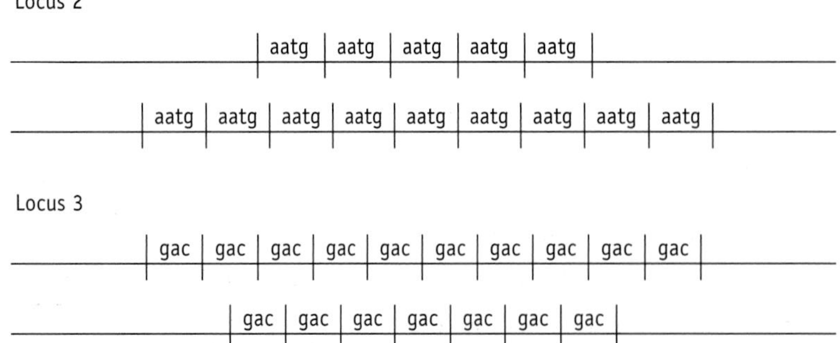

Locus	Person A	Person B
1	15, 14	12, 13
2	5, 9	11, 11
3	10, 7	10, 8
4	7, 7	5, 13
5	8, 10	7, 6

Table 6.1 Number of repeats two of the same numbers at five loci for two individuals. If the two numbers at a locus are the same, it means that the person has the same number of repeats on both chromosomes.

How is a DNA profile made?

DNA profiling is a multi-stage process. A tissue sample must be obtained and the DNA extracted. Sufficient DNA to manipulate in the lab must be available. Fragments of different lengths are created by cutting up the DNA. The fragments are then separated and visualised in some way; finally the profile created is compared with another. A DNA profile on its own is useless – there has to be a reference profile for comparison. The reference DNA profile might come from a suspect in a murder investigation, a relative in the case of identifying a corpse, or from the parents when establishing paternity. The same principles of separating DNA fragments are used when screening for genetic conditions like cystic fibrosis (see Topic 2), measuring genetic diversity (Topic 4), or studying evolutionary relationships (Topics 4 and 5).

Obtaining the DNA

A DNA sample can be obtained from almost all biological tissue, animal or plant. It could come from cells obtained in a cheek swab at a police station or medical clinic, from white blood cells in a blood smear found at the scene of a crime, from bone marrow in a skeleton, or from sperm left after a sexual assault. DNA extraction is relatively simple – you may have carried out Activity 2.13 to extract DNA from plant material like onions, kiwi fruit or peas. The tissue sample is physically broken down in a buffer solution that includes salt and a detergent to disrupt the cell membranes. The small suspended particles, including the DNA, are separated from the rest of the cell debris by filtering or centrifuging. Protease enzymes are incubated with the suspension to remove proteins, and then cold ethanol is added to precipitate out the DNA. Several stages of washing the DNA in a buffer solution then follow.

Creating the fragments

The original DNA profiling technique, developed by Alec Jeffreys at Leicester University in the mid 1980s, involved treating the DNA sample with **restriction enzymes** (or more correctly, restriction **endonucleases**). These enzymes are found naturally in bacteria, where their function is to cut up invading viral DNA. The enzymes will only cut DNA at specific base sequences, usually four or six base pairs long. This is why they are valuable to molecular biologists and forensic scientists. If the restriction sites are either side of a short tandem repeat sequence, that fragment of DNA will remain intact; but it will be cut away from the rest of the genome. The repeated sequences remain intact, as shown in Figure 6.8.

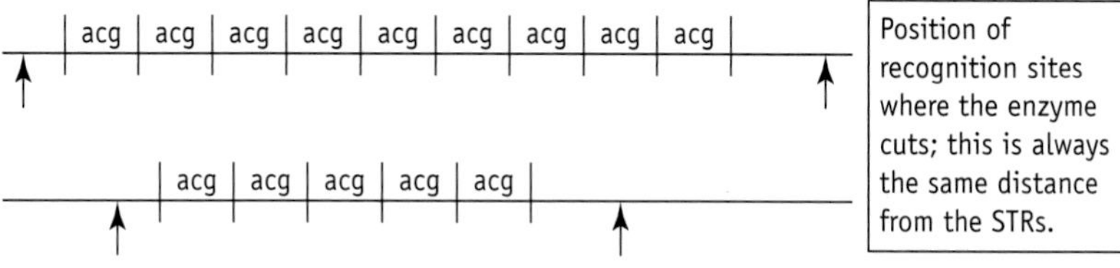

Figure 6.8 Restriction enzymes recognise specific DNA sequences on either side of the repeated sequence.

The names of restriction enzymes reflect the bacteria from which they originate. The enzyme *Eco*RI comes from the bacterium *Escherichia coli* strain RY13. It cuts DNA between the G and A bases wherever the sequence GAATTC occurs (Figure 6.9).

G|A A T T C
C T T A A|G

Figure 6.9 *Eco*RI always cuts DNA at this site. Each restriction enzyme cuts at a specific sequence.

Bacteria protect their own DNA sequences from restriction enzymes. They do this by changing the bases in the sequences that are targeted by their own restriction enzymes.

In the laboratory, restriction enzymes are used as targeted 'scissors'. They cut a DNA sample into fragments only where their specific restriction sequence occurs. If the same restriction enzyme is used to cut two identical DNA samples, identical STR fragments are produced.

Polymerase chain reaction

To allow forensic scientists to use tiny deposits of hair, skin or body fluid for identification purposes DNA is copied numerous times using the **polymerase chain reaction** (PCR). The process uses **DNA primers**. These are short DNA sequences complementary to the DNA adjacent to the STR. The DNA primers are marked with fluorescent tags. The forensic sample is placed in a reaction tube with **DNA polymerase**, DNA primers and nucleotides. Once in the PCR machine, the tube undergoes a cycle of temperature changes (Figure 6.10). The first separates the double stranded DNA. The second temperature optimises prime binding to the target DNA sequence in the sample. The polymerase attaches and replication occurs, as shown in Figure 6.11. The final temperature is the optimum temperature for the heat-stable DNA polymerase.

As the cycle continues, huge numbers of the targeted DNA fragments are produced.

The UK Forensic Science Service generally analyses DNA samples for the presence of ten short tandem repeat sequences, also known as micro-satellites. Each target STR is four bases in length. An additional primer is used to determine gender; it targets a sequence on the sex chromosome.

Activity

The animation in **Activity 6.4** lets you visualise what is happening in the PCR reaction, or you might be able to carry out PCR practically for yourself. **A6.04S**

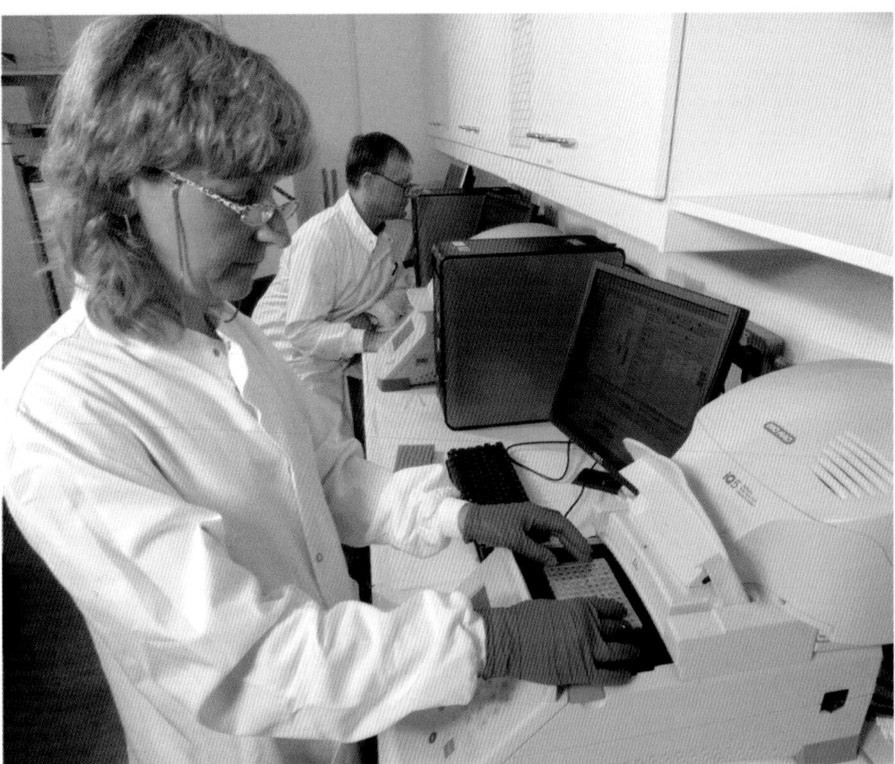

Figure 6.10 The reaction tubes are placed in the PCR machine.

A

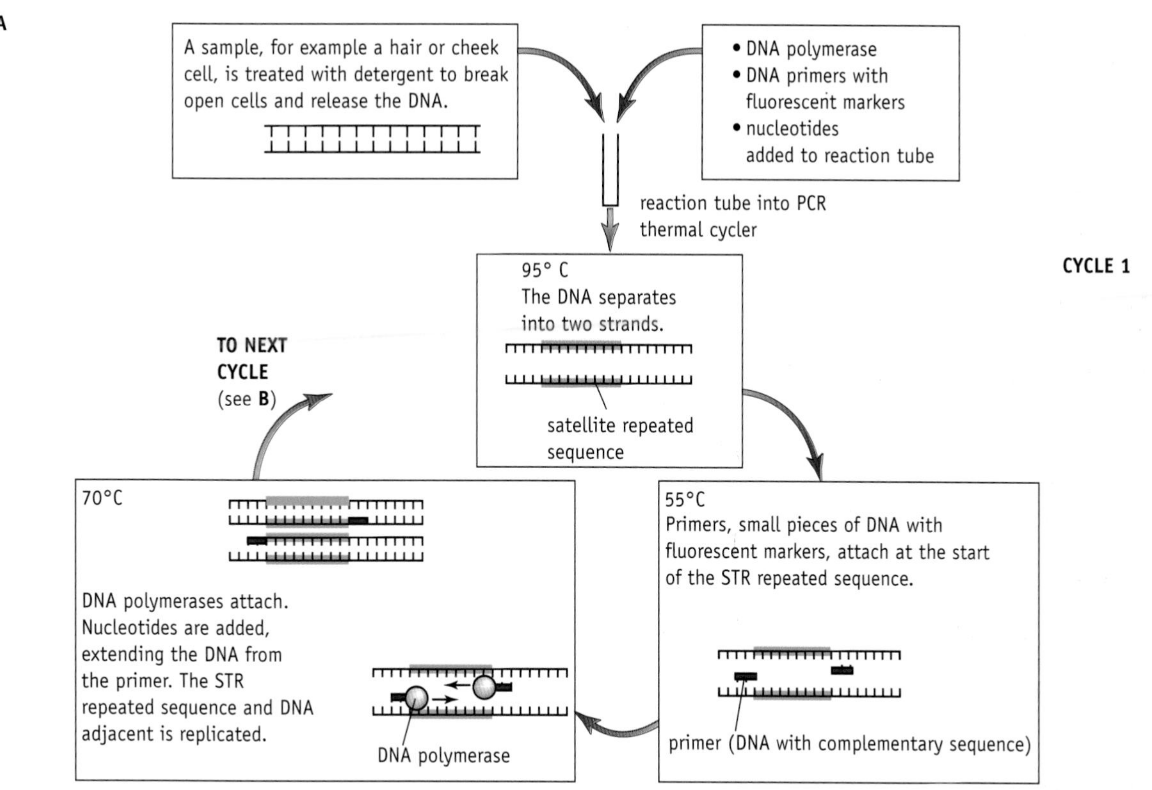

B

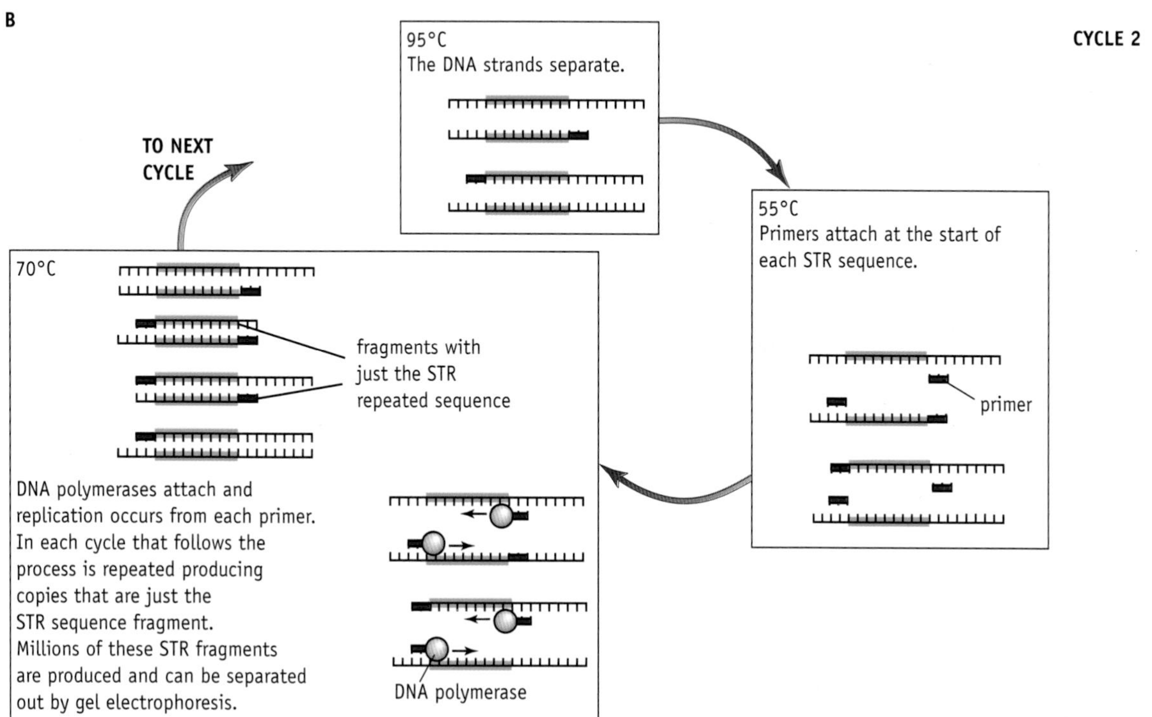

Figure 6.11 The polymerase chain reaction is used to multiply DNA samples to produce sufficient DNA for separation by gel electrophoresis. The sample goes through many cycles of temperature change, and **A** and **B** show what happens in the first two cycles. The exact temperatures used will depend on which species the DNA comes from. The whole cycle is repeated 25–30 times, taking about three hours and producing about 30 million copies of the DNA. 28 cycles are used for standard DNA profiling, 32 cycles are used when the DNA is not visible to the naked eye.

Separating the fragments

DNA fragments produced by restriction enzymes or PCR can be separated by **gel electrophoresis** according to their size (Figure 6.12). The DNA is placed on a gel of agarose or polyacrylamide, which both provide a stable medium through which the fragments can move. The gel is submerged in a buffer solution, and connected to electrodes that produce a potential difference across the gel. The negatively charged DNA fragments migrate through the gel according to their overall charge and size. Smaller fragments, with smaller numbers of repeat sequences, travel faster. In a given time, smaller fragments end up closer to the positive electrode. A reference sample with fragments of known length may be added to the gel. This is known as a DNA ladder or marker. The fragment lengths are measured in base pairs.

Visualising the fragments

The gel is quite fragile, and the DNA fragments are double-stranded after electrophoresis. Southern blotting (named after Ed Southern who developed the technique) is used to transfer the fragments to a more resilient nylon or nitrocellulose membrane. The membrane is placed directly onto the gel and a wad of dry absorbent paper placed on top. This acts as a wick to draw buffer solution up through the gel, carrying DNA fragments onto the membrane. During this process, the fragments maintain their positions relative to each other and are denatured into single strands, exposing the base sequences.

The membrane is then incubated with an excess of a labelled DNA probe. This is a short section of DNA with a base sequence complementary to the target DNA sequence needing to be located. After allowing time for the probe to bind to any complementary sequences (hybridise), any unbound probe is washed away. Probes may be radioactive, labelled with radioactive phosphorus (^{32}P), or labelled with a fluorescent marker. With a radioactive probe, the membrane is dried and placed next to X-ray film. The film blackens wherever the probe has bound with the DNA to form double-stranded fragments. If the probe is fluorescent, its position on the membrane can be visualised under ultraviolet light.

A single band occurs on the profile where a person's maternal and paternal chromosomes have the same number of repeats at a particular locus. Two bands occur on the profile if the two chromosomes have a different number of repeats at a locus.

Activity

Use the simulation in **Activity 6.1** to try out gel electrophoresis using restriction enzymes. **A6.01S**

Or try this practically in **Activity 6.2. A6.02S**

Q6.5 A restriction enzyme is used to cut out the STR sequences shown in Figure 6.7. The fragments are then separated using gel electrophoresis. How many bands would be produced on a DNA profile for **a** person 1 **b** person 2?

A unique banding pattern for each person is created if probes for many different repeated sequences are used. The pattern produced is the familiar banded DNA profile such as that shown in Figure 6.12 opposite. Although the combination of lengths of STR sequences is unique to an individual's DNA, the DNA profile may not be unique if only a few probes are used (see page 83).

Activity

Summarise the processes involved in making a DNA profile by completing **Activity 6.3. A6.03S**

Q6.6 Explain what determines the distance travelled by the DNA fragments.

Q6.7 When DNA from person 1 in Figure 6.7 was analysed for the acg repeat sequence, the banding pattern shown in Figure 6.11 was produced. Copy the pattern. Sketch where the bands for person 2 would be located relative to these bands.

A

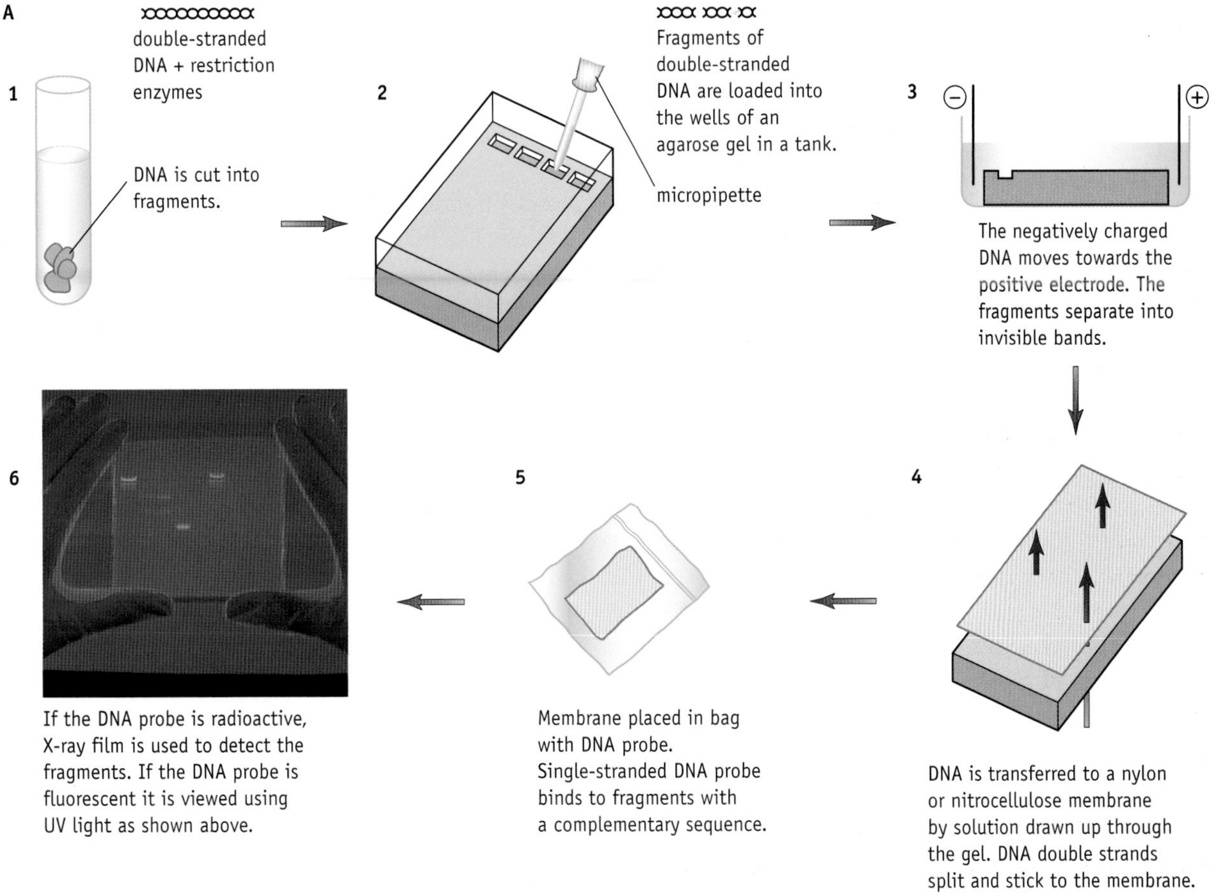

1 double-stranded DNA + restriction enzymes

DNA is cut into fragments.

2 Fragments of double-stranded DNA are loaded into the wells of an agarose gel in a tank.

micropipette

3 The negatively charged DNA moves towards the positive electrode. The fragments separate into invisible bands.

4 DNA is transferred to a nylon or nitrocellulose membrane by solution drawn up through the gel. DNA double strands split and stick to the membrane.

5 Membrane placed in bag with DNA probe. Single-stranded DNA probe binds to fragments with a complementary sequence.

6 If the DNA probe is radioactive, X-ray film is used to detect the fragments. If the DNA probe is fluorescent it is viewed using UV light as shown above.

B

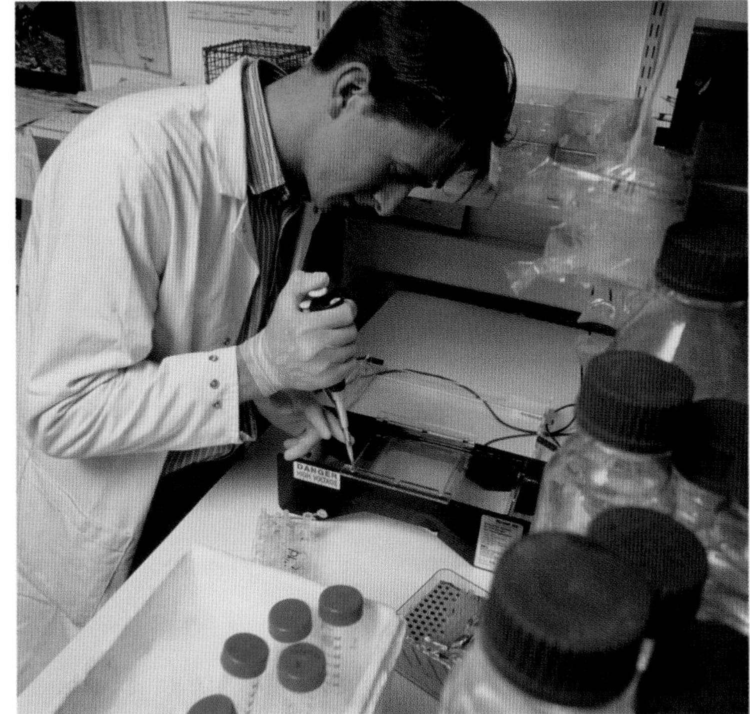

Figure 6.12 **A** The steps involved in producing a DNA profile. DNA profiles produced in this way look rather like supermarket bar codes. The Forensic Science Service now use fluorescent probes to produce a graph-style profile (Figure 6.14). In school and college laboratories the bands may be visualised on the gel by staining with a dye that attaches to the double-stranded DNA. **B** Loading the gel.

The fragments of DNA produced by PCR are analysed using gel electrophoresis. The system used by the Forensic Science Service is automated, with the position of the DNA fragments revealed as a pattern of fluorescent bands – these are caused by the fluorescent tags on the DNA primers attached to the target DNA. The fluorescent bands are detected using a camera or laser scanner.

Figure 6.13 Banded DNA profile.

The results of the gel electrophoresis procedure are produced as a graph similar to the one in Figure 6.14. The position of each peak along the x-axis corresponds to the size of the DNA fragment, and the height of the peak indicates the amount of that DNA in the sample. The colour of the peaks relate to the fluorescent-coloured tag attached to the primer. The graph is interpreted to give a series of numbers corresponding to the number of repeats in each fragment. This looks different from the 'bar code' type of profile, but the information both formats contain is the same. The graphical format is more compatible with an automated computerised system. Table 6.2 shows a digital DNA profile. There are two different figures for some of the satellites, because the number of repeated sequences is different on each chromosome in a homologous pair.

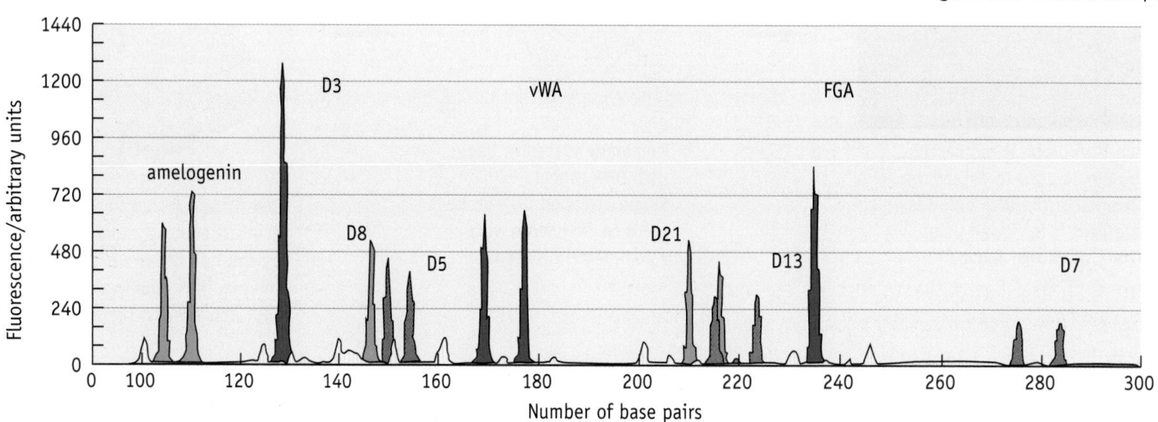

Figure 6.14 A graphical DNA profile. For each STR locus there will be one or two peaks. If two peaks occur it means the repeated sequence inherited from each parent are different lengths. The two peaks can be expressed as two numbers indicating the number of repeated units. A single peak shows that the same number of repeated units were inherited from both parents. Fluorescent peaks can be converted to a digital readout. *Source:* University of Central Lancashire.

Amg	D3	D8	D5	vWA	D21	D13	FGA	D7
XY	10, 10	16, 16	8, 9	13, 15	15, 16	18, 20	30, 30	14, 16

Table 6.2 A digital DNA profile for the nine STRs above would produce something like this. Amg indicates male or female.

Using the DNA profile

The STRs are inherited in the same way as alleles of a gene, with offspring receiving one repeated sequence randomly from each parent. This means that genetic profiling can be used for such things as identification purposes, settling paternity disputes, identifying stolen animals, and looking at variation and evolutionary relationships between organisms.

Q6.8 **a** Poppy is a racing pigeon and her chick has been stolen. Her owner, a successful breeder of racing pigeons, suspects that a fellow breeder has stolen the chick and is trying to pass it off as one of his bird Patsy's own brood. Genetic profiling for three satellites produced the results outlined in Table 6.3. Is the chick the offspring of Poppy or Patsy? Give a reason for your answer.

Checkpoint

6.1 Draw up a flow chart that outlines the steps required in DNA amplification and separation to produce a DNA profile.

Satellite	Number of repeats for each satellite		
	Chick	Poppy	Patsy
A	6,12	5,12	10,6
B	7,4	4, 4	8,5
C	9,5	9,13	6,11

Table 6.3 Digital genetic profile for three pigeons.

b Look at Figure 6.15. The bands in these DNA profiles are marked M for mother, F for father and C for child. Decide if the father is the biological parent of both children. Give reasons for your answer.

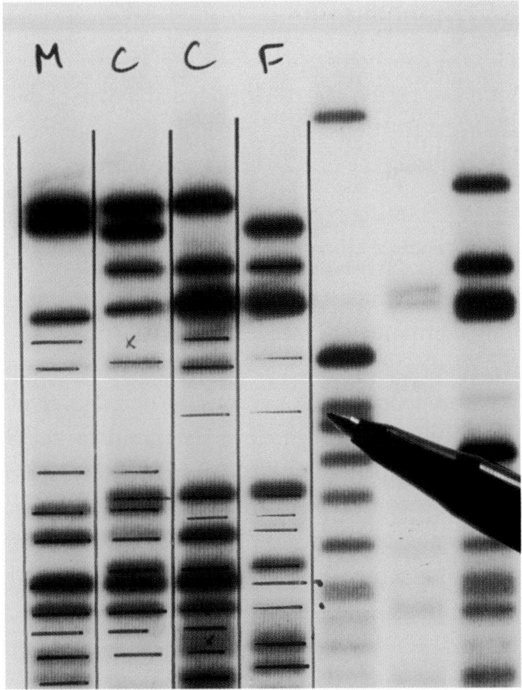

Figure 6.15 DNA profiling is frequently used for settling disputes about paternity.

c A homeowner is concerned about several large cracks running across the back wall of his house. They seem to be getting worse, and he thinks they are caused by tree roots affecting the foundations. He has done some digging, and suspects that his neighbour's tree is the culprit. The neighbour denies this, insisting that one of the trees in the homeowner's own garden is responsible. How could the neighbour establish who is right?

d Basmati rice is more expensive than other types of rice. The name Basmati can only be used for certain varieties of long grain rice grown in India and Pakistan. The Rice Association says that Basmati rice must not contain more than 7% non-Basmati rice but cheaper rice varieties are often mixed with Basmati and sold as Basmati. In tests on 382 UK Basmati samples, 46% were found to contain more than 7% non-Basmati rice; some samples were more than 60% non-Basmati. DNA profiling is used to determine if samples exceed the 7% limit. Suggest how DNA profiling can be used to check the quality of rice samples.

e Separating DNA fragments is used in screening for genetic conditions and measuring genetic diversity. Suggest how fragments might reveal the genotype of a person with the genetic condition.

Is DNA profiling infallible?

Genetic profiling has been widely used in legal proceedings to match samples from a crime scene to samples from an individual. It is generally thought to produce a result that is almost unique to the individual, and a near certain indication of guilt in criminal trials. A complete DNA profile would be unique (even for an identical twin, due to the accumulation of mutations during a person's life). But because the DNA profile analyses only a few repeated sequences, it is less likely to be completely unique. This is a particular problem if individuals being tested are closely related. In court a forensic scientist estimates the chance of seeing the same DNA profile in the general population; this is usually in excess of one in a billion.

DNA taken from the body in Canal Lane was compared with hair DNA samples collected from the homes of two people reported missing in the local area. The dead man was discovered to be George Watson, a 55-year-old computer software analyst.

Q6.9 Why would a DNA profile presented as evidence in court need to include ten or more short tandem repeats (STRs)?

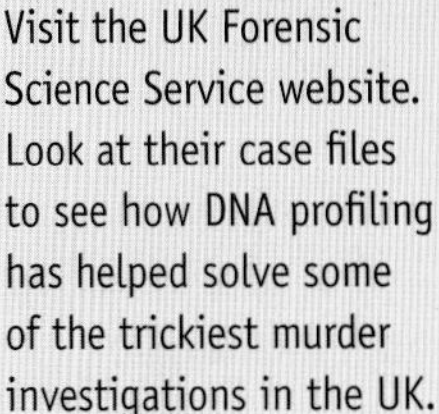

Weblink

Visit the UK Forensic Science Service website. Look at their case files to see how DNA profiling has helped solve some of the trickiest murder investigations in the UK.

Determining time of death

As soon as a person dies, a series of physical and chemical changes start to take place in their body. These changes occur in a known order, and can be used to estimate the time of death. The temperature of the body, the degree of **rigor mortis**, and the state of decomposition can be used to estimate time of death. In addition, entomological (insect) evidence may provide further clues to help identify when the person died.

Body temperature

Human core body temperature is normally in the range of 36.2 to 37.6 °C. But as soon as a person dies, their body starts to cool due to the absence of heat-producing chemical reactions. The temperature of a body can be useful for estimating time of death during the first 24 hours post-mortem.

Q6.10 Measurement of body temperature to estimate time of death is useful in cool and temperate climates. Why might it be less useful in the tropics?

Core body temperature is measured via the rectum or through an abdominal stab. A long thermometer is needed; an ordinary clinical thermometer is too short and has too small a temperature range. An electronic temperature probe can also be used. Environmental conditions must be noted, as these will affect how the body has cooled. The cooling of a body follows a **sigmoid curve**, as shown in Figure 6.16. The initial temperature plateau normally lasts between 30 and 60 minutes. The graph assumes that the person's temperature was normal, 37 °C, at the time of death. However, this may not always be the case. If a person has a fever or is suffering from hypothermia, their body temperature at the point of death will be elevated or depressed.

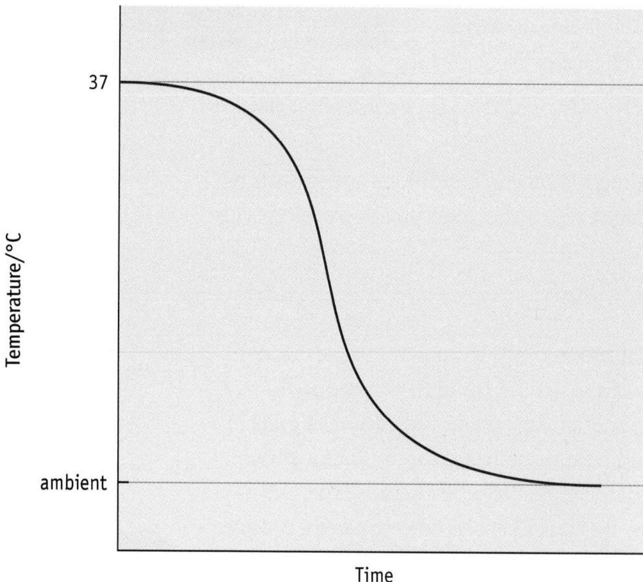

Figure 6.16 The temperature of a body will follow this sort of sigmoid curve as it cools after death. The initial plateau can last from 30 minutes to several hours depending on conditions.

Over the near linear part of the cooling curve, the temperature decline per hour can be used to give an estimate of the time of death. Many factors will affect post-mortem cooling, including:

- body size
- body position
- clothing
- air movement
- humidity
- temperature of surroundings.

If the body is immersed in water it will cool much more rapidly, as water is a better conductor of heat than air. These factors need to be taken into account when estimating time of death from temperature measurements.

Q6.11 For each of the factors mentioned in the above paragraph say how it might affect the rate of cooling.

Q6.12 The Canal Lane body was found to have a core temperature of 18 °C. Why might this measurement have been of little use to the forensic pathologists investigating the case?

Rigor mortis

After death, muscles usually totally relax and then stiffen; this stiffening is known as rigor mortis (meaning 'stiffness of death'). Joints become fixed during rigor mortis and their position, whether bent (flexed) or straight (extended), will depend on the body position at the time of death. After a further period of time, rigor mortis passes and muscles are again relaxed. The following sequence of events occurs.

1 After death, muscle cells become starved of oxygen, and oxygen-dependent reactions stop.
2 Respiration in the cells becomes anaerobic and produces lactic acid.
3 The pH of the cells falls, inhibiting enzymes and thus inhibiting anaerobic respiration.
4 The ATP needed for muscle contraction is no longer produced. As a result, bonds between the muscle proteins become fixed.
5 The proteins can no longer move over one another to shorten the muscle, fixing the muscle and joints.

There is a progression in the development of rigor mortis, with smaller muscles stiffening before larger ones. The rigor mortis passes off as muscle tissue starts to break down, in the same order in which it developed.

Most human bodies will have complete rigor mortis six to nine hours after death. (Table 6.4 shows a simple rule of thumb.) However, rigor mortis will set in more quickly and last for a shorter period if environmental temperature is high, or if the person has been physically active before death.

Temperature of body	Stiffness of body	Approximate time since death
Warm	Not stiff	Not dead more than 3 hours
Warm	Stiff	Dead 3–8 hours
Cold	Stiff	Dead 8–36 hours
Cold	Not stiff	Dead more than 36–48 hours

Table 6.4 A rough guide to the onset and loss of rigor mortis in temperate climates.

Decomposition

After death, tissues start to break down due to the action of enzymes. **Autolysis** occurs first. This is when the body's own enzymes, from the digestive tract and from lysosomes, break down cells. (See AS Topic 3 and the interactive cell in Activity 3.1 for the function of lysosomes.)

Bacteria from the gut and gaseous exchange system rapidly invade the tissues after death, releasing enzymes that result in decomposition. The loss of oxygen in the tissues favours the growth of anaerobic bacteria.

Signs of decomposition

The first sign of decomposition (also known as putrefaction) in humans is a greenish discoloration of the skin of the lower abdomen. This discoloration is due to the formation of sulphaemoglobin in the blood. It will spread across the rest of the body, darken to reddish-green, and then turn a purple-black colour. Gas or liquid blisters may appear on the skin. Due to the action of bacteria, gases including hydrogen sulphide, methane, carbon dioxide, ammonia and hydrogen form in the intestines and tissues. This causes the body to smell and become bloated. As the tissues further decompose, the gas is released and the body deflates. When the fluid associated with putrefaction drains away, the soft tissues shrink and the decay rate of the dry body is reduced.

There is variation in the time taken for decomposition. In average conditions in a temperate climate, the discoloration of the abdominal wall will occur between 36 and 72 hours after death. Gas formation occurs after about a week. The temperature of the body will determine the rate of decomposition. If the body remains above 26 °C, gas formation may occur within about three days.

Environmental temperatures have a major influence on the rate of decomposition. Low temperatures slow down decomposition. Warm temperatures speed it up. The rate of decomposition is highest between 21 and 38 °C. Intense heat denatures the enzymes involved in autolysis, delaying the start of decay. Injuries to the body allow the entry of bacteria that aid decomposition.

Q6.13 Which of the following conditions will speed up decomposition? Which will slow the process? Give a reason for your answer in each case.

a a well-heated room

b injuries to the body surface

c intense heat.

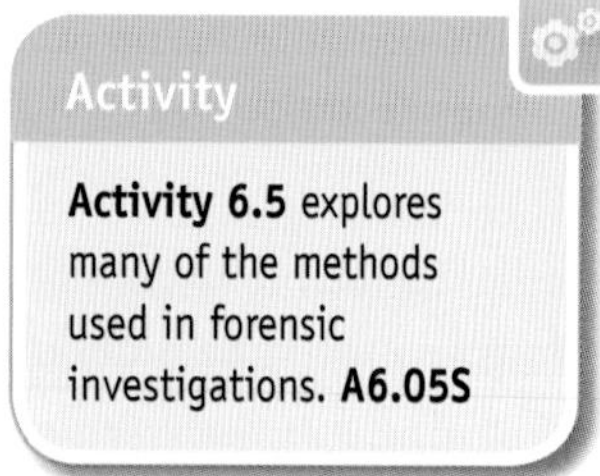

Activity
Activity 6.5 explores many of the methods used in forensic investigations. **A6.05S**

Forensic entomology

When the dog walker found the body on Canal Lane it was already discoloured over much of the torso, and a gash on the body was infested with maggots. This was a shocking experience for the person discovering the body, but provided valuable clues for the pathologists trying to determine the time of death.

Q6.14 Using the signs of decomposition described earlier, estimate how long ago the man found on Canal Lane died.

The presence of insects allows a forensic entomologist to make an estimate of how much time has elapsed since death. Forensic entomologists record information about the location and condition of the body. They take samples of any insects found on, near or under the body, noting exactly where and when they were found. The temperature of the air, ground, body and 'maggot mass' are measured, in order that the rate of maggot development can be determined. The temperature history of where the body was found may be established from local meteorological records. Sometimes a local weather station is set up at the site. Its temperature readings are correlated with the nearest weather station readings, to help calculate a more accurate temperature history for the body before its discovery.

Q6.15 Suggest how setting up a local weather station would allow a more accurate temperature history to be calculated.

Some of the maggots will be killed at the time of collection so that their age can be determined later, and some will be taken back to the lab alive. The live maggots are fed on meat, allowing them to complete their development (Figure 6.17); this is useful for species identification, and to establish when they pupate.

Q6.16 Explain how an entomologist would identify the species of fly found on a body.

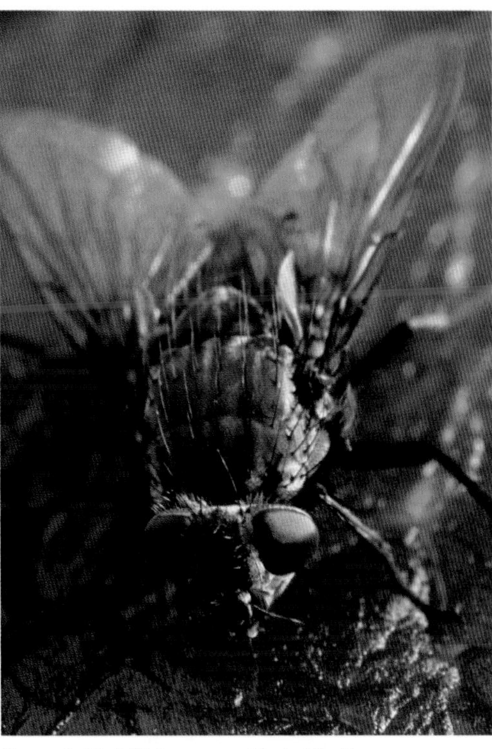

Figure 6.17 *Calliphora* sp., a bluebottle fly.

Estimating time since death

There are several ways in which the forensic entomologist can determine the age of maggots and so estimate the time when the eggs were laid. This provides a minimum time since death.

For the commonest bluebottle species found on bodies, *Calliphora vicina*, the age of the maggot can be read from a graph like the one shown in Figure 6.18. This method can only be used if the temperature conditions of the body have remained fairly constant.

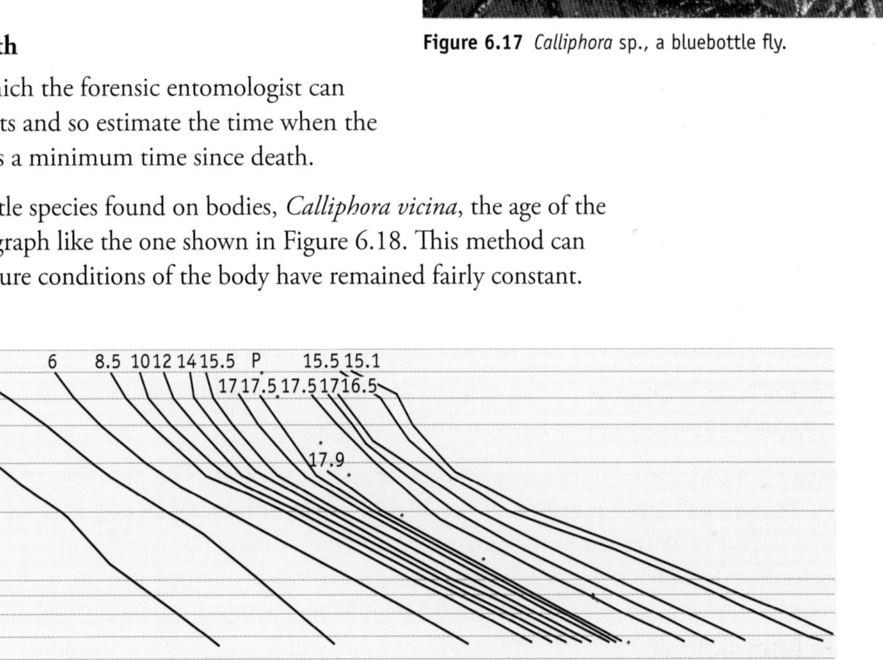

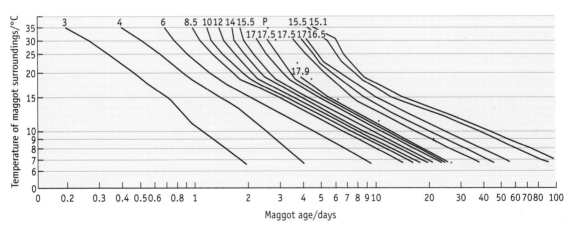

Figure 6.18 This graph was produced from a detailed laboratory study of the bluebottle *Calliphora vicina*. Each line is for a maggot of a particular length – the length being given in mm at the top of the line. The maggot age can be found by referring to the appropriate maggot length line and looking up the temperature of the place where the maggot was found. For example, a maggot 3 mm in length found on a body at 28 °C will be 0.3 of a day (i.e. 8 hours) old. Lines on the right of the dotted line labelled P are for post-feeding maggots prior to pupation.

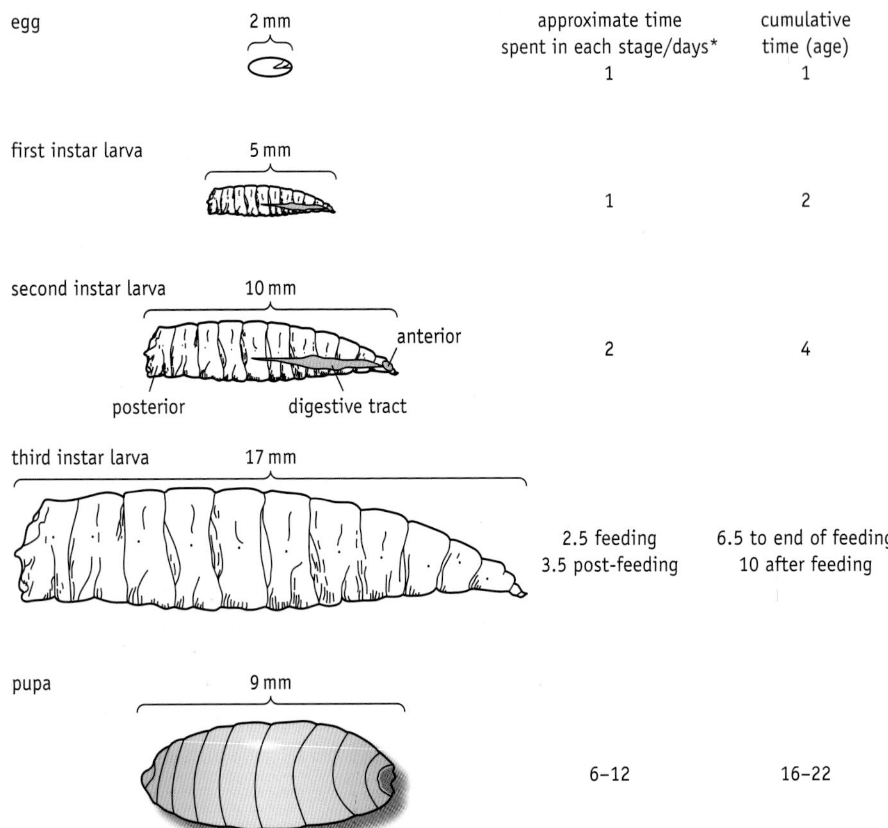

Figure 6.19 Blowflies (family Calliphoridae) have four life stages: egg, larva (maggot), pupa and adult. There are three larval stages, known as instars. At the end of each instar, the maggot sheds its skin to allow more growth in the next instar. The pupa forms inside a dark hard case, the cuticle of the third instar larva.

Identifying the maggot's stage of development with reference to the life cycle of the fly can also give an estimate of age (see Figure 6.19).

If the maggot is not *C. vicina* and its stage of development is unclear, it can be allowed to mature. This gives a date of pupation. If the normal length of time for an egg to develop and pupate is subtracted from the date of pupation, it should be possible to work out the date on which the eggs were laid. The time of egg-laying may give an underestimate of time of death, because there is no knowing how long it took the flies to find the body. Generally, however, in the UK blowflies will lay eggs within one day of death on a body outdoors in the summer. Other factors, such as any toxins in the body, will affect the results. For example, cocaine would accelerate development.

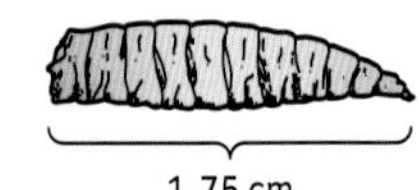

Figure 6.20 A maggot collected from the Canal Lane body.

Q6.17 The post-feeding bluebottle maggot in Figure 6.20 was collected from the Canal Lane body. Determine its approximate age using the graph in Figure 6.18 (see Q6.12 for the body temperature to use).

Q6.18 Bluebottle eggs from a dead body hatched ten hours after they were collected. The body they came from had a temperature of 15 °C. Using the graph in Figure 6.21, decide how long the eggs had been on the body before they were collected.

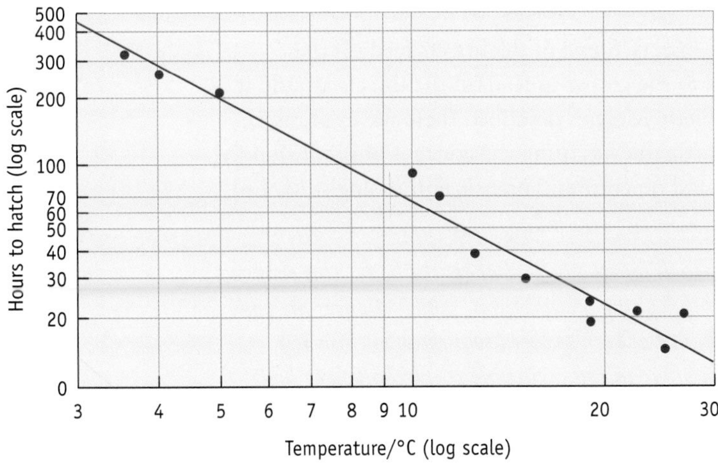

Figure 6.21 Time taken for eggs of *Calliphora vicina* to hatch at different temperatures.
Source: The Natural History Museum.

Q6.19 The body of Kathleen McClung was found in Guildford in June 1969. The stage of development of *Calliphora vicina* maggots on the body could not be clearly identified, so they were allowed to develop on raw beef. Pupae were produced between 4 and 8 July. With reference to the information in Figure 6.19, decide when the bluebottle eggs were laid and estimate the date of death.

Weblink

Visit the Natural History Museum website and read about the forensic work undertaken by their Entomology Department.

Did you know?

Maggots in medicine

Maggots are increasingly being used to clean wounds. Maggots of the fly *Lucilia sericata* are specially bred in sterile conditions. When placed on a wound, they eat dead and dying flesh and leave the healthy tissue (Figure 6.22). Maggot saliva has antiseptic properties.

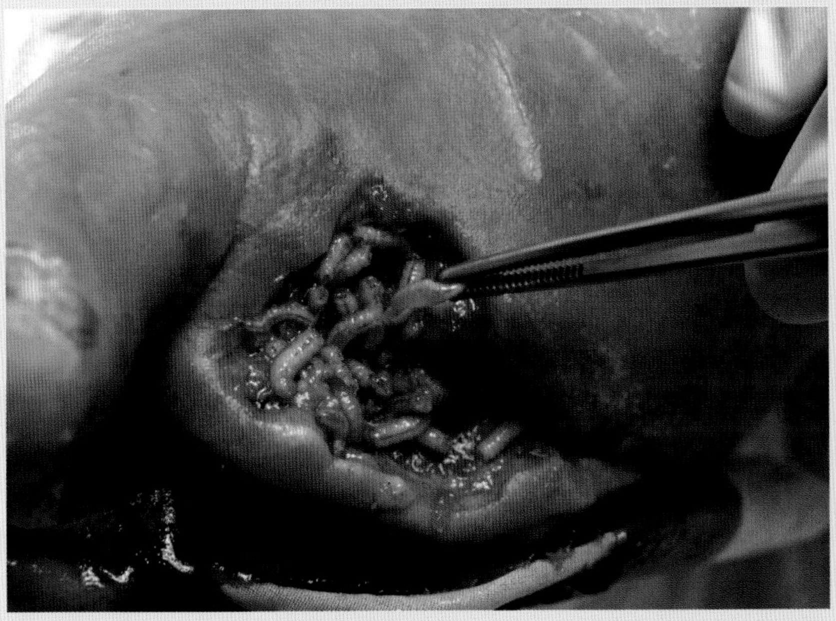

Figure 6.22 Maggots being used to clean a wound.

Succession on corpses

A corpse attracts many different types of insects. Some of the insects feed off the decaying corpse, and others are attracted to the corpse to feed on the insects around it. In 1894, the French entomologist Jean Pierre Megnin described these waves of insects. Although he did not use the term succession, he was using the concept of succession to make sense of the range of organisms found on corpses. Forensic entomologist M. Lee Goff explained such succession like this:

> *As each organism or group of organisms feeds on a body, it changes the body. This change in turn makes the body attractive to another group of organisms, which changes the body for the next group, and so on until the body has been reduced to a skeleton. This is a predictable process, with different groups of organisms occupying the decomposing body at different times.*

Table 6.5 gives some examples of the species that occur in the succession on a decomposing body. Normally eggs are laid in wounds or at the openings to the body, for example the mouth or nose. Prevailing conditions determine the community of species that first occupy the body, and how this changes with each stage in the succession. The season, weather conditions, size and location of the body will all influence the type and number of species present. Some species like *Calliphora vicina* are found in more urban situations, whereas *Calliphora vomitoria* occurs in rural locations, and is rarely found on bodies indoors. Bluebottles lay their eggs in more shaded situations; some greenbottle species prefer sunny spots.

	Organism	State of body
First wave	bluebottles: *Calliphora vicina* *Calliphora vomitoria* greenbottle *Lucilia sericata* (sheep blowfly) house fly *Mucsa domestica* cow face fly *Mucsa autumnalis*	fresh
Second wave	flesh flies *Sarcophaga* species	bloated by gases
Third wave	*Dermestes* beetle larvae *Aglossa* tabby moth maggot	active decomposition (fatty acids have turned to a waxy substance)
Fourth wave	cheese skippers *Piophilia* spp. lesser house fly *Fannia canicularis*	active decomposition (fermentation)

Table 6.5 Species succession on a human body.

Table 6.6 shows how the number of species increases through the stages of the succession. The length of each stage in the succession depends on the condition of the body, which in turn depends on environmental conditions. Note, though, that unlike plant succession, where many of the early species are replaced as conditions change, most of the early insects *remain* on the body until the advanced stage of decay.

Q6.20 **a** Explain how temperature might affect the processes of succession on a human body.

b What other factors might affect succession?

Stage of decomposition	Total number of species attracted to each stage in the decomposition	Percentage of species attracted to another stage of decomposition				
		Fresh	Bloated	Active decay	Advanced decay	Dry
fresh	17	100	94	94	76	0
bloated	48	33	100	100	90	2
active decay	255	6	19	100	98	13
advanced decay	426	3	10	59	100	38
dry	211	0	1	16	76	100

Table 6.6 Total number of and percentage of species attracted to the different stages of decay.

Insects can also help determine if a body has been moved. There may be species of insect found on a body that would not naturally occur in that location. For example, insects normally found in woods occurring on a body discovered indoors would suggest that it has been moved some time after the time of death.

Other decomposers

Insects are not the only organisms involved in the decomposition of a body. Bacteria from the gut quickly invade the tissues after death. Other bacteria and fungi from the surroundings colonise the corpse, contributing to decay and changing conditions on the decomposing body. There doesn't seem to be a set succession of bacteria and fungi that colonise in a particular sequence, but genera often found on corpses in the early stages include *Bacillus, Staphylococcus, Candida* and *Streptococcus*, followed by *Salmonella, Cytophaga* and *Agrobacterium*. These microorganisms are collectively referred to as decomposers.

A corpse is a great source of energy for the decomposers. The organic carbohydrates, proteins, fats and nucleic acids of the corpse are used as a food source, with energy being released through aerobic and anaerobic respiration. This energy enables the bacteria and fungi to grow and multiply rapidly, ensuring yet more decomposition. Carbon dioxide is released into the atmosphere by the respiring decomposers. This recycles the carbon back into a form that can be used in photosynthesis by plants, to synthesise more organic molecules. Of course, all dead organic material – plant, animal, microbial – will provide energy in this way. Decomposition is a major process sustaining the carbon cycle (see Topic 5).

Checkpoint

6.2 Produce a bullet point summary of methods used to determine time of death.

6.2 Cause of death

A post-mortem examination may be performed if there is a sudden or unexpected death, or if the cause of death is unknown. The pathologist undertaking the examination will first make an external examination.

An internal examination is then undertaken. An incision is made down the front of the body and organs are taken out for detailed examination (Figure 6.23). The state of the internal organs may allow conclusions to be made about the health of the person and illnesses suffered. For example, the condition of the heart and arteries may show whether atherosclerosis and a heart attack were responsible for death. Cirrhosis of the liver, characterised by death of liver cells and formation of fibrous tissue, may suggest inadequate diet, excessive alcohol consumption, or infection.

Blood and tissue samples may be taken and tested for toxins, infection or tumours. The contents of the stomach may be analysed. This can show what was last eaten and when. This may help in determining when the person was alive and where they were.

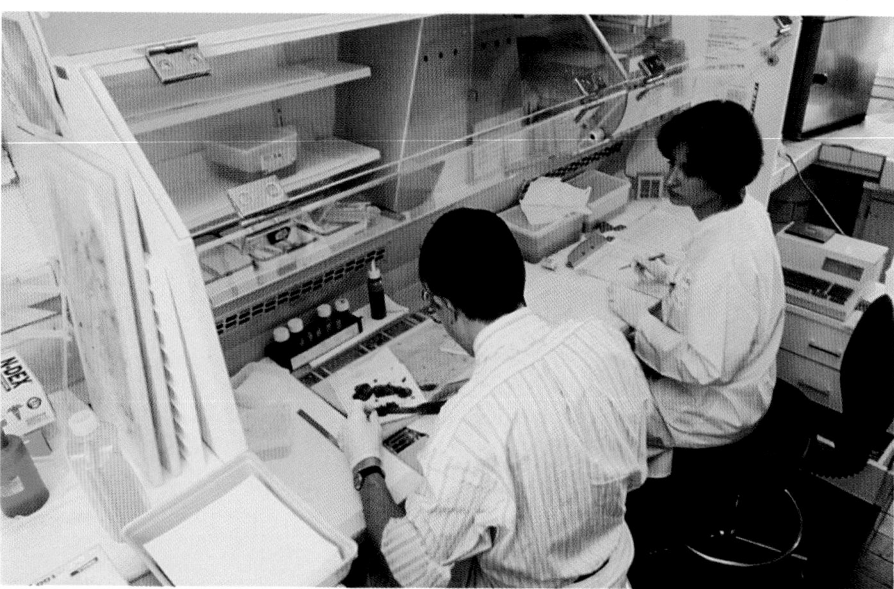

Figure 6.23 Medical laboratory scientific officers examine lung tissue.

Weblink

Visit the virtual autopsy website (it's not too gruesome!).

What killed George Watson and Nicki Overton?

The post-mortem on George Watson, the Canal Lane body, showed that he had an aneurysm in the aorta. As you saw in AS Topic 1, blood builds up behind a section of the artery, which has narrowed and become less flexible; the artery bulges as it fills with blood.

Q6.21 **a** What process is likely to have caused the narrowing of the artery?

b What major risk factors could have contributed to the development of the aneurysm?

c What may have happened to the aneurysm, and why would it have caused the man's death?

Results of Nicki Overton's post-mortem showed she was infected with two pathogenic (disease-causing) microorganisms. These were *Mycobacterium tuberculosis*, the bacterium responsible for the disease **tuberculosis** (**TB**), and the **human immunodeficiency virus** (**HIV**), which causes **AIDS**. What is the difference between these two microbes? How might she have contracted the infections? Could one or the other, or both, have caused her death? To determine whether they were the cause of her death, the extent of her illness and the immune response of the body must be ascertained.

Key biological principle: What is the difference between a virus and a bacterium?

Bacteria

As you saw in AS Topic 3, bacteria are prokaryotic cells; these are much simpler than eukaryotic cells (Figure 6.24). They do not have a nucleus, lack membrane-bound organelles, and do not produce a spindle during cell division. The average diameter of a bacterial cell is between 0.5 and 5 μm. Bacteria reproduce asexually by binary fission; after replication of their DNA they divide into two identical cells.

Q6.22 Which of the following features would be found in a prokaryote?

a cytoplasm
b a long, circular strand of DNA
c plasmids
d cell surface membrane
e mitochondria
f mesosomes (infolding of the cell surface membrane)

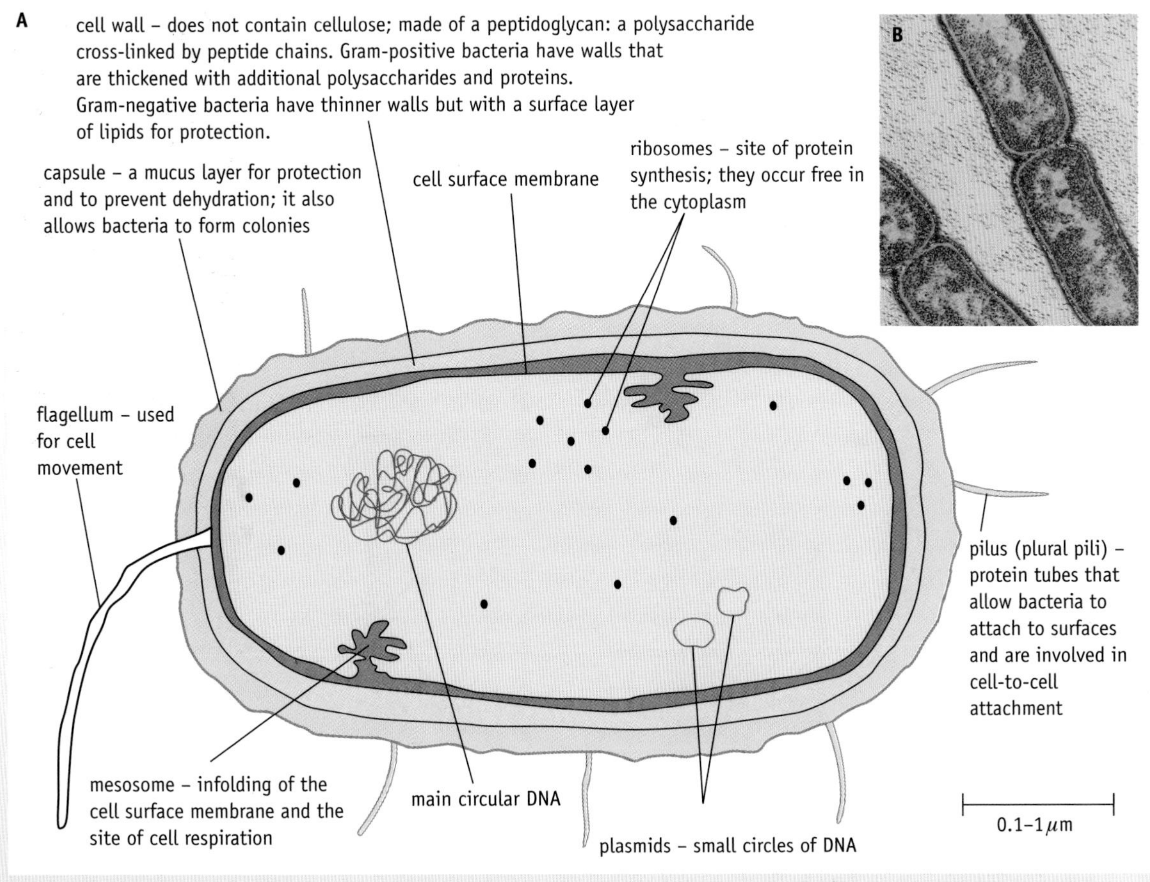
A
cell wall – does not contain cellulose; made of a peptidoglycan: a polysaccharide cross-linked by peptide chains. Gram-positive bacteria have walls that are thickened with additional polysaccharides and proteins. Gram-negative bacteria have thinner walls but with a surface layer of lipids for protection.
capsule – a mucus layer for protection and to prevent dehydration; it also allows bacteria to form colonies
cell surface membrane
ribosomes – site of protein synthesis; they occur free in the cytoplasm
B
flagellum – used for cell movement
pilus (plural pili) – protein tubes that allow bacteria to attach to surfaces and are involved in cell-to-cell attachment
mesosome – infolding of the cell surface membrane and the site of cell respiration
main circular DNA
plasmids – small circles of DNA
0.1–1 μm

Figure 6.24 A The basic structure of a bacterium. **B** *Mycobacterium tuberculosis*, first identified by Robert Koch in 1882. Magnification × 4800.

CONTINUED ▶

Viruses

Viruses are small organic particles with a structure that is quite different from that of bacteria and very much simpler. They basically consist of a strand of nucleic acid (RNA or DNA) enclosed within a protein coat (Figure 6.25). Viral DNA can be single or double stranded.

Some viruses also have an outer envelope taken from the **host** cell's surface membrane; the envelope therefore contains lipids and proteins. Viral envelopes also have glycoproteins from the virus itself. These are **antigens**, molecules recognised by the host's immune system as not being its own self. The envelope helps the virus attach to the cell and penetrate the surface membrane. The human immunodeficiency virus (HIV) is an example of an enveloped virus (Figure 6.48).

Viruses come in a wide variety of sizes and shapes of different complexity (see Figures 6.25 and 6.26). Plants can also be infected by viruses, as shown in Figure 6.27.

Viruses lack some of the internal structures required for growth and reproduction. This means they have to enter the cells of the organisms they infect (the host), and use the host's metabolic systems to make more viruses (see Figure 6.28). When viruses hijack the host cell's biochemistry, the normal working of the cell is disrupted. After reproducing inside the host cell, new virus particles may bud from the cell surface or burst out of the cell, splitting it open. This splitting kills the cell, and is called **lysis**. It results in the cell contents being released into the surrounding tissues; the many enzymes and other chemicals released can damage neighbouring cells. These processes cause the disease symptoms produced by the virus infection.

Activity

In **Activity 6.6** you produce an animation brief for a new prokaryote section of the virtual cell.
A6.06S

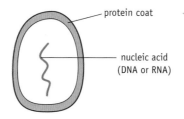

A Basic structure

B Viruses come in a wide variety of shapes and sizes (10–300 nm diameter).

spherical/icosahedral protein coat

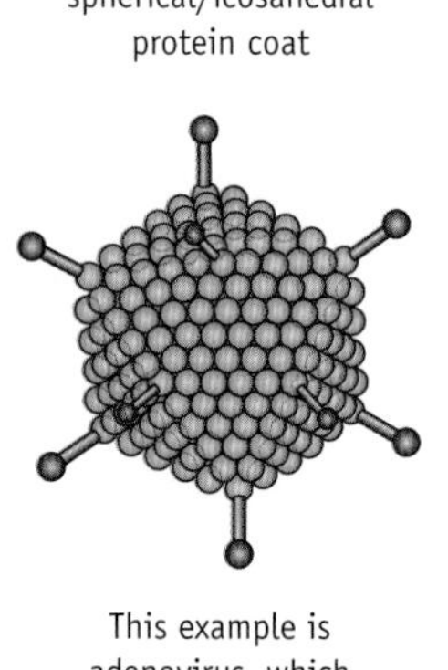

This example is adenovirus, which causes sore throats.

protein subunits around a nucleic acid strand

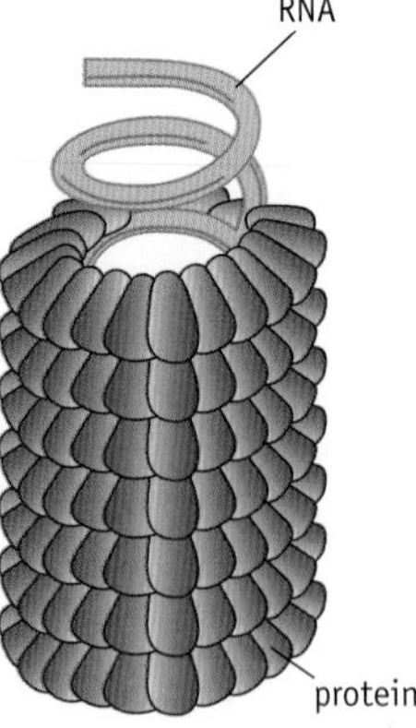

This is tobacco mosaic virus.

complex protein coat

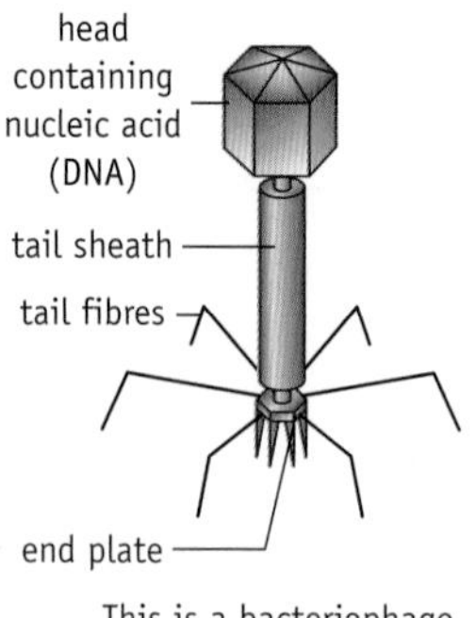

This is a bacteriophage, a virus that infects bacteria.

Figure 6.25 **A** The basic structure of a virus. **B** Some examples of different viruses.

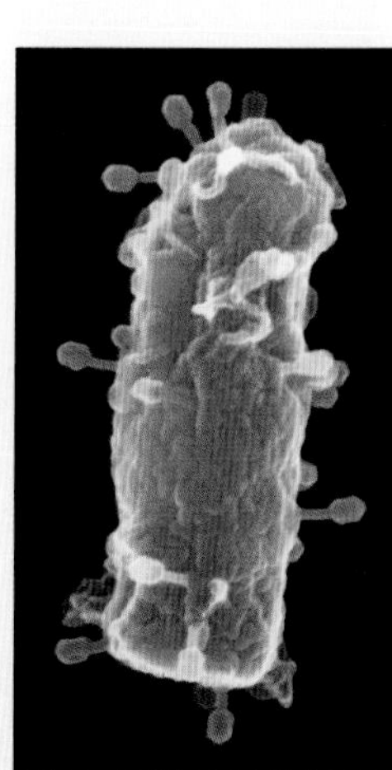

Figure 6.26 Electron micrograph of bacteriophage viruses attacking an *Escherichia coli* bacterium. Magnification × 19 000.

CONTINUED ▶

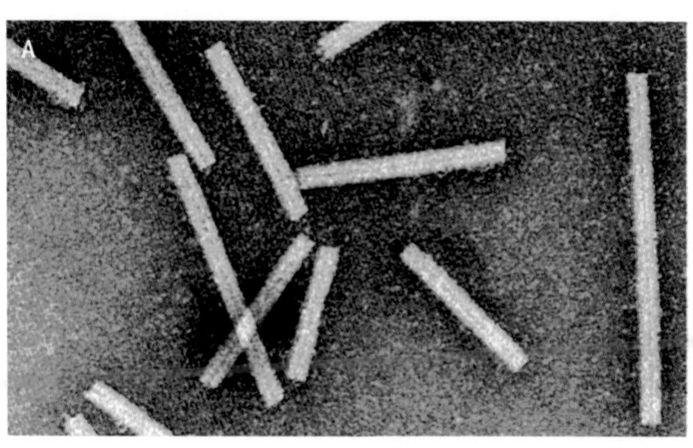

Figure 6.27 **A** Coloured election micrograph of the potato mop-top virus. **B** Potato plants infected by the virus are dwarfed, and tissues become blotched and die.

1 Virus attaches to the host cell.

2 Virus inserts nucleic acid.

3 Viral nucleic acids replicate.

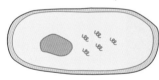

4 Viral protein coats synthesised.

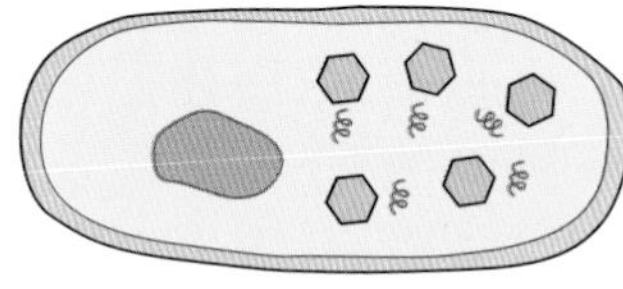

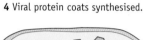

5 New virus particles formed.

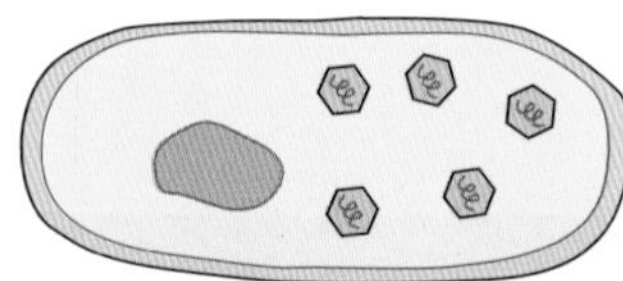

6 Virus particles released due to cell lysis.

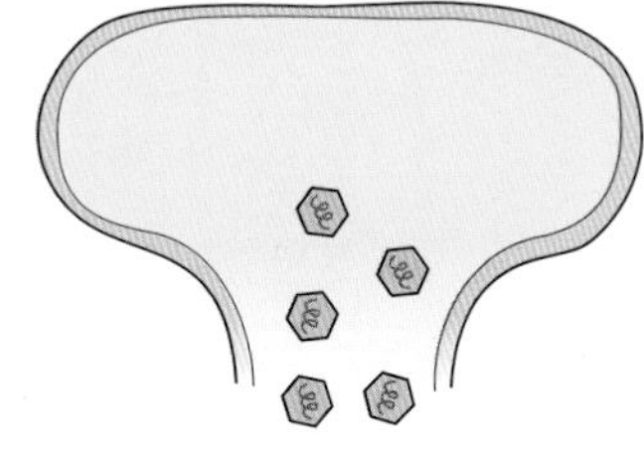

Figure 6.28 Viruses use the host's protein synthesis machinery to manufacture new virus particles.

Bacteria and viruses that cause diseases are known as **pathogens**. Common human diseases caused by bacteria include *Salmonella* food poisoning, gonorrhoea and cholera. Viruses cause many human diseases including flu, measles, chicken pox and cold sores. Many plant diseases are also the result of viral infection. For example, tobacco mosaic virus, as its name suggests, causes mosaic disease in tobacco plants.

Checkpoint

6.3 Draw up a table of comparison to allow you to distinguish between the structures of bacteria and viruses.

Q6.23 Which of the following diseases are caused by bacteria, and which by viruses?

a common cold

b food poisoning

c cold sores

d cholera

e rubella

Q6.24 Are viruses living or non-living?

Q6.25 Look at the virus shown in Figure 6.26, and *Mycobacterium tuberculosis*, the TB bacterium shown in Figure 6.24. Calculate approximately how much smaller the virus is than the bacterium.

How might Nicki have become infected?

Transmission of the TB bacterium

Mycobacterium tuberculosis is carried in the droplets of mucus and saliva released into the air when an infected person talks, coughs or sneezes (Figure 6.29). Others then inhale the droplets. This is known as droplet infection. The droplets can remain suspended for several hours in poorly ventilated areas. Close contact with an infected person increases the risk of developing the disease, as do poor health, poor diet, and overcrowded living conditions. *M. tuberculosis* is a tough bacterium and can survive as dust from dried droplets for several weeks, making the bedclothes and room of a TB patient potentially infectious.

Figure 6.29 TB bacteria are carried in the droplets released by an infected person, for example when they sneeze.

Transmission of HIV

HIV is not a very tough virus and cannot survive outside the body for any significant time. It can be passed on only in body fluids, such as blood, vaginal secretions and semen, but not saliva or urine. Infection can occur if you have unprotected sex with someone who is already infected, or if blood from such a person enters your bloodstream. For infection to occur, the body fluids have to be transferred directly into the body of the next host. This can occur in the following ways:

- Infection can result from sharing needles, whether used illegally for drugs or legally.
- Through unprotected sex. Worldwide, this is the most frequent route of infection. The virus can enter the bloodstream of a partner through breaks in the skin or lesions caused by other infections – usually other **sexually transmitted infections (STIs)**. The use of a condom can prevent this transmission. Infection can also occur via oral sex, though this is rare.
- Direct blood-to-blood transfer can occur through cuts and grazes. Police, paramedics and medical staff are particularly at risk from this method of transmission, and precautions are taken to minimise the risk.
- Maternal transmission from mother to unborn child or in breast milk. The risk of the virus being passed to the baby occurs in the last few weeks of pregnancy, mostly around the birth itself, when mingling of infant and maternal blood is likely to happen. Taking anti-HIV drugs during the last three months of pregnancy and giving birth by Caesarean section greatly reduces this risk, from about 20% to 5%. This option is only realistically available in countries with advanced medical care.

Q6.26 Suggest what prevents HIV from infecting a baby during most of the pregnancy.

6.3 The body's response to infection

When someone is infected by a disease-causing organism, several mechanisms in their body try to destroy the invading pathogen. This task is performed by the **immune system**, and is known as the **immune response**.

Non-specific responses help to destroy any invading pathogen, whereas specific immunity is always directed at a specific pathogen. The immune response is summarised in Figure 6.30.

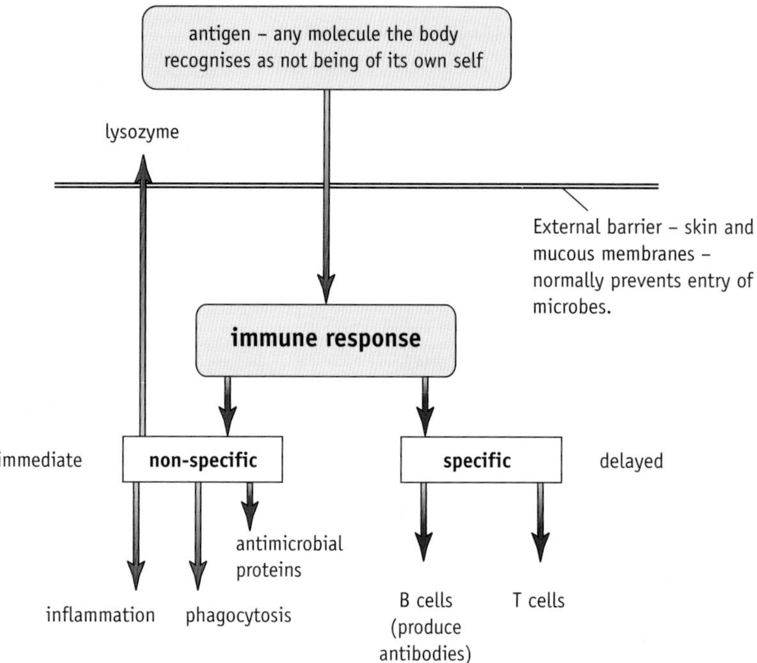

Figure 6.30 The immune response.

Non-specific responses to infection

Lysozyme

If a speck of dirt lands in your eye, your eye starts to weep. The stream of tears helps to wash out the foreign material and attacks any bacteria on the surface of the eye. Tears contain an enzyme called lysozyme (Figure 6.31) that kills bacteria by breaking down their cell walls. The same enzyme is found in saliva and nasal secretions, protecting the body from harmful bacteria in the air we breathe or the food we eat.

Q6.27 Suggest how the enzyme lysozyme might break down bacterial cell walls. (Think about the structure of bacterial cell walls before answering.)

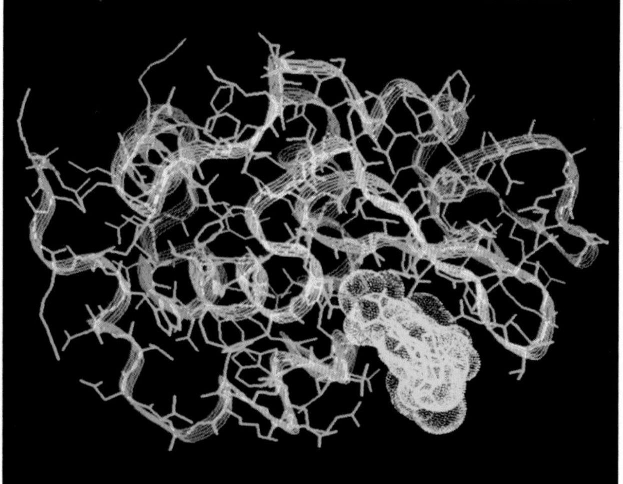

Figure 6.31 The molecular structure of lysozyme was determined in 1965. This was the first time the structure of an enzyme had ever been worked out. Lysozyme breaks down the cell walls of bacteria, and is important in defending the body against infection.

Inflammation

An injury, cut or graze enables microbes and other foreign material to enter the body. A blood clot will rapidly seal the wound. But inflammation at the site, known as the **inflammatory response**, helps to destroy invading microbes. Damaged white blood cells and mast cells, found in the connective tissue below the skin and around blood vessels, release special chemicals such as **histamine**. These chemicals cause the arterioles in the area to dilate, increasing blood flow in the capillaries at the infected site. Histamines also increase the permeability of the capillaries: cells in the capillary walls separate slightly, so the vessels leak. Plasma fluid, white blood cells and antibodies leak from the blood into the tissue causing **oedema** (swelling). The infecting microbes can now be attacked by these intact white cells.

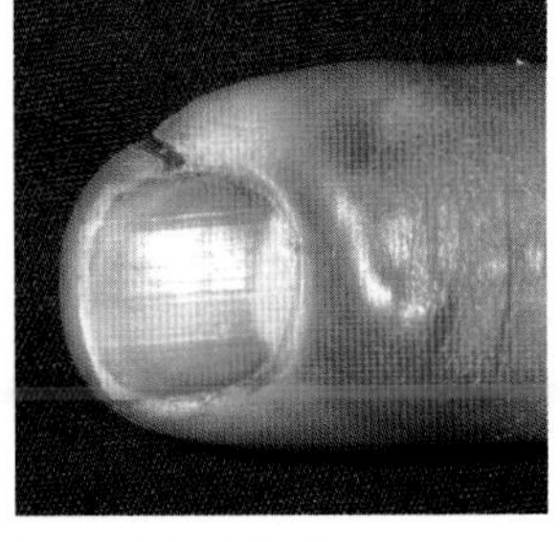

Figure 6.32 This finger became infected after the man cut himself with a bread knife.

Q6.28 Explain why the finger shown in Figure 6.32 became hot, red and swollen after the man cut it.

? Did you know?

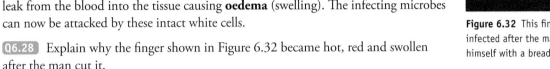

Why do people take antihistamine tablets?

People who get hay fever have an allergic reaction to certain types of pollen. The nasal tissues respond to the presence of pollen grains by releasing large amounts of histamine. This causes the nasal lining to swell and itch. It also becomes leaky, giving the characteristic runny nose. Taking antihistamine tablets greatly reduces this allergic response.

Antihistamines and histamines have similar shapes. Antihistamines bind to histamine receptors and block the binding of histamines.

Phagocytosis

Phagocytes are white blood cells that engulf bacteria and other foreign matter in the blood and tissues. Phagocytes include both **neutrophils** and **macrophages**. You can see the differences between the different types of blood cells in Figure 6.33.

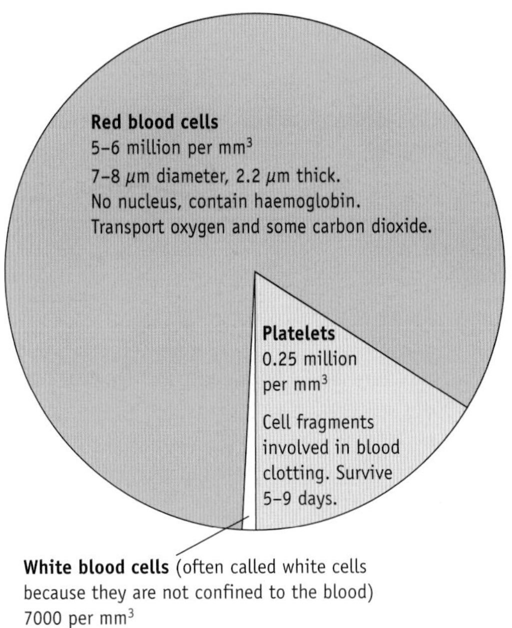

Red blood cells
5–6 million per mm^3
7–8 μm diameter, 2.2 μm thick.
No nucleus, contain haemoglobin.
Transport oxygen and some carbon dioxide.
Platelets
0.25 million per mm^3
Cell fragments involved in blood clotting. Survive 5–9 days.
White blood cells (often called white cells because they are not confined to the blood)
7000 per mm^3
All contain a nucleus.

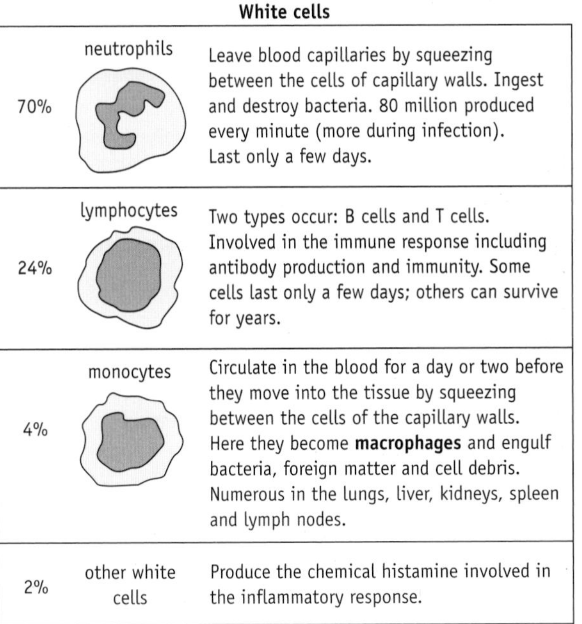

White cells

%	Cell type	Description
70%	neutrophils	Leave blood capillaries by squeezing between the cells of capillary walls. Ingest and destroy bacteria. 80 million produced every minute (more during infection). Last only a few days.
24%	lymphocytes	Two types occur: B cells and T cells. Involved in the immune response including antibody production and immunity. Some cells last only a few days; others can survive for years.
4%	monocytes	Circulate in the blood for a day or two before they move into the tissue by squeezing between the cells of the capillary walls. Here they become **macrophages** and engulf bacteria, foreign matter and cell debris. Numerous in the lungs, liver, kidneys, spleen and lymph nodes.
2%	other white cells	Produce the chemical histamine involved in the inflammatory response.

Figure 6.33 Structures and functions of blood cells.

Action at the infected site

Chemicals released by bacteria and the cells damaged at the site of infection attract phagocytic white cells. Neutrophils are the first to arrive; they engulf between 5 and 20 bacteria before they become inactive and die (Figure 6.34). The neutrophils are followed by macrophages. These larger, longer-lived cells each have the potential to destroy as many as 100 bacteria. They will also ingest debris from damaged cells, and foreign matter such as particles of carbon and dust in the lungs. The ingested material is enclosed within a vacuole (Figure 6.35). Lysosomes containing digestive enzymes fuse with the vacuole, the enzymes are released, and they destroy the bacteria or other foreign material.

The large numbers of phagocytic cells that collect at the site of infection can engulf huge numbers of bacteria. After a few days, the area is full of dead cells, mainly neutrophils, which form a thick fluid called pus. The pus may break through the surface of the skin, but usually it gradually gets broken down and absorbed into the surrounding tissue.

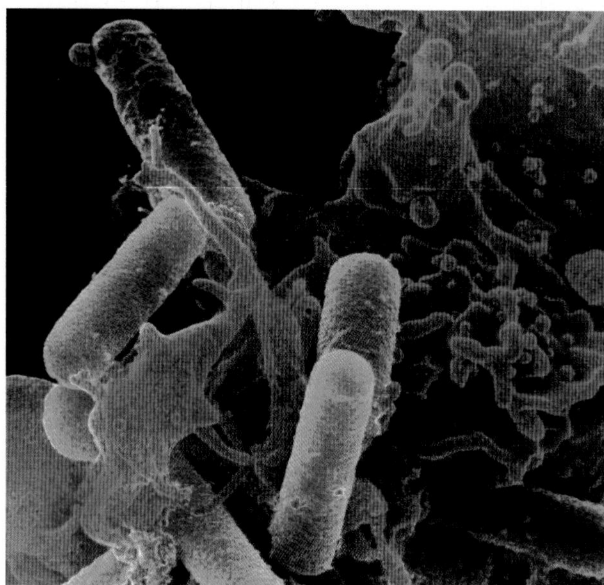

Figure 6.34 Phagocytosis. A white blood cell (coloured orange) is engulfing *Bacillus cereus* bacteria cells (coloured blue). These rod-shaped soil bacteria are a cause of food poisoning. Magnification × 9200.

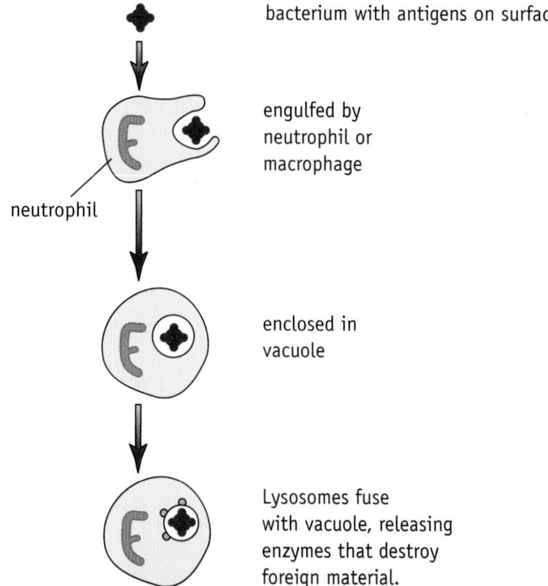
bacterium with antigens on surface
engulfed by neutrophil or macrophage
neutrophil
enclosed in vacuole
Lysosomes fuse with vacuole, releasing enzymes that destroy foreign material.

Figure 6.35 Neutrophils and macrophages engulf bacteria at the site of infection.

Did you know?

Radicals and antioxidants

After engulfing bacteria, phagocytes, especially neutrophils, produce the highly reactive radicals nitric oxide (NO•) and superoxide (O_2•). These kill the engulfed bacteria by attacking their DNA and protein.

As well as helping to defend against infections, radicals can also damage the body's own cells, thereby increasing the risk of diseases such as atherosclerosis. Certain enzymes counteract this effect by destroying any excess radicals. Antioxidants in the diet, for example vitamins C and E, have a similar protective action (see AS Topic 1).

Components of cigarette smoke cause the production of unnaturally high levels of radicals. This is partly why smokers are much more likely to have heart attacks and strokes than non-smokers.

Action to prevent the spread of infection

In spite of intense phagocytic activity at the infected site, some live bacteria usually get carried away either by the blood or in the lymph. The spread of these bacteria is hindered by the action of macrophages in the **lymph nodes**, spleen and liver. The role of the lymphatic system is outlined in Figure 6.36. Only occasionally does the system fail, leading to widespread infection known as septic shock or 'blood poisoning'.

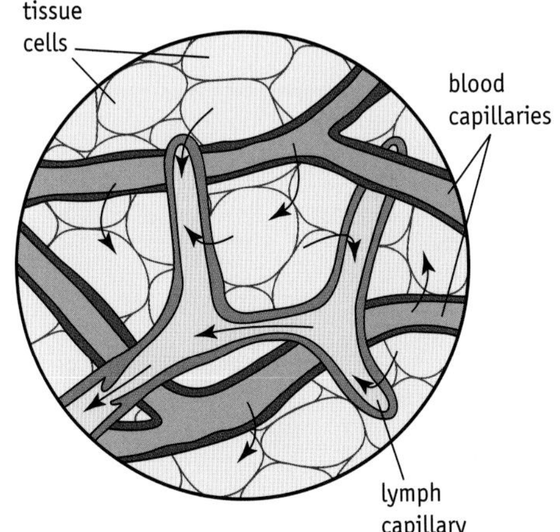

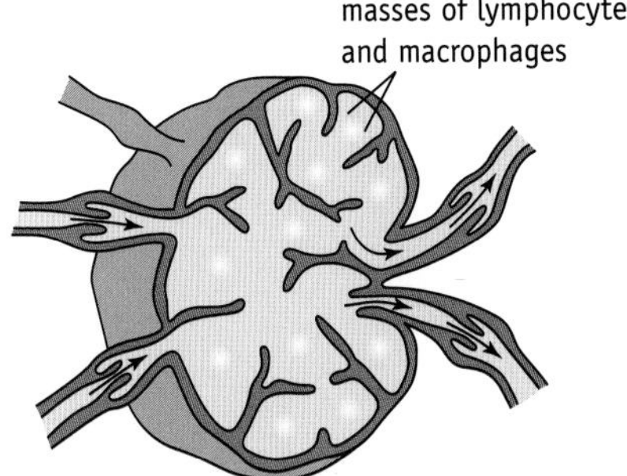

1 Tissue fluid drains into the lymphatic vessels.
tissue cells

lymph capillary

3 As lymph passes through the lymph nodes any pathogens present activate lymphocytes and macrophages, which can then destroy the microbes.
masses of lymphocytes and macrophages

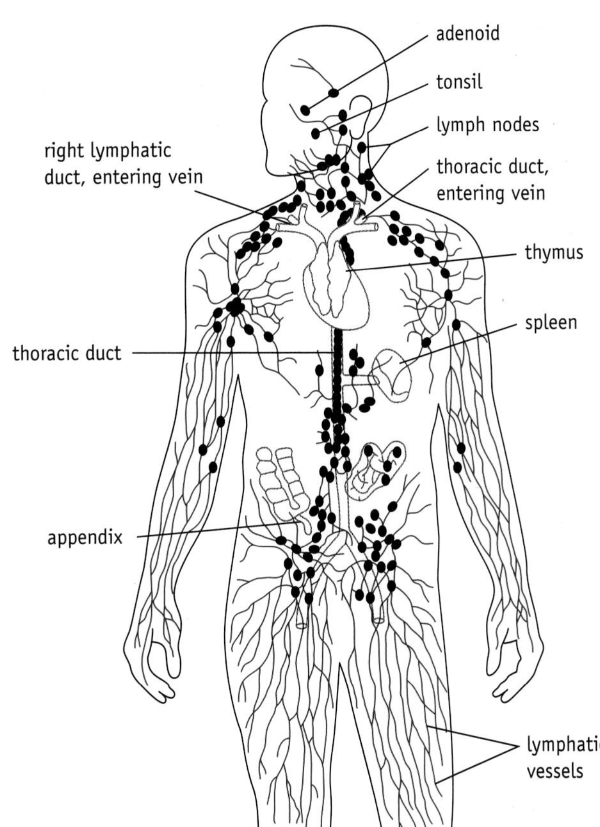

2 The fluid, called lymph, flows along the lymph vessels. It passes through lymph nodes and eventually returns to the blood via the lymphatic and thoracic ducts.

right lymphatic duct, entering vein
thoracic duct, entering vein

thoracic duct
appendix
lymphatic vessels

Figure 6.36 The location and role of the lymph nodes in the immune system.

Antimicrobial proteins – interferon

Most of the body's non-specific defences are aimed at invading bacteria. **Interferon** is the exception, as it provides non-specific defence against viruses. Virus-infected cells produce this protein; it diffuses to the surrounding cells where it prevents viruses from multiplying. It inhibits viral protein synthesis, and in this way limits the formation of new virus particles.

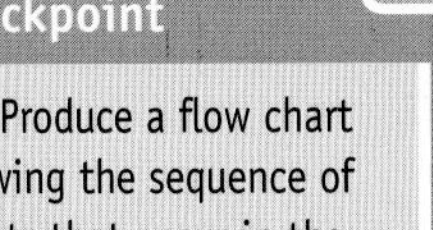

Checkpoint

6.4 Produce a flow chart showing the sequence of events that occur in the non-specific response at the site of a cut.

Did you know?

A miracle drug?

When interferon was first discovered, scientists hoped that it would become a 'miracle drug' for virus diseases and possibly cancer. Chemically, interferon is a small protein existing in a number of species-specific forms. These can be produced artificially using genetically modified bacteria. However, the production is extremely costly and when the drug is injected into patients it lasts only a short time and produces unpleasant side effects. Therefore, interferon is not widely used. But it is sometimes prescribed for the treatment of hepatitis, types of leukaemia, and certain AIDS-related cancers. It is also used in the treatment of the autoimmune disease multiple sclerosis. Recent research is trying to develop alternative drugs that can boost the natural production of interferon by the body.

Specific immunity

Lymphocytes are white blood cells that help to defend the body against specific diseases (Figure 6.33). They circulate in the blood and lymph, and gather in large numbers at the site of any infection.

Reserve supplies of lymphocytes are held in strategically-positioned lymphoid tissue (Figure 6.36). For instance, if you have an ear infection, lymphocytes in the lymph nodes of the neck go into action and you get 'swollen glands'. If you have an infected finger, it is the lymph nodes in the armpit that become swollen and active. The tonsils and adenoids are well positioned, in the throat and nose, to help deal with any upper respiratory tract infections. Patches of lymphoid tissue around the stomach and intestines protect against gut infections.

B and T cells

There are two main types of lymphocyte: **B cells** and **T cells**. Both types respond to antigens such as the ones on the surface of bacteria or viruses. Most antigens are protein molecules, and their large size and characteristic molecular shape allow the lymphocytes to identify which ones are 'foreign' (non-self). The response by lymphocytes is called the **specific immune response**.

B lymphocytes

B cells secrete **antibodies** in response to antigens. Antibodies (Figure 6.37) are special protein molecules of a class known as **immunoglobulins**. Antibodies bind to the antigens on the cell surface membrane. They act as labels, allowing phagocytes to recognise and destroy the cell (Figure 6.38).

Each B cell produces only one type of antibody, which binds to only one specific antigen. A microbe usually has several different types of antigen on its surface. Each different antigen will bind and activate different B cells.

During early human embryo development, 100 million different B cells are produced in the bone marrow. Each of these divides rapidly to produce a clone of cells, providing the baby with an immune system that can respond to a tremendous variety of antigens that might invade its body after birth. B cells have receptors on their surface; these receptors include transmembrane versions of the antibody molecules they produce.

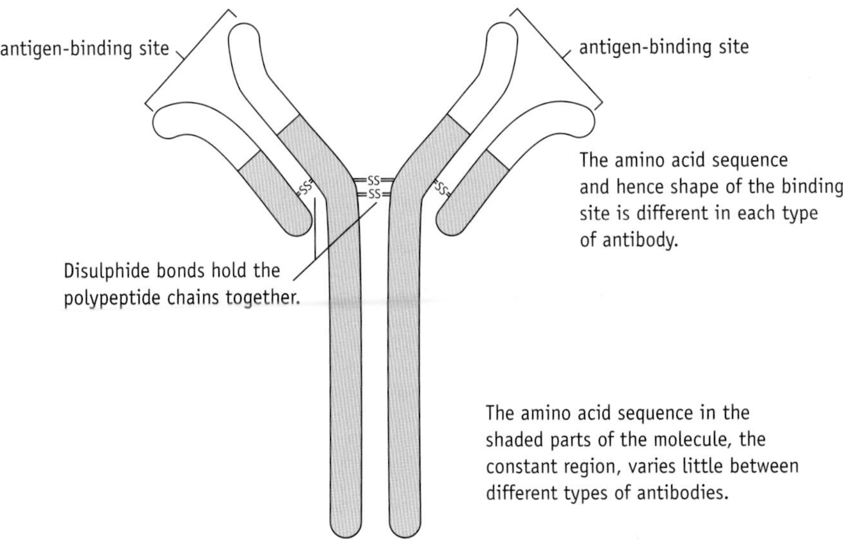

Figure 6.37 A simplified diagram of an antibody. An antigen with a complementary shape can bind to the antibody's antigen-binding site.

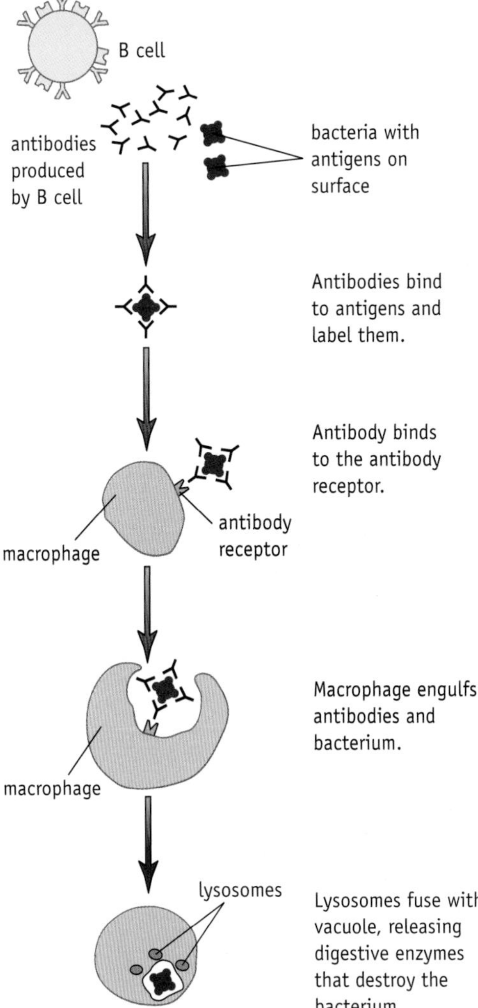

Figure 6.38 Antibodies attach to antigens, allowing phagocytes to identify them.

T lymphocytes

T lymphocytes, like B lymphocytes, are produced in the bone marrow, but unlike B lymphocytes, they mature in the thymus gland – hence T for thymus (Figure 6.39). T cells each have one specific type of antigen receptor on their surface. This only binds to an antigen with the complementary shape.

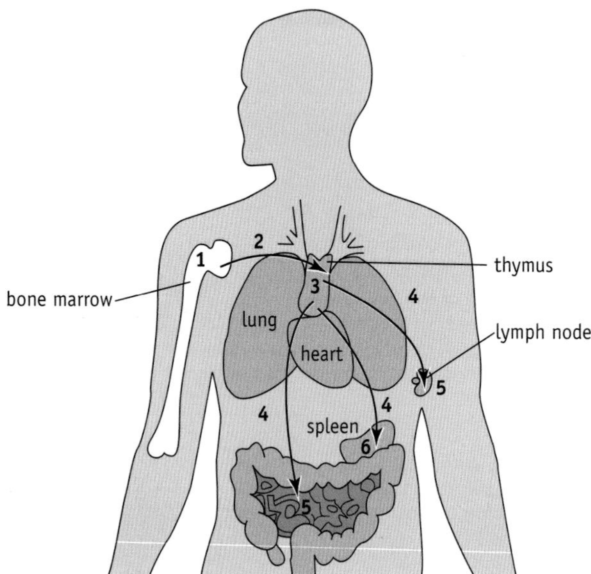

1 Immature T cells are produced by division of stem cells in the bone marrow.
2 Immature T cells move to the thymus via the blood.
3 T cells mature in the thymus.
4 Mature T cells leave the thymus in the blood and move to lymph nodes and the spleen.
5 As lymph fluid passes through a lymph node, T cells are activated by any pathogens present.
6 As blood passes through the spleen, T cells are activated by any pathogens present.

Figure 6.39 T cells move from the bone marrow to the thymus, where they mature before passing to lymph tissue via the blood.

There are two types of T cells:

- **T helper cells** – when activated, these stimulate the B cells to divide and become cells capable of producing antibodies. They also enhance the activity of phagocytes.
- **T killer cells** – these destroy any cells with antigens on their surface membrane that are recognised as foreign or 'non-self'. This includes body cells infected with pathogens. Unfortunately, it also includes tissues received as a transplant from another person.

The primary immune response

Activation of T cells

Proteins produced by cells are continually added to and removed from the 'fluid mosaic' cell surface membrane. When a piece of biological material is engulfed by a macrophage, protein fragments (peptides) from the material become attached to proteins in the cell. These are added to the macrophage's cell surface membrane, where they are displayed as 'non-self' antigens. These antigens presented on the surface of macrophages act as a signal to alert the immune system to the presence of foreign antigens in the body. Macrophages displaying non-self peptides are **antigen-presenting cells** (**APCs**) as shown in Figure 6.40A.

A T helper cell with complementary-shaped receptors, called CD4 receptors, on its surface binds to the antigen on the surface of the antigen-presenting cells. Once activated by this binding, each T helper cell divides to produce a clone of active T helper cells and a clone of **T memory cells**. This second clone of cells remains for months or years in the body. This means that, if an individual is exposed to the same antigen in the future, their immune system can respond more quickly.

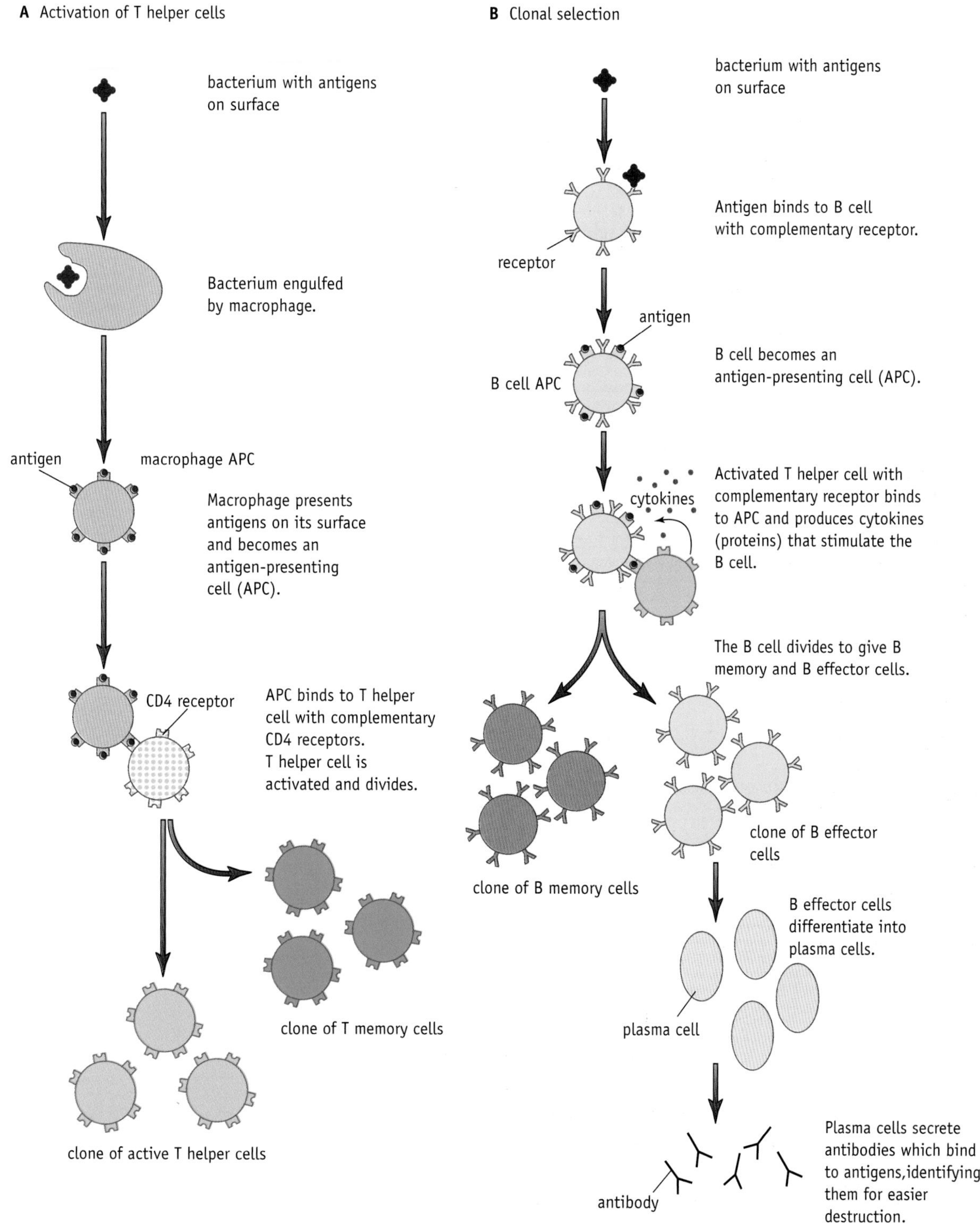

Figure 6.40 **A** Activation of T cells. **B** Cell clonal selection.

Cloning of B cells

Complementary receptors on the surface of B cells bind to non-self antigens and become antigen-presenting cells in the same way as macrophages do (Figure 6.40B). Antigen-presenting B cells bind with active, cloned T helper cells that are presenting the same antigen. Once attached, the T helper cells release chemicals called **cytokines**, which stimulate division and differentiation of the B cells.

Under the influence of cytokines, the B cells divide to produce two clones of cells:

- **B effector cells** – these differentiate to produce **plasma cells**, which release antibodies into the blood and lymph. These cells are relatively short-lived, lasting only a few days.
- **B memory cells** – like T memory cells, these cells are longer-lived. They remain for months or years in the body, enabling an individual to respond more quickly to the same antigen in the future.

The process of B cell division is called **clonal selection**. The first time a B cell comes across a non-self antigen that is complementary to its cell surface receptors, the production of sufficient antibody-producing cells takes about 10–17 days; this is the **primary immune response**. During the time it takes to produce the antibodies, the person is likely to suffer the symptoms of the infection.

Key biological principle: Complementary protein shapes

You will be familiar with lock-and-key mechanisms for enzyme specificity. This specificity of binding occurs in many other situations. The immune response relies on the recognition of specific protein antigens by T helper and B cells using receptors with the complementary shape. The specific shape of the binding sites found on antibodies is crucial to their function.

In Topic 7, 'Run for your life', you will learn that the action of signal proteins within cells depends on the binding of signal molecules to receptors with complementary shapes.

The role of killer T cells

If an intracellular bacterium or virus infects a body cell, a fragment of the antigen is presented on the cell surface membrane in the same way as occurs in macrophages. T killer cells with complementary receptors bind to the antigen presented on the body cell (Figure 6.41). The T killer cells divide to form an active clone; this division is stimulated by the cytokines from T helper cells. Without cytokines, there would not be enough T killer cells produced to fight a viral infection. The T killer cells release enzymes that create pores in the membrane of the infected cell. This enables ions and water to flow into the infected cell, which swells and bursts (undergoes lysis). The pathogens within the cell are released. Once out of the cell they can be labelled by antibodies from B cells as targets for destruction by macrophages.

Activity

View the animation in **Activity 6.7** to see phagocytosis occurring. **A6.07S**

In **Activity 6.8** an animation lets you test your overall understanding of the body's specific immune response to infection and provides you with a summary sheet. **A6.08S**

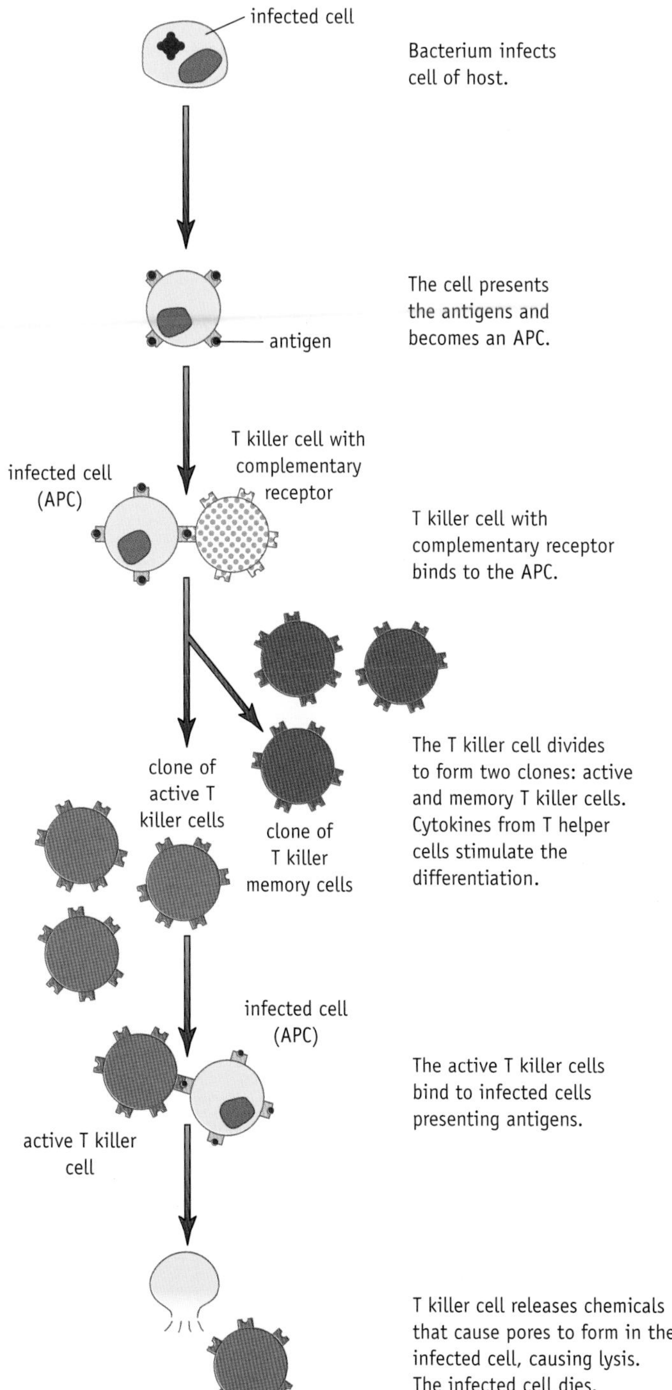

infected cell
Bacterium infects cell of host.
The cell presents the antigens and becomes an APC.
antigen
T killer cell with complementary receptor
infected cell (APC)
T killer cell with complementary receptor binds to the APC.
clone of active T killer cells
clone of T killer memory cells
The T killer cell divides to form two clones: active and memory T killer cells. Cytokines from T helper cells stimulate the differentiation.
infected cell (APC)
The active T killer cells bind to infected cells presenting antigens.
active T killer cell
T killer cell releases chemicals that cause pores to form in the infected cell, causing lysis. The infected cell dies.

Figure 6.41 The role of T killer cells.

The secondary immune response

If infected by the same bacterium or virus again, the immune system responds much faster. The **secondary immune response** involves memory cells and only takes about two to seven days. The B memory cells produced in the primary response can differentiate immediately to produce plasma cells and release antibodies. There is greater production of antibodies, and the response lasts longer as the second peak in the graph in Figure 6.42 shows. The invading viruses or bacteria are often destroyed so rapidly that the person is unaware of any symptoms. The person is said to be **immune** to the disease.

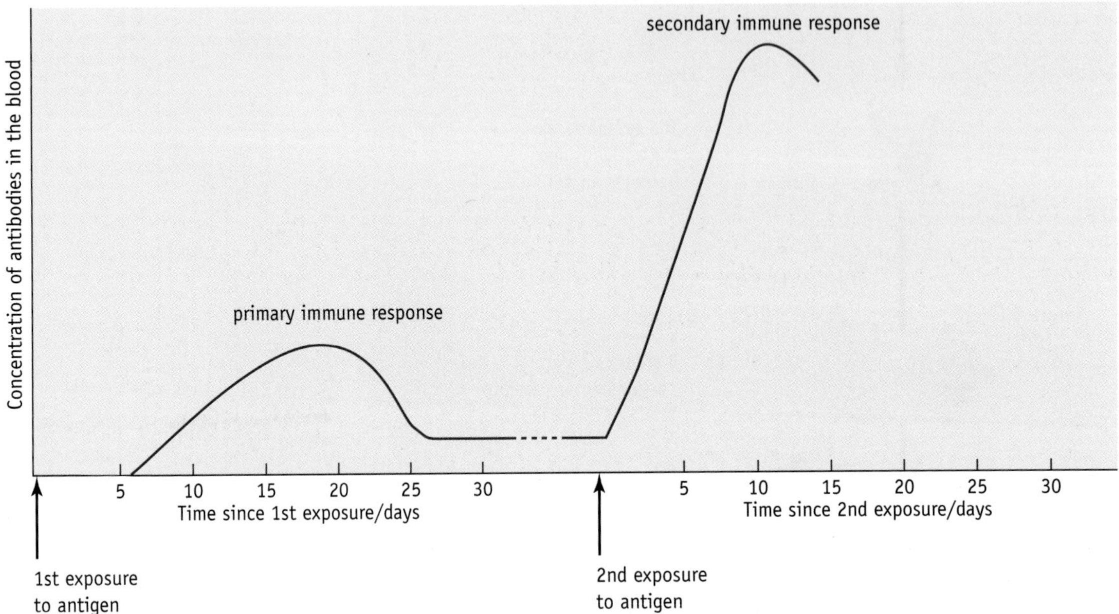

Figure 6.42 Changes in the concentration of antibodies after infection with an antigen.

Q6.29 Which of the B cells will be selected by the antigens in Figure 6.43?

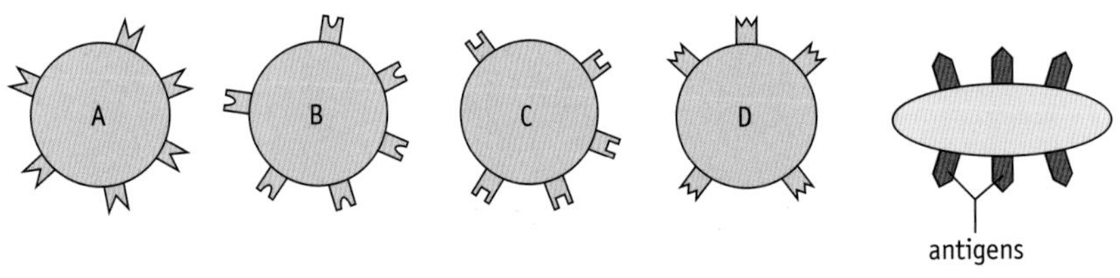

Figure 6.43 Which B lymphocyte will be selected?

Q6.30 The graph in Figure 6.42 shows the antibody concentration in the blood after the first, and then after the second, exposure to a foreign antigen. Use the graph to suggest three reasons why people often have no symptoms the second time they encounter a disease.

> **Checkpoint**
>
> **6.5** Sketch a diagram or use downloaded figures from the mediabank to show how the artworks in Figures 6.38, 6.40A, 6.40B and 6.41 are interconnected.

Avoiding attack by our own immune system

Some of the membrane proteins on the surface of our cells act as 'bar codes'. These proteins mark the body cell as 'self'. They allow us to distinguish between our own cells and those of 'foreign invaders'. There are hundreds of alleles for these proteins, so the combination of proteins on our cell surfaces is unique to each individual, and is not found on the cells of anyone else.

As B and T cells mature in the bone marrow and thymus, any lymphocytes for 'self' membrane proteins are destroyed by **apoptosis** (programmed cell death). Only lymphocytes with receptors for foreign, 'non-self' antigens remain.

Just occasionally the body attacks itself. Particular cells may alter in some way so that they appear 'foreign' and get destroyed by the immune system. An example of this is the auto-destruction of insulin-secreting cells in the pancreas, leading to insulin-dependent diabetes. Other autoimmune diseases are rheumatoid arthritis and multiple sclerosis.

Did you know?

Rhys gets a genetically modified immune system

In 2001 gene therapy transformed the life of baby Rhys Evans. The little boy from Cardiff was born with a faulty immune response. He was constantly ill and spent much of his time living in a sterile 'bubble' in London's Great Ormond Street Hospital (Figure 6.44). Rhys had SCID (severe combined immunodeficiency), a rare sex-linked condition caused by a single mutated gene on the X chromosome. Without treatment he would have died within months.

Rhys's defective gene prevented him from producing T cells, so he was unable to fight off infections. Even chicken pox or cold sores were life-threatening.

At first Rhys's doctors hoped to give him a bone marrow transplant. This would have allowed healthy T cells to develop from the donated stem cells. However, they were unable to find a donor whose 'self' membrane proteins were sufficiently similar to prevent rejection of the donated cells.

The only alternative treatment was gene therapy (see AS Topic 2). Doctors removed bone marrow from Rhys and inserted working versions of his defective gene into the stem cells, using a gibbon virus as the vector. The genetically modified (GM) stem cells were then re-implanted into Rhys, where they multiplied and started to generate healthy immune cells. By the time he was a toddler Rhys was out of his 'bubble' and playing with other children. His new GM stem cells should last a lifetime, producing healthy lymphocytes to protect him against infections. In the words of Dr Adrian Thrasher, who led the medical team at Great Ormond Street Hospital: 'We're very excited by this – he was incredibly sick, with a nasty pneumonia, a life-threatening infection. After his gene therapy, he was running around at home – he's a normal little boy now.'

In 2008, Rhys was continuing to do very well, was attending a normal school, and was happy and healthy.

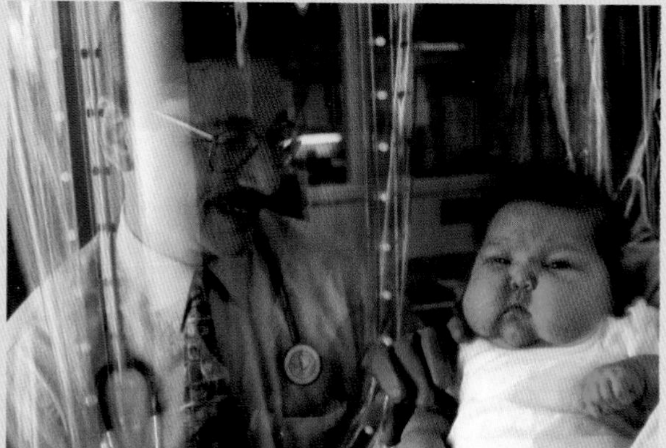

Figure 6.44 Children with SCID usually have to live their whole lives within a sterile bubble to protect them from infection.

6.4 The body's response to TB

Nicki Overton was infected with TB and HIV. What symptoms might she have had? How might her immune system have responded to each of the pathogens? Could the diseases have been the cause of her death?

What is tuberculosis?

Tuberculosis (TB) is a contagious disease caused by the bacterium *Mycobacterium tuberculosis*. It is an ancient disease: evidence of infection has been found in the bones of 3000-year-old Egyptian mummies. In Western Europe and North America, TB became truly rampant during the nineteenth and early twentieth centuries. The poverty and cramped living conditions of city slums promoted rapid spread of infection, resulting in very large numbers of TB cases. Two hundred years ago one UK death in four was caused by TB. The disease was so active in the population that it was known as the 'white plague', or 'consumption', because it 'consumed' the body of the patient.

The composer Chopin, poet John Keats and the Brontë sisters were among many famous figures affected by the disease in the nineteenth century. In 1880 Robert Koch said 'If the number of victims which a disease claims is the measure of its significance, then all diseases ... must rank far behind tuberculosis.'

Today, it is estimated that nearly two billion people – almost one third of the world's population – is infected. There are estimated to be over nine million new cases each year, and around two million people will die from the disease every year. However, most infected people are perfectly healthy. Table 6.7 shows the incidence and mortality figures for 2006. Respiratory or pulmonary TB is the most common form. It affects the lungs and is highly contagious.

WHO region	Number of cases	Deaths from TB	Deaths per 100 000 population
Africa	2 975 000	674 000	80
The Americas	331 000	41 000	6
Eastern Mediterranean	570 000	108 000	17
Europe	430 000	62 000	6
South-east Asia	3 100 000	515 000	27
Western Pacific	1 911 000	291 000	16
Global	9 317 000	1 691 000	25

Table 6.7 World Health Organization estimates of TB incidence and mortality in 2006.

Improved housing and living conditions, coupled with the development of **antibiotics**, saw a decline in the number of TB cases in the UK and other industrialised nations (Table 6.8) during the twentieth century. However, there has recently been a resurgence of the disease, and although it is still relatively rare in the UK, there are approximately 7000 new cases each year with about 500 deaths, mostly in major cities.

Year	Respiratory TB	Non-respiratory TB	Total	Population
1915	68 309	22 283	90 592	35 284 000
1920	57 844	15 488	73 332	37 247 000
1925	58 545	19 228	77 773	38 935 000
1930	50 583	16 818	67 401	39 801 000
1935	39 635	12 435	52 070	40 645 000
1940	36 151	10 421	46 572	39 889 000
1945	42 165	9944	52 109	37 916 000
1950	42 435	6923	49 358	43 830 000
1955	33 580	4554	38 134	44 441 000
1960	20 799	2806	23 605	45 775 000
1965	13 552	2551	16 103	47 671 200
1970	9475	2426	11 901	48 891 300
1975	8208	2610	10 818	49 469 800
1980	6670	2472	9142	49 603 000
1985	4660	1197	5857	49 990 500
1990	3942	1262	5204	50 869 500
1995	4123	1483	5608	51 820 200
2000	4825	1742	6572	52 943 284
2001	4835	1861	6714	52 041 916
2002	4802	1951	6753	52 480 500
2003	4585	1933	6518	52 793 700
2004	4555	2168	6723	53 046 000
2005	5077	2551	7628	53 390 300
2006	5110	2511	7621	53 728 800

Table 6.8 Number of TB cases in England and Wales notified to the Communicable Disease Surveillance Centre.

Q6.31 Use the data in Table 6.8 to work out how the risk of any person in England and Wales contracting TB changed between 1915 and 2005.

Symptoms of the disease

Only 30% of people closely exposed to TB will become infected, and only 5–10% of those infected will develop the symptoms of the disease. Infection may occur when *M. tuberculosis* bacteria are inhaled and lodge in the lungs. Here they start to multiply. There are two phases to the disease, primary infection (the first phase) and active tuberculosis (the second phase).

Primary infection with TB

The immune system responds

The first phase (primary infection) can last for several months, and may have no symptoms.

The immune system of a person infected with TB responds to and deals with the infection. The *M. tuberculosis* causes an inflammatory response from the host's immune system. In an individual with a healthy immune system, macrophages engulf the bacteria. A mass of tissue known as a granuloma forms; this is produced in response to infection. In TB these tissue masses are anaerobic and have dead bacteria and macrophages in the middle (Figure 6.45). They are called tubercules and give the disease its name. After three to eight weeks, the infection is controlled and the infected region of the lung heals. Most primary infection happens during childhood, and over 90% of infections heal without ever being noticed.

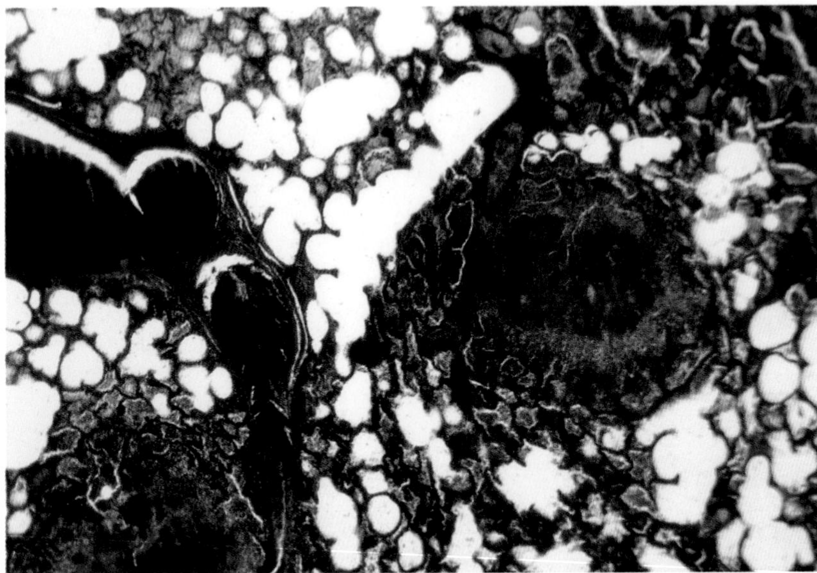

Figure 6.45 Section from a lung of a TB patient. The blue and green areas in the centre and bottom left are tubercules with dead tissue at their centre; the white areas are alveoli. Magnification × 26.

Q6.32 *M. tuberculosis* bacteria are obligate aerobes (need oxygen to survive). Why do they die within tubercules?

Activity

Activity 6.9 is a comprehension exercise on TB. **A6.09S**

Bacteria evade the immune system

In the normal course of events, macrophages engulf and destroy bacteria. However, the *M. tuberculosis* bacteria can survive inside macrophages. The bacteria are taken up by phagocytosis, but once inside they resist the killing mechanisms used by these cells. The bacteria have very thick waxy cell walls, making them very difficult to break down. They can lie dormant for years, and if the immune system is weakened the infection can become active again.

Not only can TB bacteria survive and breed inside macrophages, they are also known to target the cells of the immune system. TB bacteria can suppress T cells. This reduces antibody production and attack by killer T cells.

Weblink

Go to the Wellcome Trust's Big Picture on Epidemics and read about TB and other infectious diseases.

Active tuberculosis

The second phase (active tuberculosis) occurs if the patient's immune system cannot contain the disease when it first arrives in the lungs. This may be because the number of bacteria is too great. Alternatively, an old infection may break out if the immune system is no longer working properly. About 80% of active TB cases are reactivations of previously controlled infections.

The activity of the immune system may be reduced for several reasons. In old age or in the very young (0–5 years) it is less able to respond quickly to pathogens. Malnutrition and poor living conditions also adversely affect the immune system. But the most significant factor in many recent infections is AIDS. HIV, the virus

that causes AIDS, directly targets white blood cells and greatly reduces a patient's ability to fight any infection. TB is such an aggressive disease that it kills many people with AIDS, especially in sub-Saharan African countries where millions of people are infected with HIV.

With active tuberculosis in the lungs (known as respiratory pulmonary tuberculosis), the bacteria multiply rapidly and destroy the lung tissue, creating holes or cavities. The lung damage will eventually kill the sufferer if they are not treated with an appropriate antibiotic.

Q6.33 What effect will the damage shown in Figure 6.45 have on:

a gas exchange in the lungs

b the breathing rate of the patient?

Explain your answers.

Symptoms of active TB

A patient with active tuberculosis will experience a range of symptoms including:

- coughing: the patient may cough up blood
- shortness of breath
- loss of appetite and weight loss
- fever and extreme fatigue.

The role of fever

A person infected with TB (or many other pathogens) experiences fever and night sweats. These occur because, as part of the inflammatory response, fever-causing substances are released from neutrophils and macrophages. These chemicals affect the **hypothalamus** and alter the set point for the core body temperature to a higher temperature. Effectors act to warm the body up to the new set point (see 'Key biological principle: Homeostasis' and 'Temperature control' in Topic 7, Section 7.4). The patient has a high fever, with a temperature in the region of 40.5 °C.

It is thought that the raised temperature enhances immune function and phagocytosis. In addition, bacteria and viruses may reproduce more slowly at the higher temperature. TB itself is temperature-sensitive and will stop reproducing at temperatures above 42 °C. However, this temperature is harmful to the patient. Above 40 °C human enzymes are increasingly denatured; a fever of 42–43 °C is life-threatening.

Q6.34 Why is the denaturing of enzymes life-threatening?

Glandular TB

TB bacteria can also move to infect other parts of the patient's body. Common sites include the bones, lymph nodes and central nervous system. Occasionally one of these is the only site of disease, but more often these infections follow an initial pulmonary infection. In glandular TB the main symptom is enlarged lymph glands, usually in the neck or armpits. Sometimes only lymph glands in the chest are affected, and these can only be seen on an X-ray. Asian people are more likely to get glandular TB, whereas Caucasians get more pulmonary TB.

Did you know?

Scrofula – the King's evil

The most common type of glandular TB is a swollen infection of the lymph nodes in the neck (cervical nodes). If you have these swellings you have scrofula.

Scrofula was known as the 'King's evil' because it was believed that the touch of a king could cure the disease. This superstition goes back at least to Edward the Confessor (1002–1066). Edward I (1239–1307) touched as many as 533 sufferers in one month, but the record must go to Charles II (1630–1685) who 'cured' nearly 100 000 people during his reign. References to the royal miracle cure are found in Shakespeare's *Macbeth*.

Before the development of anti-TB drugs in the twentieth century, surgical removal of all the infected lymph nodes was the only alternative treatment. This would have been done without any real anaesthetic, so you can see why the royal touch was such an attractive option. There are many written accounts of the time supporting the success of kings (and queens) in curing scrofula, but the superstition died out in the nineteenth century.

Q6.35 Suggest a possible explanation for the reported success of kings and queens in curing scrofula.

How is TB diagnosed?

To check if someone suspected of having TB really has the disease, a 'history' is taken from the patient – they describe their symptoms to the doctor. These might include feeling generally tired, a persistent cough, weight loss, fever and night sweats. A range of tests can then be used to check for TB.

Skin and blood tests

A skin test may be done, similar to the test you might have had before your BCG. A small amount of tuberculin is injected under the skin of the forearm. Tuberculin is the name given to extracts from several species of *Mycobacteria*, and is usually composed of a mixture of purified proteins. A positive result shows as an inflamed area of skin around the site of the test injection. Antibodies in the blood cause this inflammation, indicating that TB antigens are already present. Unfortunately the test can give a negative result if the person has latent TB, i.e. the disease is not active. It can also give a false positive result if the person has had a BCG anti-TB vaccination. To overcome these problems, blood tests have been developed that analyse blood samples for T cells specific to antigens only occurring on *Mycobacterium tuberculosis*.

Identification of bacteria

To confirm a positive skin test, a sample of sputum coughed up by the patient is taken and cultured to see what bacteria are present in it. Different bacteria can be identified from the culture using staining techniques. A variety of stains are used. Only some types of bacteria take up particular stains, depending on the make-up of their cell wall.

Activity

Activity 6.10 allows you to try out a staining technique for bacteria. **A6.10S**

Chest X-rays

Chest X-rays are usually taken to discover the extent of the damage and disease in the lungs. The patient whose X-ray is shown in Figure 6.46 has extensive infection. X-rays of other organs may also be taken if the disease is thought to have spread outside the lungs.

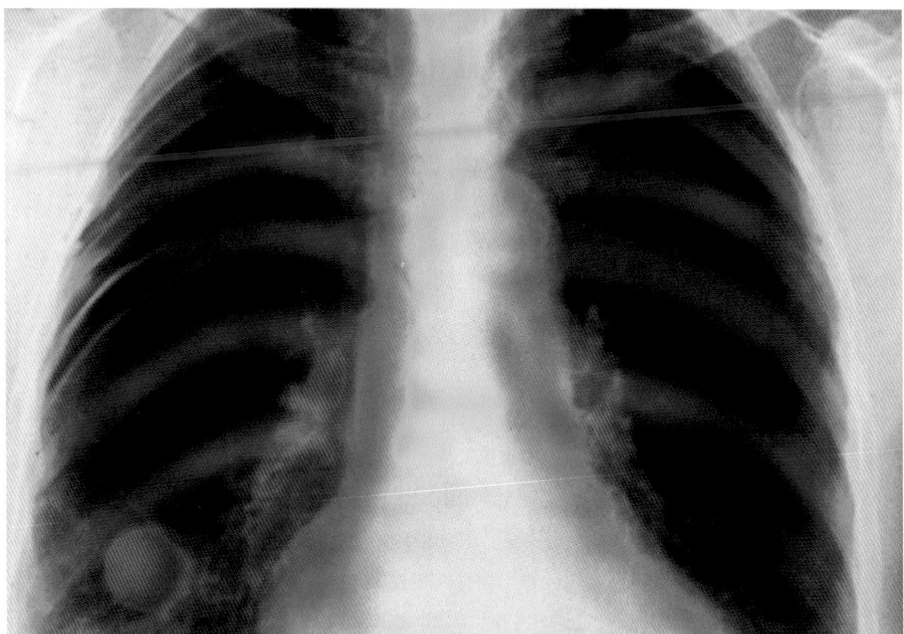

Figure 6.46 Chest X-ray of a patient with pulmonary TB. Damaged tissue (red areas) can be seen in the lungs (black).

Put together, these pieces of information will give the doctor a full picture of the condition of the patient. This can then be used to design the best treatment, usually a combination of antibiotics and improved lifestyle.

If the TB sufferer is contagious, all their recent contacts – friends, family and so on – should be tested as well, to check whether they have been infected.

Q6.36 What lifestyle changes might help the patient avoid any future infection with the disease?

6.5 The body's response to HIV and AIDS

What are HIV and AIDS?

Worldwide there is an AIDS epidemic, which started in the early 1980s. Globally more than 25 million people have died of AIDS since 1981. In 2007, an estimated 33 million people were infected with the virus, with 2.5 million new infections in 2007 (Figure 6.47). These figures are lower than those previously estimated by the World Health Organization and UNAIDS. Reasons for the lower estimates include

> **Weblink**
>
> You can find out the latest figures by visiting the UNAIDS or World Health Organization websites.

Adults and children estimated to be living with HIV/AIDS in 2007

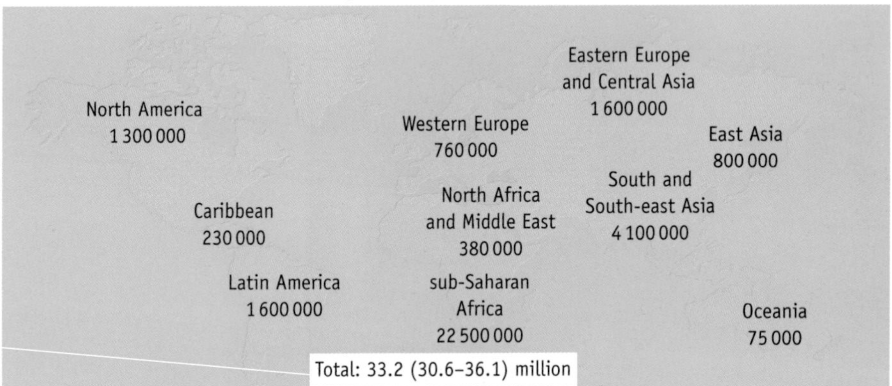

Estimated number of adults and children newly infected with HIV during 2007

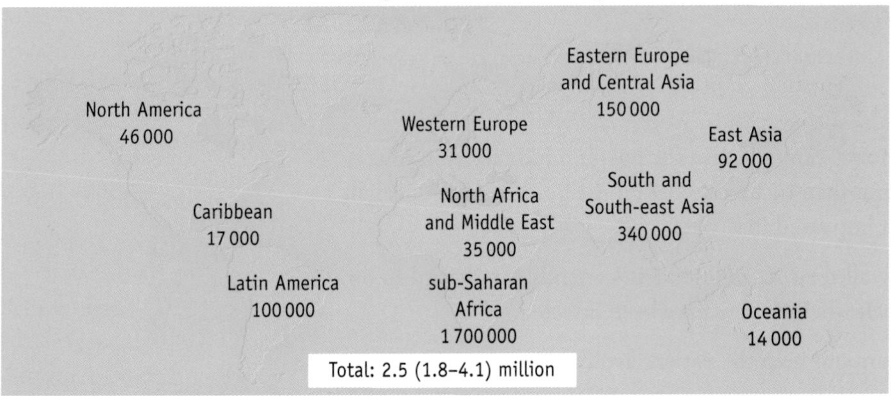

Estimated adult and child deaths due to AIDS during 2007

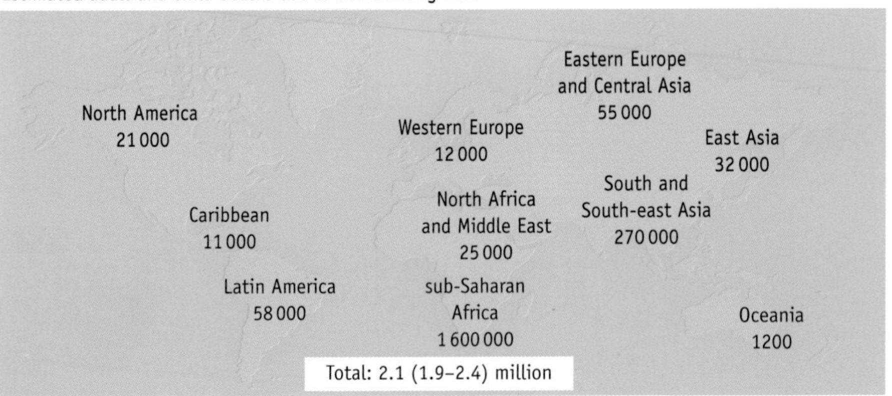

Figure 6.47 HIV infection is not distributed evenly across the globe. The figures are only estimates; the possible range of values is indicated alongside each total figure.
Source: UNAIDS.

improved methodology, better HIV surveillance by countries, and changes in key epidemiological assumptions used to calculate the estimates.

Things are getting better as a result of more investment in HIV prevention programmes and greater access to drugs. In 2008 UNAIDS reported significant gains in preventing new HIV infections and AIDS-related deaths; new HIV infections declined from 3 million in 2001 to 2.7 million in 2007. But although the number of new HIV infections has fallen in several countries, UNAIDS do not consider the AIDS epidemic to be over. Rates of new HIV infections are rising in many countries, with an estimated 7500 new infections every day. At the end of 2006, approximately 73 000 people in the UK were living with HIV infection, of whom about a third were unaware of their infection.

AIDS, **acquired immune deficiency syndrome**, is caused by infection with the human immunodeficiency virus, HIV (Figure 6.48). A syndrome is a collection

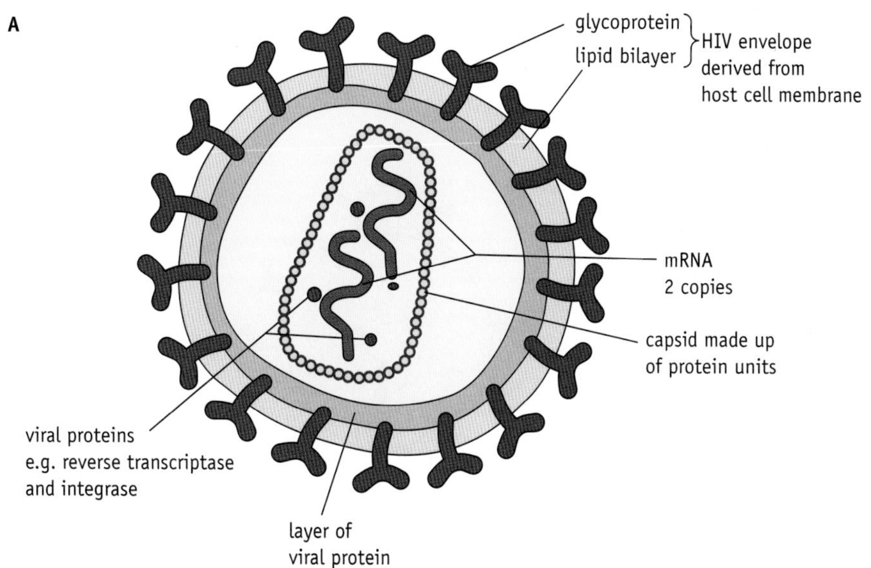

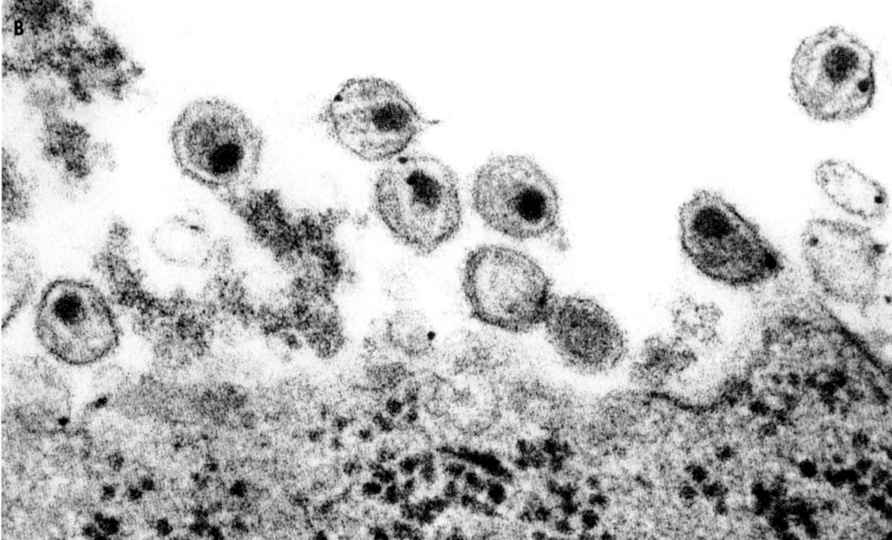

Figure 6.48 **A** The structure of the human immunodeficiency virus. **B** The transmission electron micrograph shows the surface of a T cell with HIV particles.

of symptoms related to the same cause, in this case the action of HIV, which gradually destroys part of the immune system. The symptoms of AIDS are those of opportunistic infections to which the patient becomes susceptible as their immune system is weakened.

In Figure 6.48 you can see the internal structure of HIV. HIV is a structurally complex virus, an example of an enveloped virus. The lipid envelope is formed from the host cell membrane as the new virus particles emerge from the cell cytoplasm. Sticking through the envelope are viral glycoprotein (gp) molecules.

Activity

Activity 6.11 looks at the structure of the HIV virus in detail. **A6.11S**

HIV invades T helper cells

HIV invades T helper cells within the immune system. Particular glycoprotein molecules, called gp120 and located on the virus surface, bind to the CD4 receptors on the surface of the T helper cells. They then combine with a second receptor. This allows the envelope surrounding the virus to fuse with the T helper cell membrane, enabling the viral RNA to enter the cell (Figure 6.49). Macrophages also have CD4 receptors, so the virus can also infect them.

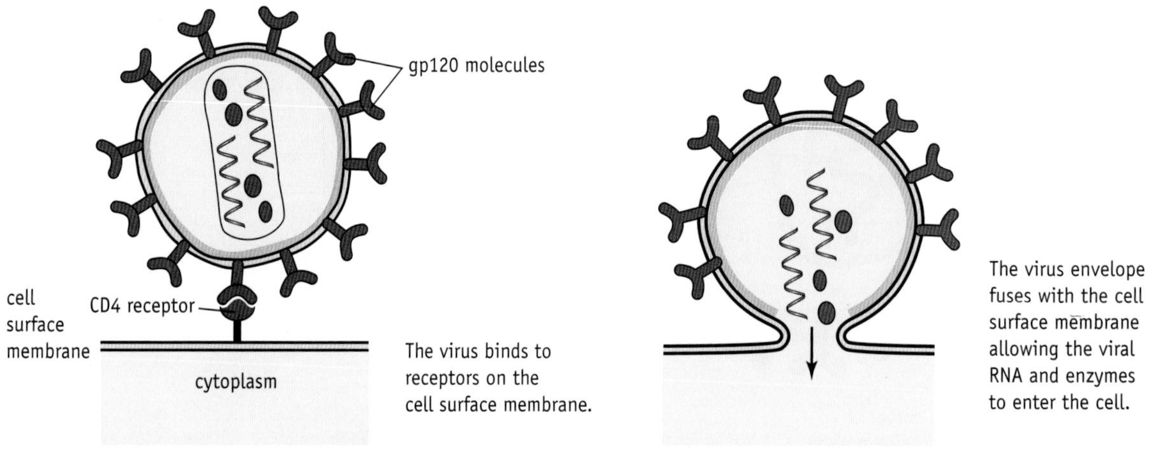

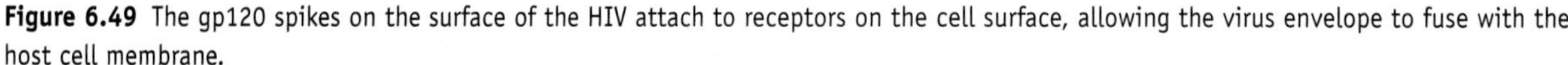

Figure 6.49 The gp120 spikes on the surface of the HIV attach to receptors on the cell surface, allowing the virus envelope to fuse with the host cell membrane.

HIV hijacks the cell's protein synthesis

Once inside the host T helper cell, the virus needs to make the host cell replicate new viral components. HIV nuclear material is in the form of RNA not DNA. Therefore the first step is to reverse normal transcription and manufacture DNA from the RNA template. To do this, the virus uses an enzyme called **reverse transcriptase** (see Figure 6.50). Viruses that contain RNA and use reverse transcriptase in this way are known as retroviruses.

Q6.37 Copy Figure 6.50 and complete the gaps on the DNA. Do this by recalling the principles of base pairing.

Once the HIV DNA strand is produced, it is integrated into the host's DNA by another HIV enzyme, **integrase**. Once the HIV genome is integrated into the host cell's genome, it can be transcribed and translated to produce new viral proteins.

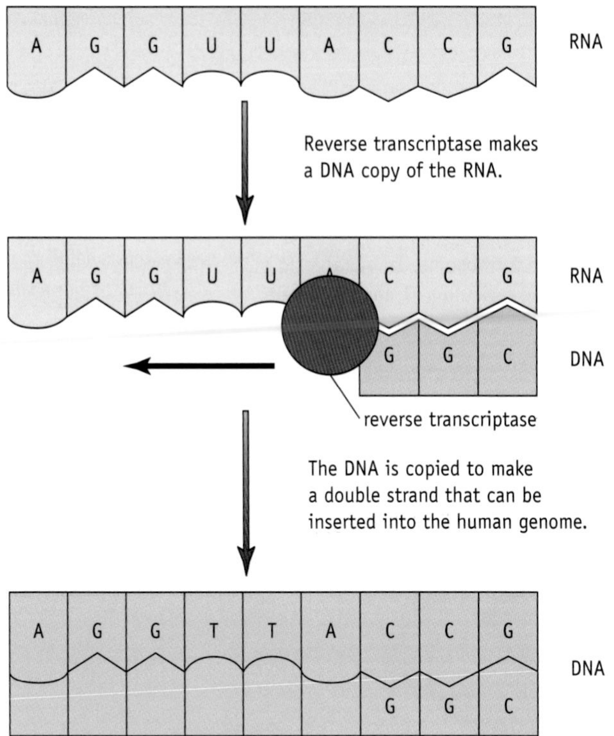

Figure 6.50 The action of reverse transcriptase.

Key biological principle: Protein synthesis revisited

Protein synthesis was outlined in AS Topic 2. Here we look at it in more detail, and see how the HIV DNA integrated into the host DNA is transcribed and translated exactly like the host DNA.

Look at Figure 6.51 to remind yourself about transcription, translation, and the relationship between the base codes on DNA, mRNA and tRNA.

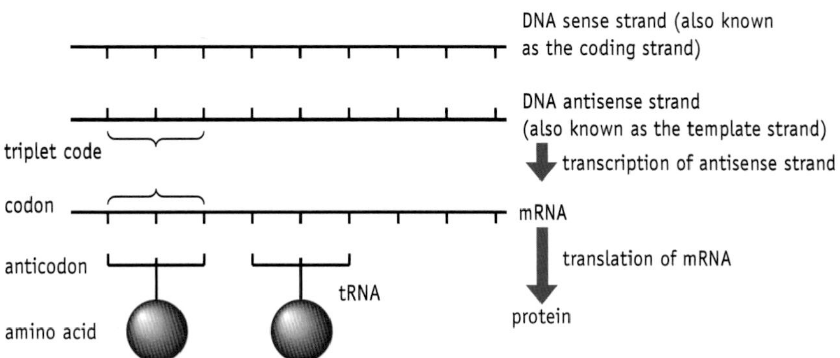

Figure 6.51 Codons, anticodons and complementary base pairing. Remember that the template strand is also known as the antisense strand. This is because, once transcribed, it makes an mRNA molecule with the same base sequence as the DNA coding strand. The coding strand is known as the sense strand.

CONTINUED ▶

Protein synthesis – transcription

In **transcription**, an enzyme called **RNA polymerase** attaches to the DNA (Figure 6.52). The hydrogen bonds between paired bases break, and the DNA molecule unwinds. RNA nucleotides with bases complementary to those on the template strand of the DNA pair up and bond to form an mRNA molecule. When complete, the mRNA molecule leaves the nucleus through a pore in the nuclear envelope.

Once the mRNA passes out of the nucleus it attaches to a ribosome. In eukaryotic cells, ribosomes are usually attached to the endoplasmic reticulum. Translation can now begin.

Activity

Use the computer simulation in **Activity 6.12** to add the protein synthesis detail to what you learned in AS Topic 2. **A6.12S**

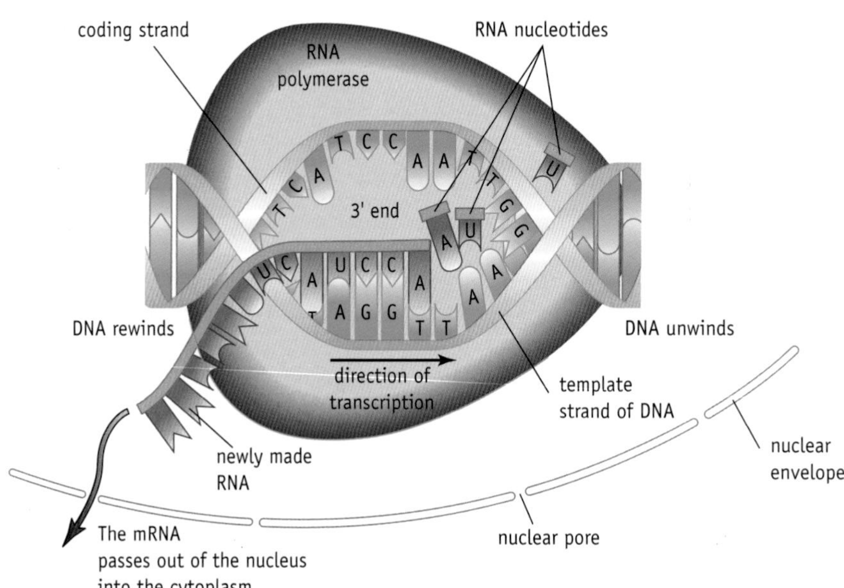

Figure 6.52 Protein synthesis – transcription.

Protein synthesis – translation

Translation starts once the mRNA attaches to the surface of a ribosome. Ribosomes are made up of two subunits. The mRNA attaches to the smaller subunit, so that two mRNA **codons** face the two binding sites of the larger subunit (Figure 6.53).

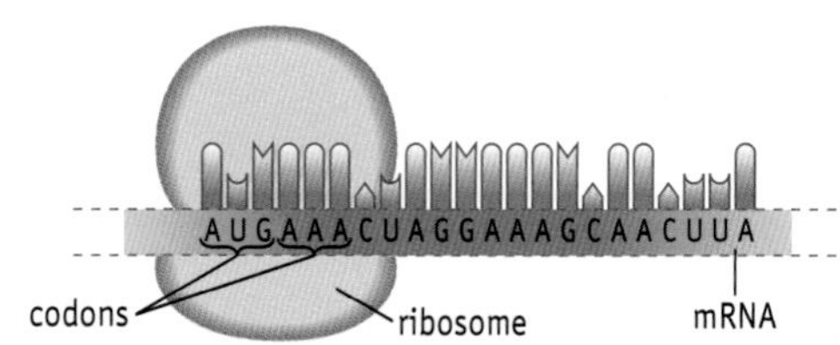

Figure 6.53 mRNA attached to a ribosome.

At one side of a tRNA molecule is a triplet base sequence called an **anticodon.** The three bases of the anticodon are complementary to the mRNA codon for an amino acid (Figure 6.54). For example, the mRNA codons for the amino acid lysine are AAA and AAG. The complementary anticodons are UUU and UUC. Within the cytoplasm, free amino acids become attached to the correct tRNA molecules. Each amino acid has its own specific tRNA that carries it to the ribosome. See Table 6.9 for the full genetic code.

The first codon exposed on the ribosome is always the start code, AUG. This codes for the amino acid methionine. The tRNA molecule with the complementary anticodon, UAC, hydrogen bonds to the codon. The next codon is facing the next binding site. This codon attracts the tRNA–amino acid complex with the complementary anticodon, and it binds to it.

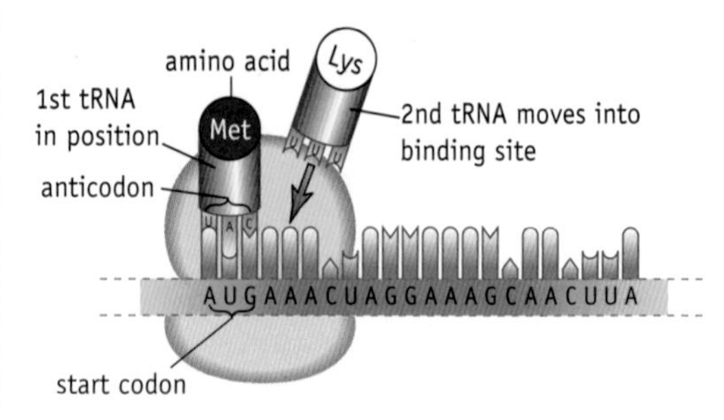

Figure 6.54 tRNA molecules bind.

CONTINUED ▶

The ribosome holds the mRNA, tRNAs, amino acids and associated enzyme in place while a **peptide bond** forms between the two amino acids (Figure 6.55). The peptide bond is a condensation reaction between the amine group of one amino acid and the carboxylic acid group of the next, and forms a dipeptide.

Once the peptide bond has formed, the ribosome moves along the mRNA to reveal a new codon at the binding site. The first tRNA returns to the cytoplasm (Figure 6.56). The whole process is repeated, and translation continues until the ribosome reaches a stop signal: UAA, UAC or UGA.

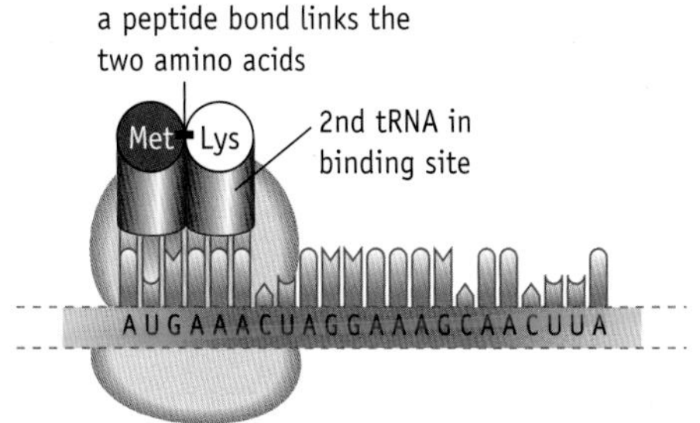

Figure 6.55 A peptide bond forms between the two amino acid molecules.

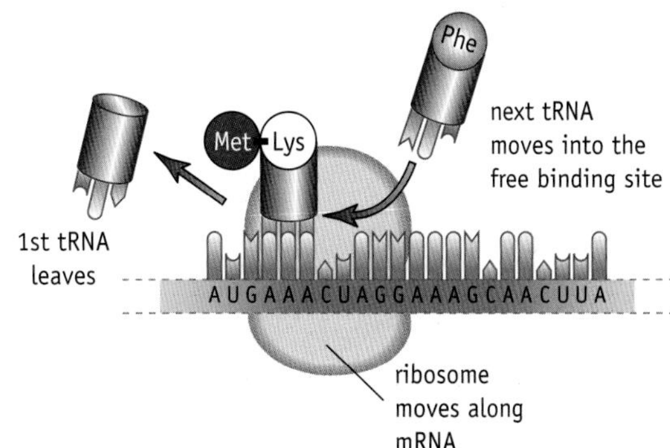

Figure 6.56 The ribosome moves along the mRNA, allowing another complementary tRNA molecule to bind.

The nature of the genetic code

We have seen that the genetic code is read as sequences of three bases – triplet codes on DNA and codons on mRNA. The code is non-overlapping. No base of one triplet contributes to part of the next triplet. There are 64 possible combinations of the four bases in DNA if they are grouped into triplets. One triplet sequence is a start code and three are stop codes. This leaves on average three triplet codes for each of the 20 naturally occurring amino acids in protein molecules. There are thousands of them in a single protein molecule.

As several triplets can code for the same amino acid, the genetic code is described as a degenerative code. In some cases, all the codes with the same first two letters code for the same amino acid. For instance, all triplet codes starting with GU code for valine. Some amino acids such as tryptophan are coded for by a single triplet code. Table 6.9 shows all the mRNA codons and their corresponding amino acids. You do not need to learn the triplet codes or codons of the genetic code.

mRNA codons

First base	Second base U		Second base C		Second base A		Second base G		Third base
U	UUU	Phe	UCU	Ser	UAU	Tyr	UGU	Cys	U
U	UUC	Phe	UCC	Ser	UAC	Tyr	UGC	Cys	C
U	UUA	Leu	UCA	Ser	UAA	Stop	UGA	Stop	A
U	UUG	Leu	UCG	Ser	UAG	Stop	UGG	Trp	G
C	CUU	Leu	CCU	Pro	CAU	His	CGU	Arg	U
C	CUC	Leu	CCC	Pro	CAC	His	CGC	Arg	C
C	CUA	Leu	CCA	Pro	CAA	Gln	CGA	Arg	A
C	CUG	Leu	CCG	Pro	CAG	Gln	CGG	Arg	G
A	AUU	Ile	ACU	Thr	AAU	Asn	AGU	Ser	U
A	AUC	Ile	ACC	Thr	AAC	Asn	AGC	Ser	C
A	AUA	Ile	ACA	Thr	AAA	Lys	AGA	Arg	A
A	AUG	Met	ACG	Thr	AAG	Lys	AGG	Arg	G
G	GUU	Val	GCU	Ala	GAU	Asp	GGU	Gly	U
G	GUC	Val	GCC	Ala	GAC	Asp	GGC	Gly	C
G	GUA	Val	GCA	Ala	GAA	Glu	GGA	Gly	A
G	GUG	Val	GCG	Ala	GAG	Glu	GGG	Gly	G

Key:
Ala = alanine
Arg = arginine
Asn = asparagine
Asp = aspartic acid
Cys = cysteine
Gln = glutamine
Glu = glutamic acid
Gly = glycine
His = histidine
Ile = isoleucine
Leu = leucine
Lys = lysine
Met = methionine
Phe = phyenylalanine
Pro = proline
Ser = serine
Thr = threonine
Trp = tryptophan
Tyr = tyrosine
Val = valine

Table 6.9 The dictionary of mRNA codons for amino acids. Rather confusingly, the amino acids' names are shortened to three letters.

CONTINUED ▶

mRNA splicing

Between transcription and translation, messenger RNA is often edited, with some sections being removed and other sections spliced together. The non-coding introns are removed. The remaining sequences, which will be expressed, are exons (for *ex*pressed regi*ons*) (Figure 6.57). This means that several proteins can be formed from one length of mRNA if it is spliced in different ways.

The discovery of mRNA splicing has modified what was once the central dogma of molecular biology, from 'one gene → one protein' to 'one gene → several related proteins'. Estimates of the number of genes in a human were initially in the region of 100 000, but it is now thought to be around 20–25 000. Some of this reduction is accounted for by these post-transcriptional changes to messenger RNA.

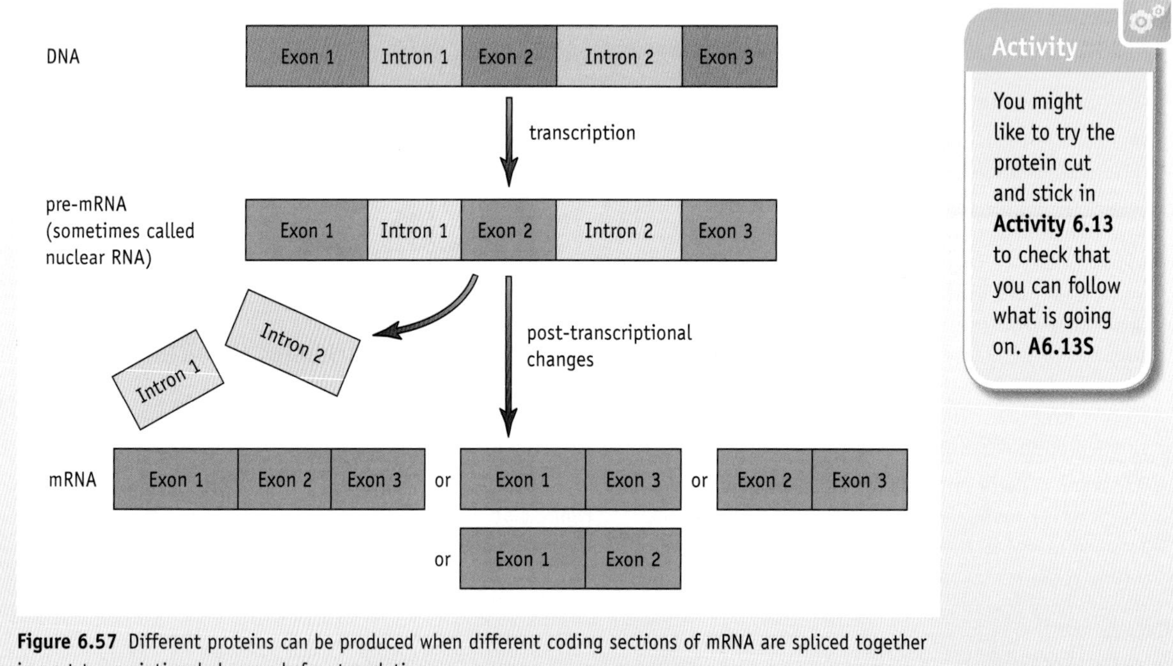

Figure 6.57 Different proteins can be produced when different coding sections of mRNA are spliced together in post-transcriptional changes before translation.

Activity

You might like to try the protein cut and stick in **Activity 6.13** to check that you can follow what is going on. **A6.13S**

New virus particles destroy T helper cells

The new viral proteins produced, together with glycoproteins and nuclear material, are assembled into new viruses (Figure 6.58 overleaf). The new viruses bud out of the T cell, taking some of the host cell surface membrane with them as their envelope, and killing the cell as they leave.

Infected T helper cells will also be destroyed by T killer cells. Thus as the number of viruses increases, the number of host T helper cells decreases. The loss of T helper cells results in macrophages, B cells and T killer cells not being successfully activated and therefore not functioning properly. This means that the infected person's immune system becomes deficient.

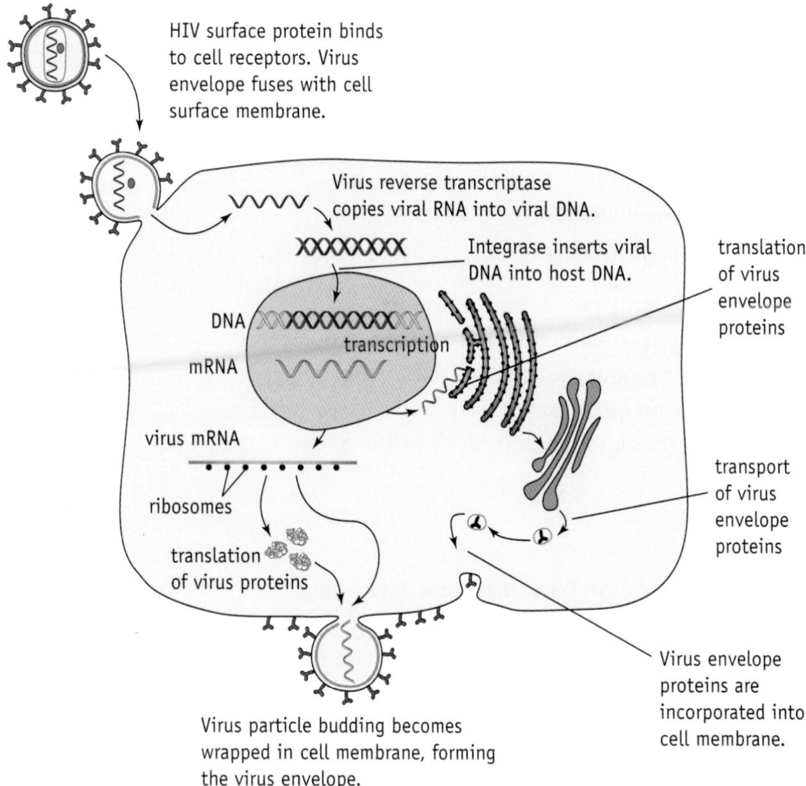

Figure 6.58 A simplified view of the HIV life cycle.

The course of the disease – AIDS

You probably know that AIDS does not always follow HIV infection straight away. There are several stages in the course of the disease once someone is infected, and this depends on many factors. The health of the host before infection, their genetic resistance to infection, the quality of their immune response to initial infection, their lifestyle and nutrition, and the availability of the currently expensive and complicated drug treatment, all contribute to an individual's disease and life expectancy. There is currently, and for the foreseeable future, no cure for AIDS. AIDS nearly always follows HIV infection eventually, and has 100% mortality unless treated by a cocktail of drugs.

The acute phase

When a person is first infected by HIV, there is an acute phase of infection. The following events occur during this phase:

- HIV antibodies appear in the blood after 3–12 weeks.
- The infected person may experience symptoms such as fever, sweats, headache, sore throat and swollen lymph nodes, or they may have no symptoms.
- There is rapid replication of the virus and loss of T helper cells.
- After a few weeks, infected T helper cells are recognised by T killer cells, which start to destroy them. This greatly reduces the rate of virus replication but does not totally eliminate it.

The chronic phase

There is now a prolonged chronic phase. The chronic phase is sometimes called the 'latent' phase. The virus continues to reproduce rapidly, but the numbers are kept in check by the immune system. The phase is characterised by the following features:

- There may be no symptoms during this phase, but there can be an increasing tendency to suffer colds or other infections, which are slow to go away.
- Dormant diseases like TB and shingles can reactivate.

The chronic phase can last for many years. In fit young people with a healthy lifestyle it can last for 20 years or more, especially if combined with drug treatment. Unfortunately, most HIV infection occurs in developing countries with little money available to provide drugs, and amongst people who often do not have access to sufficient food or clean water. These people may go on to develop AIDS within a few years of infection.

The disease phase

Eventually, the increased number of viruses in circulation (viral load) and a declining number of T helper cells indicates the onset of AIDS, the disease phase. During this phase, the decrease in the number of T helper cells leaves the immune system vulnerable to other diseases. A normal T helper cell count is over 500 per mm^3 of blood; below 200 per mm^3 of blood there is a high risk of infection by diseases that take advantage of the weakened immune system – so-called **opportunistic infections**, which can rapidly be fatal.

Opportunistic infections often found in people with AIDS include pneumonia and TB. There can be significant weight loss. Patients can also develop dementia (memory and intellect loss). People with AIDS are also susceptible to tumours such as Kaposi's sarcoma. This cancer is readily identified by purple/black patches on the skin. This is rare in the general population, and it was the sudden occurrence of several cases of people with Kaposi's sarcoma, noticed by doctors in the USA in 1981, which contributed to the identification of AIDS as a new disease.

Checkpoint

6.6 Produce a flow chart that summarises the course of the disease for TB and for AIDS.

Did you know?

Where diseases come from

The common diseases that kill or have killed huge numbers of people, such as smallpox, TB, malaria, flu, cholera and AIDS, have evolved from diseases of animals that have been transferred to humans. Molecular biologists can identify the closest relatives of these disease microbes. For instance, measles evolved from rinderpest, a disease of cattle, TB from *Mycobacterium bovis* in cattle, and HIV from a virus found in wild monkeys.

Close contact between livestock and humans increases the chances of transfer from animals to humans. We can occasionally pick up infections from wild or domestic animals, such as leptospirosis from dogs and brucellosis from cattle. These pathogens are at an early stage of evolution, and cannot yet spread between human hosts, unlike more specialised human pathogens.

Avian influenza virus does not normally infect humans, but cases of human infection have been reported as a result of direct contact with infected birds. Health authorities monitor outbreaks of human illness associated with avian influenza because of concerns about the potential for more widespread infection in the human population.

Many epidemic diseases evolved and became widespread in Europe when people lived in close contact with their animals. The diseases were then taken to the New World, where the native people of the Americas fell victim to the new diseases. The conquerors of the Aztecs and Native Americans used violence, but their microbes were the real conquerors.

6.6 Could the infections have been prevented?

Tests showed that Nicki Overton had a high HIV viral load and low T helper cell count. This suggests that she had entered the disease phase of AIDS, and she would have been very susceptible to opportunistic infections. She had been staying in hostels while travelling, and the crowded conditions could have exposed her to TB bacteria. Her weakened immune system would have been unable to deal with TB, which resulted in her death. Could the body have prevented entry of the pathogens? Could she have developed any immunity to the diseases? Could she have been vaccinated against the diseases or received treatment? Find out in the following section.

Preventing entry of pathogens

Normally the body has mechanisms to prevent the entry of pathogens. These include physical barriers and chemical defences.

The skin

The skin's keratin (hard protein) outer layer is effective in stopping entry of microorganisms. Entry can occur through any wounds, but blood clotting seals the wound and thus reduces the number of microorganisms gaining access. Large numbers of microbes (known as the **skin flora**) live on the skin surface. These are harmless and prevent colonisation by other bacteria. The bacteria that occur naturally on the skin surface are well adapted to the environment there. Other bacteria are not so well suited to the conditions created by salty sweat and excreted chemicals such as urea and fatty acids.

Mucous membranes

The mucous membranes that line the airways and gut provide easier routes into the body. This is because, lacking any keratin layer, the surface is always moist, making it a more favourable environment for bacterial growth. As described in AS Topic 2, entry of microbes to the lungs is limited by the action of **mucus** and **cilia**. The mucus, secreted by goblet cells in the trachea and bronchi, traps microbes and other particles; then beating cilia carry the mucus up to the throat where it is swallowed. Secretions in the mouth, eyes and nose contain **lysozyme**, an enzyme that breaks down bacterial cell walls, causing the cell to burst.

Digestive system

Stomach acid

Gastric juices secreted by gastric glands in the stomach walls contain hydrochloric acid, giving a pH of less than 2.0. This kills most bacteria that enter with food, but is also the optimum pH for the digestive enzyme pepsin, which is also secreted in the gastric juices.

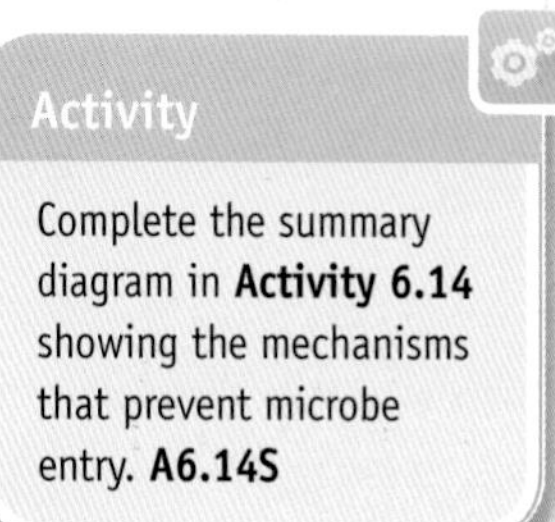

Activity

Complete the summary diagram in **Activity 6.14** showing the mechanisms that prevent microbe entry. **A6.14S**

Gut flora

Bacteria are found in the small and large intestines. There are hundreds of different species found within the intestines, including harmless strains of *Escherichia coli*. These natural flora benefit from living within the gut where conditions are ideal: warm, moist, and with plentiful food supplies. The host (each of us) also benefits from their presence; it is a mutualistic relationship. The bacteria may aid the digestive process, and competitively exclude pathogenic bacteria, competing with the pathogens for food and space. The bacteria also secrete chemicals such as lactic acid that are useful in the defence against pathogens.

Extension

Read about the development of probiotic foods, which contain useful bacteria, in **Extension 6.3. X6.03S**

Becoming immune

You can become immune to a specific disease in various different ways. Use the profiles in Table 6.10 to find out the differences between:

- **active artificial immunity**
- **passive artificial immunity**
- **active natural immunity**
- **passive natural immunity**.

Type of immunity	Example
passive natural immunity	Catherine has just been born. Her immune system is undeveloped, but antibodies that have been crossing the placenta from her mother during the past three months protect her.
passive natural immunity	Simon is one month old and his immune system is not fully developed. He is protected from infectious diseases by antibodies received from his mother's milk. The colostrum, or first-formed milk, which he had during his first few days, was particularly rich in maternal antibodies. In addition, he is still getting protection from the maternal antibodies that crossed the placenta before he was born.
active natural immunity	Owen is three years old and is recovering from chicken pox. His body's specific immune response to the foreign antigens helps to destroy the virus and make him immune to chicken pox in the future. He has a supply of antibodies and B memory and T memory cells in his system that will respond quickly if he is re-infected with the pathogen.
active artificial immunity	Sital is about to start school. She has been vaccinated against diphtheria, tetanus, whooping cough, meningitis C, polio, measles, mumps and rubella. The antigens in the vaccines did not cause the diseases, but in each case stimulated a specific immune response. This has given her immunity to all eight of these serious and potentially deadly illnesses.
active artificial immunity	Students in class 9K are being vaccinated against meningitis C. This vaccine was introduced in 1999, initially for young people. It was hugely successful. By 2001 there had been a 90% reduction in the UK in both cases and deaths. The MenC vaccine is now included as a primary immunisation given to babies when they are three or four months old. The MenC vaccine does not give immunity to all types of meningitis and anyone showing symptoms, even if they have been vaccinated, should still get urgent medical treatment.
passive artificial immunity	Chris got a long thorn in her finger while she was gardening, and is at risk from tetanus bacteria that may be in the wound. Because she has not been vaccinated against tetanus, she needs immediate protection. She is therefore being given an injection of tetanus antibodies that should stop her from getting the disease.

Table 6.10 Natural and artificial immunity.

Q6.38 Which types of immunity:

a give immediate protection

b develop after a time lag

c last for a short time, perhaps only a few weeks

d give long-lasting protection

e involve memory cells

f require medical treatment?

Checkpoint

6.7 Write a definition for each of the four types of immunity: passive natural immunity, active natural immunity, active artificial immunity and passive artificial immunity.

Being vaccinated

When you are vaccinated for a particular disease, your immune system responds to the vaccine in the same way as it responds to the disease. Antibodies are produced and memory cells ensure lasting protection. It is important to realise that vaccination doesn't stop you contracting the pathogen. But if you do come into contact with the pathogen in the future, you can rapidly destroy it before the onset of any symptoms – you are said to be immune.

A vaccine must contain one or more antigens that are also found on the pathogen or the toxin they produce. This can be achieved in several different ways. Vaccines may contain the following:

- Attenuated viruses – these viruses have been weakened so they are harmless. For example, the measles vaccine contains attenuated measles viruses.
- Killed bacteria – one of the commonly used whooping cough vaccines contains whooping cough bacteria that have been killed.
- A toxin that has been altered into a harmless form – the diphtheria vaccine contains a harmless form of the toxin produced by *Corynebacterium diphtheriae,* a bacterium that causes diphtheria.
- An antigen-bearing fragment of the pathogen – some of the newer vaccines, like that for meningitis C, are of this type.

After the initial vaccination it is usually necessary to have one or more 'boosters' to ensure long-lasting immunity. In the case of influenza, vaccination with a completely new vaccine is required every year in order to protect against new strains of the virus.

Vaccination for infectious diseases protects not just the individual but also the community. When enough people are immunised, the disease is less likely to be transferred from one person to another, and so there is less disease in the community as a whole. This means that anyone who did not respond to the vaccine or could not be given it for medical reasons is still, in effect, protected. This group protection is called **herd immunity**. For measles, which is extremely infectious, 95% of the population need to be vaccinated to achieve herd immunity.

Q6.39 **a** Use Figure 6.59 to find out the year in which herd immunity to measles was first achieved in England and Wales.

b Suggest why, before 1968, there was a measles epidemic every two years.

c Why were there still epidemics between 1968 and 1988?

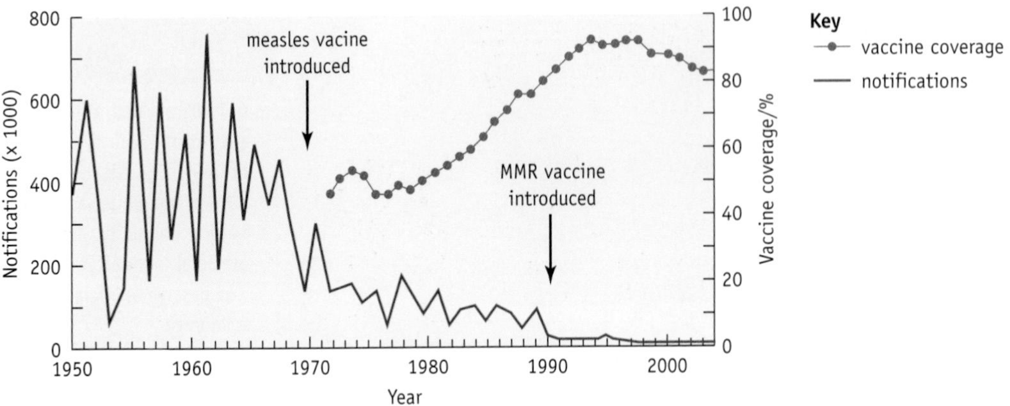

Figure 6.59 Annual measles notifications and vaccine coverage in England and Wales, 1950–2003. *Source:* Office for National Statistics and Health Protection Agency.

There is currently no vaccine against AIDS, but the BCG vaccine against TB has been widely used in the UK since the 1950s. The BCG in schools programme meant all young people were given the BCG vaccine around the time they started secondary school, providing they showed no reaction to the Heaf skin test. An extract from TB bacteria is injected just below the skin using a ring of small needles. If the area of skin becomes red and swollen after about three days, the test is positive, and the individual has immunity to the disease. About 8% of UK children test positive. If the test is negative, a BCG vaccination is given. This contains live, chemically attenuated bacteria. (BCG stands for Bacille (bacteria) of Calmette and Guerin, after the two French doctors who developed the vaccine.)

By the beginning of the twenty-first century, the epidemiology of tuberculosis had changed from a disease of the general population to one predominantly affecting high risk groups. TB in Britain is now largely concentrated in the major cities, with over 40% of cases in London. The highest rates are in particular risk groups; 60% of reported cases are in people born abroad, the rate being higher in certain ethnic groups in the first few years after they enter Britain. Rates remain high in the children of these immigrants, wherever they were born. Other risk groups include those in contact with existing TB cases, those frequently visiting high risk areas abroad, the homeless, and those with HIV infection.

Reflecting this epidemiological shift, major changes were made to the BCG vaccination policy in 2005. The new approach involves vaccinating the following groups:

- All newborn infants living in areas where the incidence of TB is 40/100 000 or greater.
- Infants whose parents or grandparents were born in a country with a TB incidence of 40/100 000 or higher.
- All previously unvaccinated new immigrants from countries with a high incidence of TB.

At the same time, the Heaf skin test was replaced by the Mantoux test (Figure 6.60). Newborns are generally vaccinated without a skin test.

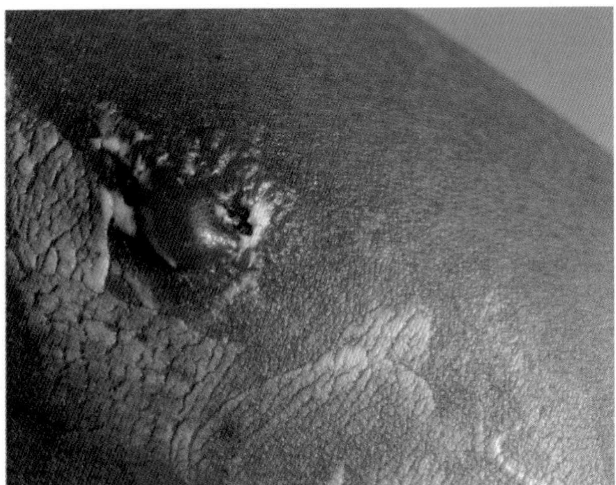

Figure 6.60 A positive result for a Mantoux skin test. A small amount of tuberculin is injected just under the skin. The reaction to the test is viewed 48–72 hours after injection. If there is no reaction or only a small hard lump forms, a BCG is given. The Mantoux test does not measure immunity – it is used to detect latent TB infection.

Did you know?

Edward Jenner and vaccination

Edward Jenner (1749–1823), a British physician and naturalist, investigated the observation that milkmaids who had suffered from cowpox did not get the more serious and often fatal smallpox. In 1796 he inoculated a healthy boy, James Phipps, with fluid from a cowpox blister on a milkmaid's finger. After the boy had recovered from cowpox, Jenner inoculated him with smallpox. The boy did not develop smallpox. Jenner named the immunising process 'vaccination' after the cowpox virus (vaccinia, Figure 6.61). Vaccination against smallpox was made compulsory in Britain in 1853, and smallpox was declared extinct worldwide outside laboratories in 1980.

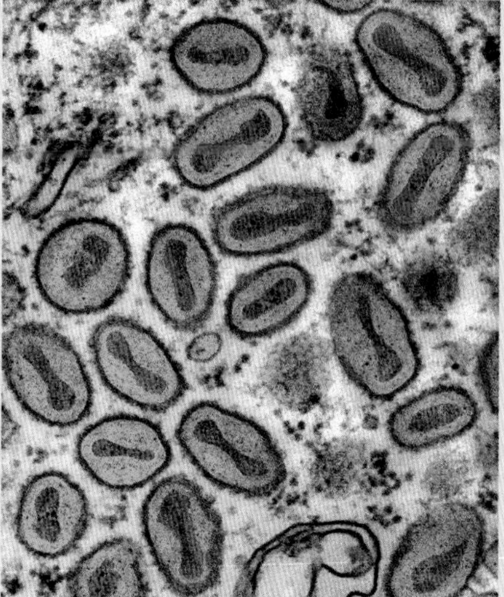

Figure 6.61 Coloured transmission electron micrograph (TEM) of sectioned vaccinia virus particles. The virus is covered by membrane layers (green) taken from the host cell that replicated the virus.

Long before Jenner came up with the idea of vaccination in 1796, people in the Middle East, Africa and Asia were inoculating against smallpox. Lady Mary Wortley Montague, the wife of the British Ambassador in Constantinople, wrote in 1717 describing the process:

> *The small pox, so fatal, and so general amongst us, is here entirely harmless, by the invention of engrafting ... the old woman comes with a nutshell of the matter [pus] of the best sort of small pox, and asks which vein you please to have opened. She immediately rips open that you offer her, with a large needle (which gives you no more pain than a common scratch) and puts into the vein as much matter as can lie on the head of her needle ... Every year, thousands undergo this operation ... There is no example of any one that has died in it.*

Lady Mary was so impressed that she had her son inoculated in Constantinople and, on her return to England, tried hard to spread the practice in Britain. She did manage to convince the King and Queen. After an ethically questionable set of tests on prisoners and orphans, they had their own children inoculated in 1721. But despite royal approval, inoculation never became very widespread in Europe. This may have been because sometimes those who had been inoculated did develop full-blown smallpox.

Are vaccinations dangerous?

Some vaccinations cause mild soreness at the site of the injection, fever or a general feeling of being unwell.

Research studies have indicated that, very occasionally, much more serious and long-term damage can be linked to certain vaccines, some of which have now been withdrawn. Two brands of MMR vaccine were withdrawn from use in the UK because they contained a particular strain of mumps virus that caused a very few children to develop mild meningitis. The whooping cough vaccine can, on very rare occasions, cause brain damage. However, a child who gets whooping cough is at *more* risk of brain damage and even death. The *balance of risk and benefit is crucial.* It is always necessary to weigh up the probability of getting the disease, together with its risks and complications, against the effectiveness of the vaccine and the risk of adverse (i.e. harmful) reactions.

6.7 Are there treatments for AIDS and TB?

Treating AIDS

There is no treatment to get rid of HIV in someone who is infected with the virus because the virus is hidden inside the T helper cells. But there are drugs available that reduce the production of more viruses. These are known as antiretroviral drugs. There are two main types:

- Reverse transcriptase inhibitors, which prevent the viral RNA from making DNA for integration into the host's genome.
- Protease inhibitors, which inhibit the proteases that catalyse the cutting of larger proteins into small polypeptides for use in the construction of new viruses.

An integrase inhibitor and fusion inhibitor drugs have also been approved for use.

Q6.40 Suggest mechanisms for how integrase inhibitor and fusion inhibitor might work in reducing the effects of HIV infection.

HIV can develop resistance to anti-HIV drugs and therefore these drugs are often given in combination. If the virus becomes resistant to one drug, it may still be susceptible to the other drugs being taken. Access to antiretrovirals is increasing across the world and with it comes declines in the number of AIDS-related deaths.

Treating tuberculosis

Active TB bacteria can be killed by antibiotics. Usually a combination of four antibiotic drugs are given for two months, with two continued for a further four months. This treatment ensures that any dormant bacteria are destroyed.

What are antibiotics?

More than 3000 years ago, the Egyptians, the Chinese, and Central American Indians used moulds to treat rashes and infected wounds. They did not understand what caused the diseases or how the mould helped to treat them; in some cases they believed that the moulds drove away the evil spirits that caused the disease.

In 1928, a research scientist by the name of Alexander Fleming was working in a hospital in London. He was studying the bacterium *Staphylococcus aureus*. One day Fleming found a spot of green growing in one of the agar plates. He noticed there was a clear, bacteria-free ring of agar gel around the mould. He thought this might mean that the mould had killed the bacteria. Fleming watched the dish over the next couple of days and saw that, the more the mould spread, the more bacteria were killed. He noticed that the mould produced tiny droplets of fluid on its surface (Figure 6.62), and wondered if this chemical was destroying the bacteria.

Figure 6.62 The fungus *Penicillium notatum* growing on an agar plate.

Fleming drew off the liquid. He found that this liquid could kill bacteria in a test tube. The name of the mould was *Penicillium notatum*, so he decided to call the liquid penicillin. Later, other scientists discovered that penicillin could cure certain infections in mice and rabbits without harming the animals in any way. Fleming was working on the development of vaccines, so did not take his discovery any further.

Q6.41 One explanation for the effects of penicillin was that it is a toxic chemical that kills bacteria. Another was that it is an enzyme that digests the bacteria. Describe how you could carry out an investigation to distinguish between these two hypotheses.

? Did you know?

Penicillin production

In 1939 a scientific team led by Howard Florey and Ernst Chain was set up in Oxford to produce usable quantities of penicillin. Ernst Chain developed a freeze-drying technique for purifying penicillin, but mass production was difficult. They started to grow *Penicillium notatum* in containers in their labs. They used this penicillin for the first clinical trials, but they could not produce enough.

Howard Florey knew that penicillin would be useful to treat infected wounds in soldiers during the Second World War, but British chemical companies were producing explosives and were unable to mass-produce penicillin as well. When the USA joined the war in 1941, American chemical companies started to produce the drug. However, it was not until 1944 that there was enough penicillin to meet all the needs of the allies.

The Nobel Prize for Medicine in 1945 was awarded to Fleming, Florey and Chain for the discovery and production of penicillin.

Penicillin is now produced on a large scale. It can be made by microbial fermentation, or by a semi-synthetic process in which one antibiotic is converted into another. The fungus *Penicillium chrysogenum* is grown in stainless steel fermenters containing up to 200 000 dm^3 of nutrient medium (Figure 6.63). The penicillin is released into the fluid. Adjusting the pH at the end of the fermentation process causes the penicillin to crystallise and it can then be extracted. Then, using immobilised enzyme technology (see Activity 1.15), the penicillin can be modified to give a range of novel penicillins.

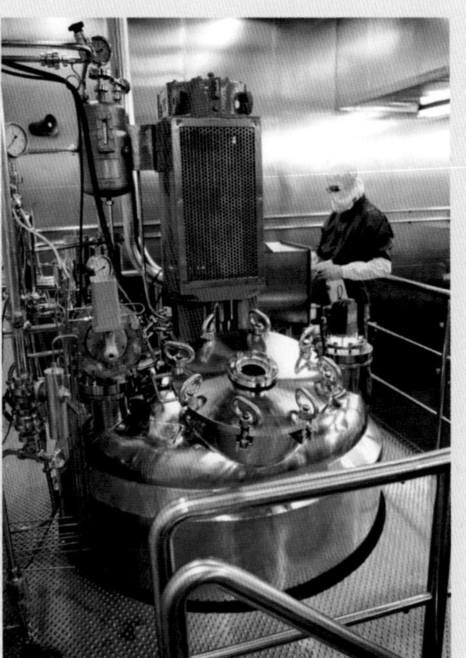

Figure 6.63 Large-scale fermenters being used to produce penicillin.

Searching for new antibiotics

Penicillin had been proven to work against pneumonia, scarlet fever, and several other diseases. However, it had no effect on the bacteria that caused typhoid, TB, or many other diseases.

In 1932, an American called Selman Waksman showed that when *Mycobacterium tuberculosis*, the cause of tuberculosis, was added to certain types of soil, the bacteria died. In 1943 Waksman and a research student called Albert Schatz tested hundreds of soil organisms for antibacterial activity. After three months they isolated an organism called *Streptomyces griseus* from a sick chicken. They found it produced an antibiotic that they called streptomycin, which they then purified. Streptomycin was the first drug found that could cure tuberculosis.

In 1953, Waksman decided to call these antibacterial chemicals antibiotics, and he defined the term as '... a chemical substance, produced by microorganisms, which has the capacity to inhibit the growth and even to destroy bacteria and other microorganisms, in dilute solutions'. In 1952 Selman Waksman was awarded the Nobel Prize for the discovery of streptomycin.

Q6.42 Imagine that you are a scientist working in the 1940s. You have a limited amount of streptomycin and you have been asked to carry out a test to see whether it is an effective treatment for tuberculosis, caused by *Mycobacterium tuberculosis.*

You have three possible options:

1 Treat the next one hundred patients with TB coming into the hospital, and compare them with previous TB patients.
2 Give streptomycin to every second TB patient who comes into the hospital, and compare them with the untreated ones.
3 Select TB patients at random, and give them the streptomycin. Compare their progress with that of untreated patients.

Which is the best way of testing how effective streptomycin is at treating TB?

During the 1950s and 1960s a large number of new antibiotics were developed. However, even with the discovery of a large range of antibiotics, there are still bacterial infections that don't respond to them, or become resistant to them. The search goes on.

Why do some microorganisms make antibiotics? The usual answer is that antibiotics help microorganisms to compete in the environment. For example, a soil-living fungus may secrete an antibiotic so that other fungi or bacteria are unable to grow near it. However, some scientists are not convinced that microorganisms make antibiotics for this reason. Antibiotics are not produced in large amounts until the cells are ageing. If antibiotics were produced to help a microorganism compete, it would be expected that they would be produced mainly in young cells.

One feature of antibiotics is that they are effective against bacterial cells but leave eukaryotic cells unharmed. They are also useless against viruses, which is why diseases such as colds and flu should not be treated with antibiotics.

How antibiotics work

Classifying antibiotics

Antibiotics are classified according to their method of action. There are two types:

- **Bactericidal** antibiotics destroy bacteria.
- **Bacteriostatic** antibiotics prevent the multiplication of bacteria. The host's own immune system can then destroy the pathogens.

Q6.43 Clear zones around test discs impregnated with various antibiotics provide a method of comparing the effectiveness of different antibiotics (Figure 6.64). This method can also be used to compare disinfectants. The wider the zone that remains clear due to no bacterial growth, the more effective the antibiotic is at preventing growth. Look at the plate in Figure 6.64 and decide which of the six antibiotics being tested is the most effective.

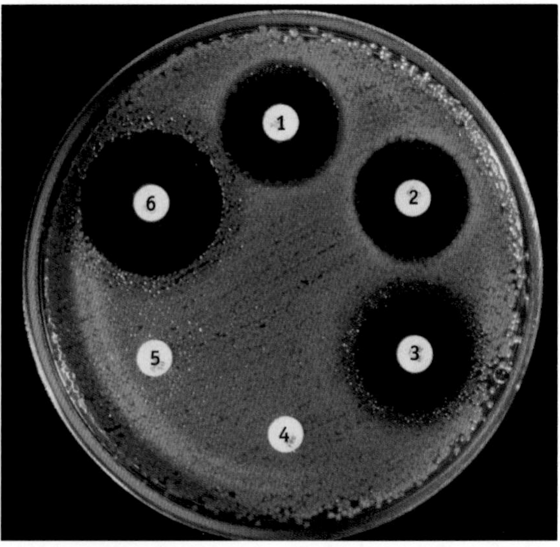

Figure 6.64 Each disc contains a different antibiotic.

How antibiotics disrupt bacterial cell growth and division

There are several ways in which antibiotics can interfere with bacterial cell growth and division. These include the following:

- Inhibition of bacterial cell wall synthesis. If a weak wall forms, this can lead to lysis (bursting) of the cell.
- Disruption of the cell membrane, causing changes in permeability that lead to cell lysis.
- Inhibition of nucleic acid synthesis, replication and transcription. This prevents cell division and/or synthesis of enzymes.
- Inhibition of protein synthesis, meaning enzymes and other essential proteins are not produced.
- Inhibition of specific enzymes found in the bacterial cell but not in the host.

Q6.44 The antibiotic vancomycin blocks bacterial cell wall synthesis. Why would it not affect the cells of a person taking the antibiotic?

Activity

In **Activity 6.15** you can compare the effectiveness of different antibiotics. **A6.15S**

Activity 6.16 will test your understanding of bactericidal and bacteriostatic antibiotics. **A6.16S**

Why do we still have diseases like TB?

Disease has been a feature of human life since we first evolved. Pathogens and hosts have always lived with each other. In the thousands of years of coexistence, the selective pressure exerted by the pathogens has resulted in selection for mutations in the human genome to make us more resistant. As you saw in AS Topic 2, carriers of the sickle cell allele are more resistant to malaria, while the CF allele may provide resistance against cholera. The Tay-Sachs gene that causes progressive degeneration of brain function is common among Ashkenazi Jews, but carriers are more resistant to TB.

Activity

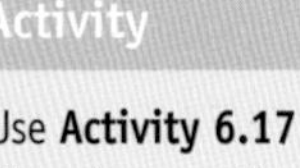

Use **Activity 6.17** to link TB and some other ideas covered in this topic. **A6.17S**

So why are we still dying of diseases that have been around for centuries? The problem is that the pathogens and host are locked in an evolutionary arms race. As quickly as we evolve mechanisms to combat pathogens, they evolve new methods of overcoming our immune system. Sometimes pathogens have the edge in this race because their reproductive cycle is faster than ours and their population size is greater.

Bacterial populations evolve very quickly. There are several reasons for this:

1 Bacteria reproduce very fast. For example, one *E. coli* bacterium can divide every 20 minutes, producing two million new cells in just seven hours. These cells contain a total of 8000 million genes, with, on average, 800 mutations.

2 Bacterial population sizes are usually in billions, so the number of cells containing mutations is vast.

3 Some of these random mutations will be advantageous to the cell containing them. They may allow the cell to use different food resources, reproduce more quickly, infect other cells more successfully, or produce symptoms in the host, such as coughing and sneezing, which aid the spread of the disease. Bacteria with a useful mutation are more likely to survive, reproduce and spread.

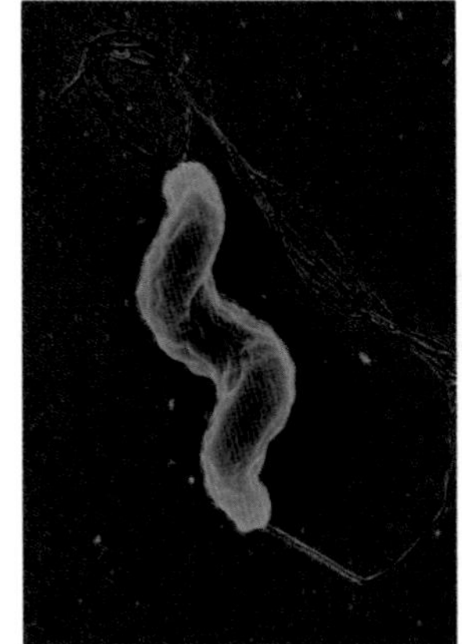

Figure 6.65 Scientists working on this bacterium, *Campylobacter*, which causes serious food poisoning, have found that it has a poor DNA repair system. This means that the genes coding for its surface proteins mutate frequently. This has given the bacterium an effective way of evading the host's immune system.

Since the human immune system is one of the main selection pressures acting on bacteria that infect humans, it is not surprising that there has been an evolution of bacterial strategies for evading or disabling the immune system. For example, slight changes in the pathogen's antigens mean that any reservoir of antibodies, and B and T memory cells from a previous infection, will be useless in combating a second infection (Figure 6.65).

We are not winning the race

It was thought that the battle against disease would be won with the use of antibiotics, and for a time it seemed as if it would be. Eradication of TB in Britain by the late 1970s was considered possible. But over 30 years later TB remains a major killer, and we are far from eradicating it.

Q6.45 Explain how *M tuberculosis* and HIV evade the host's immune system.

Antibiotics provide another selection pressure

Mutations arise in pathogenic bacteria that can make them resistant to antibiotics. The bacteria may produce an enzyme that enables the cell to break down the antibiotic, or they may use a different metabolic pathway for the reactions inhibited by the antibiotic.

In the absence of the antibiotic, bacteria with the mutation may be at a disadvantage. They may reproduce more slowly, using resources to produce enzymes that under 'normal' circumstances (no antibiotic present) are not required.

However, the presence of the antibiotic produces a selection pressure. Those bacteria that do not possess the gene for resistance are selected against and are more likely to be destroyed. Those that have the gene are selected for; they survive, grow and reproduce. The frequency of the gene for resistance within the bacterial population will increase. Since the advantageous gene is passed vertically from one generation to the next, this is sometimes termed vertical evolution.

In bacteria there is also horizontal evolution, when the gene is passed from one bacterium to another, which may be of the same or a different bacterial species. Bacteria do not undergo the sort of sexual reproduction that most animals do, but they do have cell-to-cell contact in a process called **conjugation** (Figure 6.66).

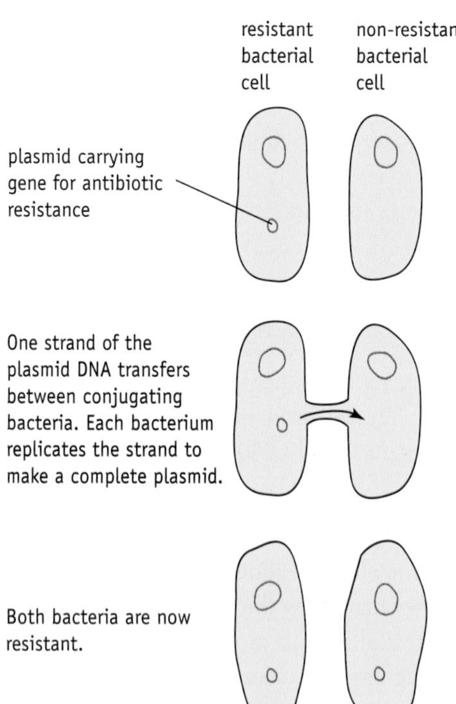

Figure 6.66 During conjugation, a copy of a DNA plasmid can be transferred between the cells. Antibiotic resistance genes are usually carried on these small loops of DNA, so when the plasmid is passed on the resistance is transferred.

Antibiotic-resistant TB

The antibiotic streptomycin was the major weapon in the treatment of TB. However, there is now widespread resistance to this and other antibiotics. In the presence of the antibiotic any resistant bacteria will be at an advantage, because they survive, and with less competition reproduce rapidly. This explains why some people with the disease may appear to improve at first in response to a single drug, and then worsen as the drug-resistant mutants multiply unchecked. A different antibiotic is then needed to clear up the infection. For many years this phenomenon did not present a problem, as new antibiotics were being developed faster than the bacteria developed resistance. In recent years, though, the bacteria seem to be catching up, with fewer new drugs being successfully developed.

Multiple-resistant bacteria

Some bacteria have evolved resistance to several antibiotics. These multiple-resistant strains are creating particular problems for the treatment of infected patients. Ironically these bacteria are common in hospitals. For example, *Staphylococcus aureus*, a bacterium that normally causes few problems, has become resistant to most antibiotics including methicillin. This strain, known as methicillin-resistant *Staphylococcus aureus* (MRSA) can cause a dangerous infection. This highly resistant bacterium can be controlled using the antibiotic vancomycin, but there are cases of vancomycin resistance and the bacterium is continually evolving resistance to any new drugs. For example, when a new drug, Linezolid, was introduced in January 2001, some MRSA bacteria had developed resistance to it by December 2002, even though the drug had not been extensively used in an attempt to prevent bacteria from evolving resistance.

Multiple-resistant TB is still relatively rare in the UK, but internationally there is growing concern about its spread.

Antibiotic resistance and hospital acquired infections

In response to the increasing incidence of patients becoming infected with healthcare associated infections (HCAIs), hospitals worldwide have had to tighten up their infection control and antibiotic prescribing practices. Infection control is now regarded by the public as more important than waiting times in the annual audits of hospital performance. Hand wash stations at the entrance to every ward, signs reminding all hospital personnel and visitors to wash their hands, and rules preventing doctors and nurses wearing ties, watches or long sleeves are all measures introduced to reduce the spread of infection. Antibiotics are only used when the patient has definitely been diagnosed as suffering from a bacterial infection. In the past doctors sometimes prescribed antibiotics even when a patient was suffering from a virus, either because the patient expected it or 'just in case'. Patients are now strongly advised to complete the whole course of their antibiotics. The consequences of stopping taking the antibiotics as soon as the patient feels better should be stressed by all those prescribing. Unfortunately the free use of these 'wonder drugs' in the past has rendered them considerably less effective now.

Preventing the development and spread of multiple-resistant bacteria is helped by a variety of methods:

- Antibiotics should only be used when needed.
- Patients should complete their treatment even when they feel better, so that all the bacteria are destroyed.
- Infection control should be used in hospitals to prevent bacteria spreading.

However, David Livermore, Director of Antibiotic Resistance Monitoring at the UK Public Health Laboratory Service, has said:

'The development of new antibiotics is essential because, however well we use antibiotics and however good [we are] at stopping the spread of infection, I don't honestly believe we can beat evolution.'

Q6.46 Nicki Overton was infected with HIV and *M. tuberculosis*. What precautions could she have taken to avoid becoming infected?

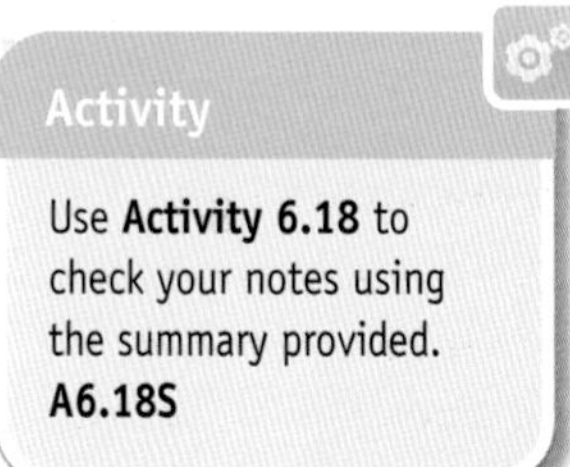

Activity

Use **Activity 6.18** to check your notes using the summary provided. **A6.18S**

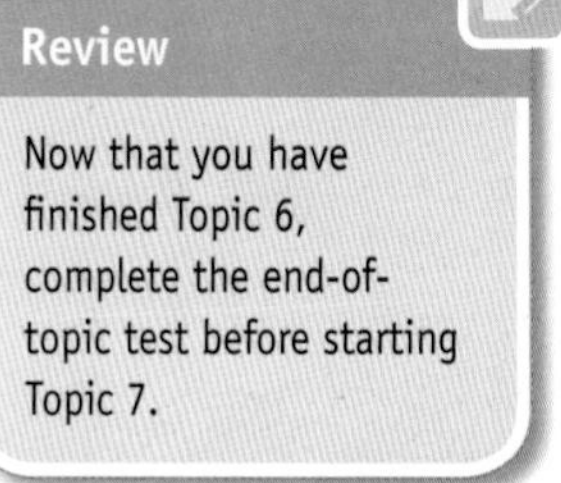

Review

Now that you have finished Topic 6, complete the end-of-topic test before starting Topic 7.

Run for your life

Why a topic called Run for your life?

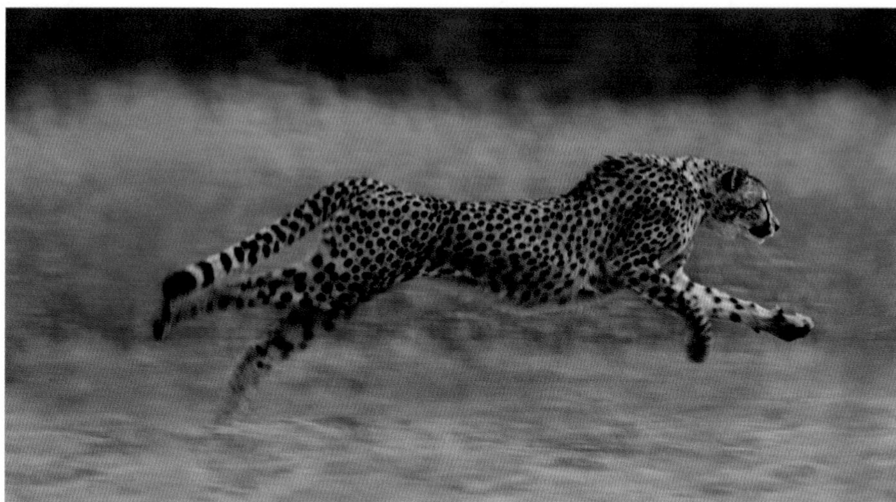

Figure 7.1 The fastest land mammal, the cheetah.

Cheetahs (Figure 7.1) can run at speeds in excess of 100 km/h, but after just a few hundred metres they must stop and rest or risk collapsing. Wildebeest (Figure 7.2) can run, though not as quickly, for many kilometres. Both animals are literally running for their lives. Whether chasing prey or seeking out new grazing pasture, their survival depends on their ability to run.

Before humans began farming just 10 000–12 000 years ago, we all lived as hunter-gatherers – working from a temporary home base, hunting wild animals and gathering plants from our natural surroundings. There would be long periods of moderate exercise while plucking shoots and berries, with occasional vigorous activity such as chasing after prey with spears.

Figure 7.2 Wildebeest travel huge distances during their annual migrations to find fresh pasture.

Although today few humans have to chase down prey or cover huge distances on foot, many of us still run. Exercise that involves running, or at least jogging, helps maintain our health, and for the professional sportsperson it also provides a living. We marvel at those who can complete a 43 km (26 mile) marathon (Figure 7.3), and are truly amazed by the ultra-marathon runners who cover 100 km, keeping going for as long as ten hours. But how is it that the marathon runner, who can complete 43 km at a pace of about 20 km/h, would fail to run 100 metres in the ten seconds or less that it takes a top class sprinter such as Usain Bolt (Figure 7.4)?

Figure 7.3 About 46 500 people start the London marathon each year with around 75% successfully completing the course. The men's marathon world record is just over two hours, and the women's is just over two and a quarter hours.

The cheetah, wildebeest, we humans and all other mammals share the same basic structures that allow us to move around. Our bones, joints and muscles are very similar, both macroscopically and microscopically; and all mammals, indeed most animals, use the same biochemical pathways to make energy available for movement. But if we are all so similar, how is it that the wildebeest can keep going for hours on end, but the cheetah must rest after just a few minutes? Why do some people excel in sprint events while others make outstanding distance runners? Why is it so rare for one person to achieve the highest level in both?

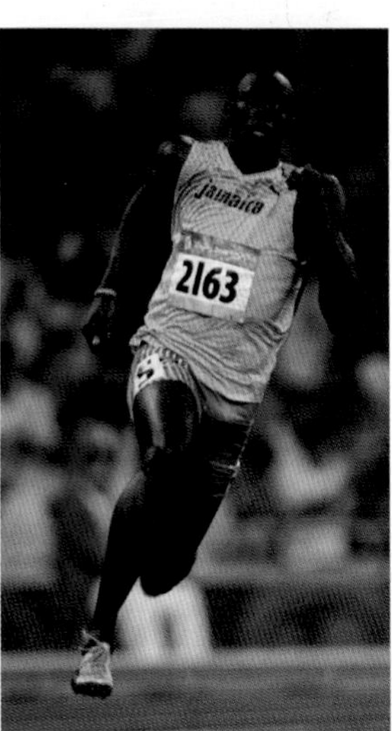

Figure 7.4 Usain Bolt of Jamaica, 100 m Olympic and world record-holder with a time of 9.69 seconds in 2008.

Exercise poses some real challenges for the body. Many changes occur without our conscious thought, such as the finely controlled adjustments that ensure oxygen and fuel are supplied in sufficient quantities to the muscles. Fit sportspeople may not be aware of this unless working very hard, when the heart starts pounding and breathing becomes laboured. Overheating may be something they are much more aware of as they cover the kilometres. How does the body ensure that muscles are well supplied with oxygen and fuel, and how does it prevent the body temperature rising too high?

We are always being encouraged to take plenty of exercise. This is good advice with our increasingly sedentary lifestyles, which are leading to an ever-higher incidence of obesity, cardiovascular disease and related diseases. But can you overdo it? What are the consequences of overtraining or trying to improve on nature?

Overview of the biological principles covered in this topic

At the start of the topic you recall ideas about joints and movement. This is then extended to provide a detailed understanding of the mechanism of contraction in skeletal muscle. You revisit ideas covered in Topic 5 about energy transfer in biological systems, ATP, redox reactions and electron transfer chains in the context of respiration. The different energy systems are considered in detail. Building on your knowledge of the cardiovascular and ventilation systems from AS Topics 1 and 2, you investigate the control of heart rate, lung volumes and breathing rate.

In this topic, homeostasis and negative feedback are discussed, using thermoregulation in a sporting context. You return to the principles of the immune response introduced in Topic 6, in the context of immune suppression as a consequence of overtraining. You investigate the use of medical technology to enable participation of those with injuries or disabilities. You will also discuss the use of performance-enhancing substances in sport, and the moral and ethical issues surrounding their use. In this context you will consider the function of hormones and transcription factors.

7.1 Getting moving

Before the starter's gun fires, the athlete in Figure 7.5 is poised for action. He is crouched on the blocks, knees and ankles flexed. When the gun fires, he pushes against the blocks to get maximum thrust and is up and running. With every step, muscles contract and relax to bend (**flex**) and straighten (**extend**) ankles, knees and hip joints, exerting a force against the ground to push his body forward. In the same way, the wildebeest in Figure 7.6 extend the joints in their hind legs to get moving.

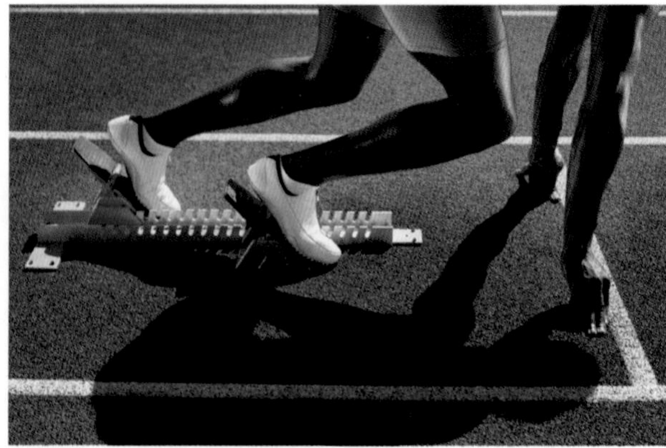

Figure 7.5 An athlete prepared for action.

Figure 7.6 A herd of blue wildebeest (*Connochaetes taurinus*) jumping into a river.

Joints and movement

Muscles bring about movement at a **joint**; most movements are produced by the co-ordinated action of several muscles. Muscles shorten, pulling on the bone and so moving the joint. Muscles can only pull, they cannot push, so at least two muscles are needed to move a bone to and fro. A pair of muscles that work in this way are described as **antagonistic**. For example, when you flex your knee by contracting the hamstring muscles at the back of the thigh, the quads at the front of the thigh relax and so are stretched. To extend the knee, the quadriceps shorten while the hamstrings relax.

A muscle that contracts to cause extension of a joint is called an **extensor**; the corresponding **flexor** muscle contracts to reverse the movement.

Q7.1 In Figure 7.7, contraction of which muscle, A or B, brings about extension of the cat knee joint?

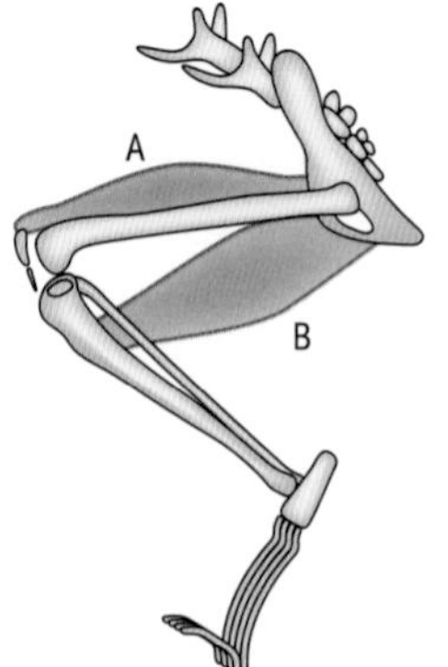

Figure 7.7 Muscles for flexion and extension of a cat knee joint.

Joint structure

The hip, knee and ankle joints are examples of **synovial joints**; the bones that articulate (move) in the joint are separated by a cavity filled with **synovial fluid**, which enables them to move freely. All synovial joints have the same basic structure as shown in Figure 7.8. The bones are held in position by **ligaments** that control and restrict the amount of movement in the joint. **Tendons** attach muscles to the bones, enabling the muscles to power joint movement. **Cartilage** protects bones within joints. The function of each part of the joint is shown in Figure 7.8.

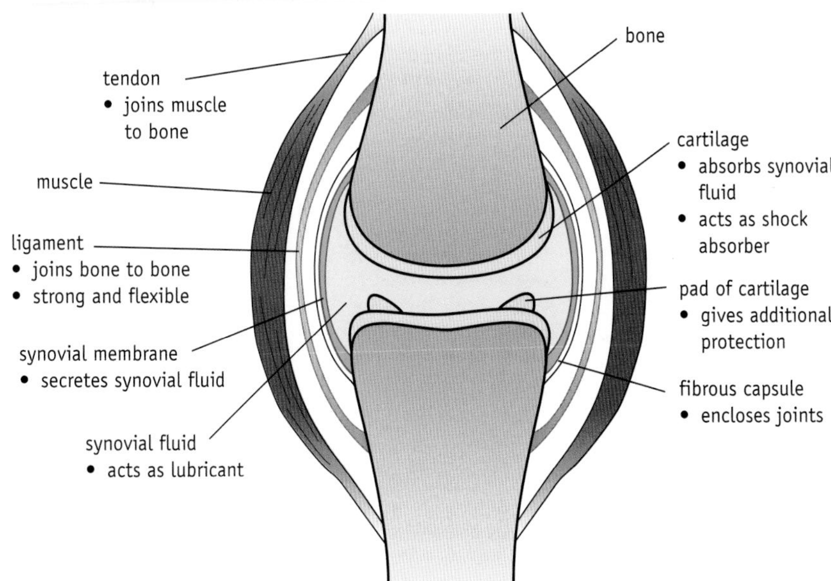

Figure 7.8 Where bones meet a joint is formed. A typical synovial joint allows bones to move freely.

Q7.2 **a** What properties of ligaments make them effective at holding the bones in place at a joint?

b What reduces the wear and tear due to friction in a mobile synovial joint?

c What properties must tendons have to make them effective at enabling movement at joints?

Q7.3 Compare the simplified synovial joint in Figure 7.8 with this diagram of the human knee joint (Figure 7.9).

a Identify the key features of the knee joint labelled A to C.

b Which features of a synovial joint are not shown in this diagram?

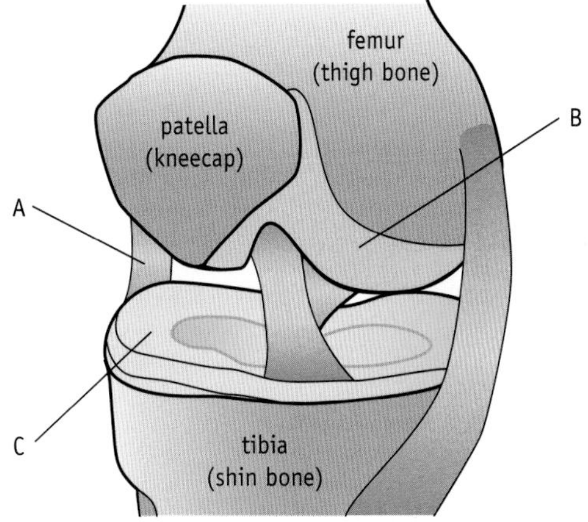

Figure 7.9 The human knee joint.

Activity

You can further your understanding of the knee joint in **Activity 7.1. A7.01S**

Did you know?

Other types of joints

Not all joints have the same structure. The plates of the skull are fixed together by fibrous tissue so very little movement occurs at these fixed joints. As a result the brain is protected inside a rigid bone casket. In the skull of a newborn baby the joints are not yet fixed; this allows the plates to move and the skull to deform (reversibly!) during birth. There are spaces between the skull bones, which fill in as the plates grow and fuse together.

Pelvic bones are joined together by cartilage, so slight movement is possible during childbirth; this is an example of a cartilaginous joint. There is also a wide variety of synovial joints allowing different degrees of extension, flexion and rotation, as shown in Figure 7.10.

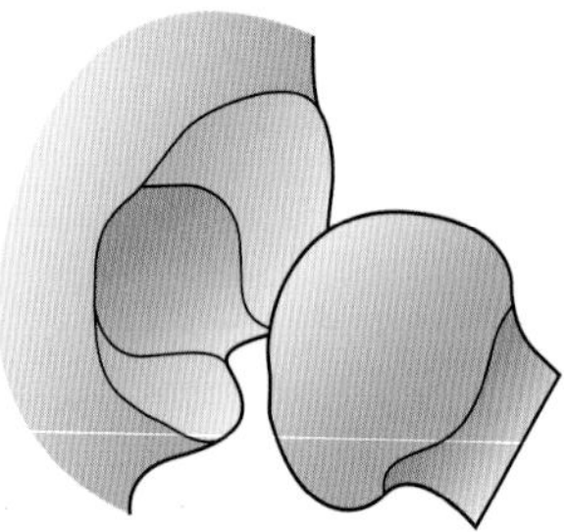

Ball-and-socket joint

A round head fits into a cup-shaped socket, e.g. the hip.

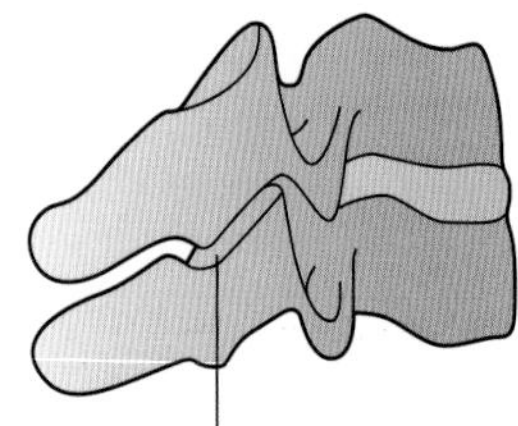

Gliding joint

Two flat surfaces slide over one another, e.g. the articulating surfaces between neighbouring vertebrae.

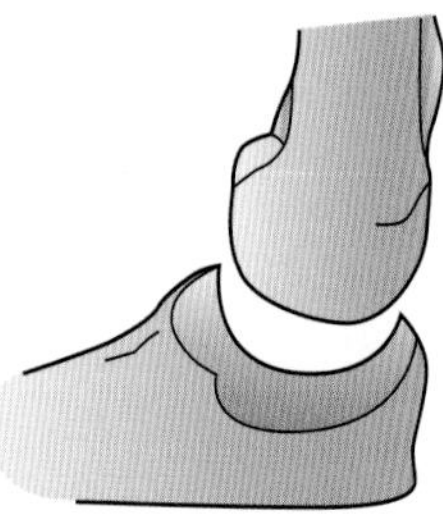

Hinge joint

A convex surface fits into a concave surface, e.g. the elbow.

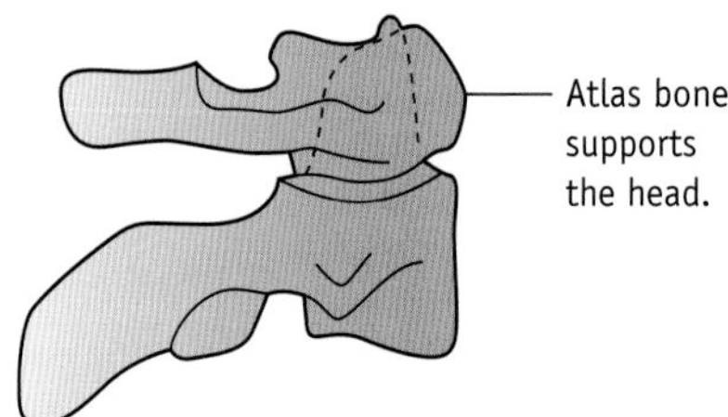
Atlas bone supports the head.

Pivot joint

Part of one bone fits into a ring-shaped structure and allows rotation, e.g. the joint at the top of the spine.

There are also saddle joints and condyloid joints in which more complex concave and convex surfaces articulate.

Figure 7.10 Different types of synovial joint.

Did you know?

What makes a cheetah so fast?

The cheetah's slender body, long legs and flexible spine allow it to run at great speed. The cheetah flexes its spine and rotates its shoulder blades forward and back, which gives it a longer stride (Figure 7.11). Unlike other cats, the cheetah does not retract its claws into protective sheaths; the exposed claws, combined with hard ridges on the feet pads, function rather like an athlete's spikes. The disadvantage of having permanently exposed claws is that they become blunt, making them less good as running spikes and for catching prey. The latter is not too much of a problem for the cheetah. It uses an enlarged claw on the inner sides of the front legs to bring down moving prey.

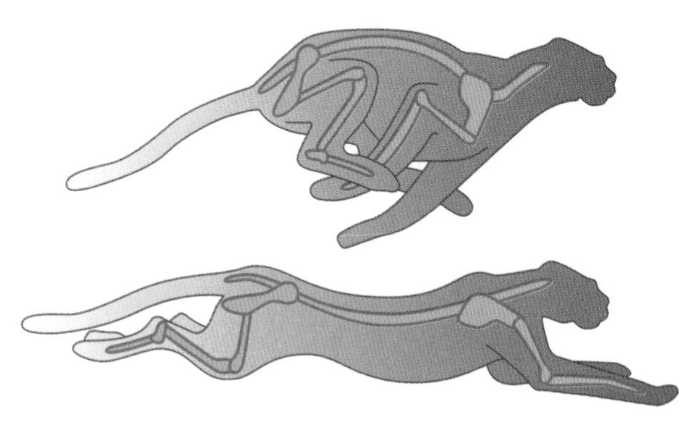

Figure 7.11 The cheetah flexes its spine.

Muscles

How do muscles work?

The key to how muscles function is in their internal structure. Muscle is made up of bundles of **muscle fibres**. Each fibre is a single muscle cell (Figure 7.12 and 7.13) and can be several centimetres in length.

Apart from the considerable length of the cells, do you notice something unusual about the muscle fibres? You should spot that each cell has several nuclei (is **multinucleate**). This occurs because a single nucleus could not effectively control the metabolism of such a long cell. During prenatal development, several cells fuse together forming an elongated muscle fibre. The muscle cells are also striped; as we will see, this is an important feature related to their ability to contract.

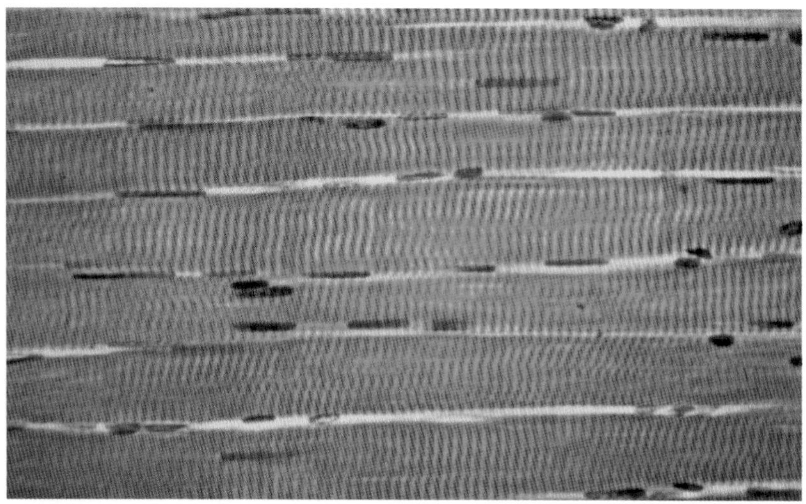

Figure 7.12 Human muscle fibres. Each muscle fibre is a single cell that is wider than this photograph.

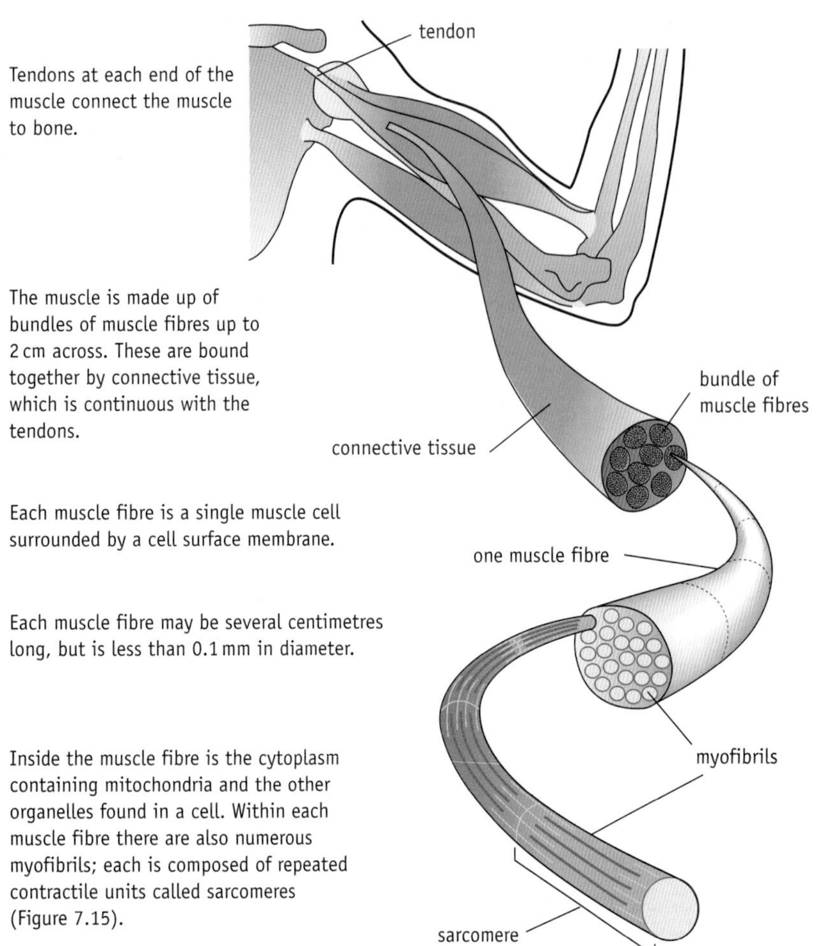

tendon
Tendons at each end of the muscle connect the muscle to bone.
The muscle is made up of bundles of muscle fibres up to 2 cm across. These are bound together by connective tissue, which is continuous with the tendons.
connective tissue
bundle of muscle fibres
Each muscle fibre is a single muscle cell surrounded by a cell surface membrane.
one muscle fibre
Each muscle fibre may be several centimetres long, but is less than 0.1 mm in diameter.
myofibrils
Inside the muscle fibre is the cytoplasm containing mitochondria and the other organelles found in a cell. Within each muscle fibre there are also numerous myofibrils; each is composed of repeated contractile units called sarcomeres (Figure 7.15).
sarcomere

Figure 7.13 The arrangement of muscle fibres within muscles.

Weblink

Compare the muscle cell with the typical animal cell in the interactive tutorial on cell structure and function in **Activity 3.1**.

Q7.4 Why would a single nucleus be unable to control the metabolism of a long, thin cell like a muscle fibre?

Inside a muscle fibre

Within each muscle fibre are numerous **myofibrils** (Figure 7.13). These are made up of a series of contractile units called **sarcomeres**, as shown in Figures 7.14 and 7.15.

The sarcomere is made up of two types of protein molecules: thin filaments made up mainly of the protein **actin**, and thicker ones made of the protein **myosin**. Contractions are brought about by co-ordinated sliding of these protein filaments within the muscle cell sarcomeres.

Figure 7.14 Electron micrograph showing the banding pattern of sarcomeres.

The arrangement of the filaments in a sarcomere is shown in Figures 7.14 and 7.15. The proteins overlap and give the muscle fibre its characteristic striped (striated) appearance under the microscope (Figure 7.14). Where actin filaments occur on their own, there is a *light* band on the sarcomere. Where both actin and myosin filaments occur, there is a *dark* band. Where only myosin filaments occur, there is an intermediate-coloured band (Figure 7.15B).

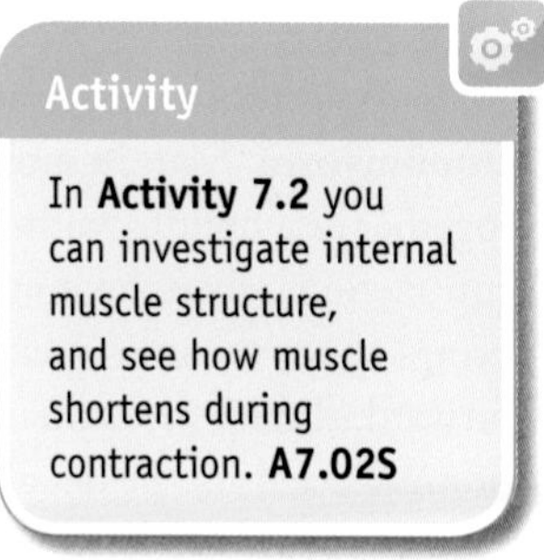

Activity

In **Activity 7.2** you can investigate internal muscle structure, and see how muscle shortens during contraction. **A7.02S**

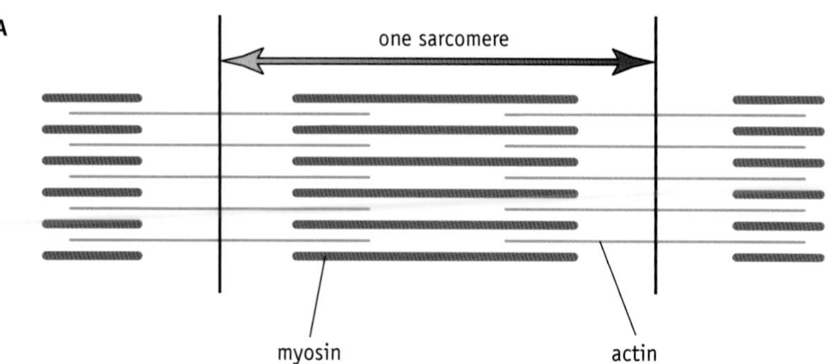

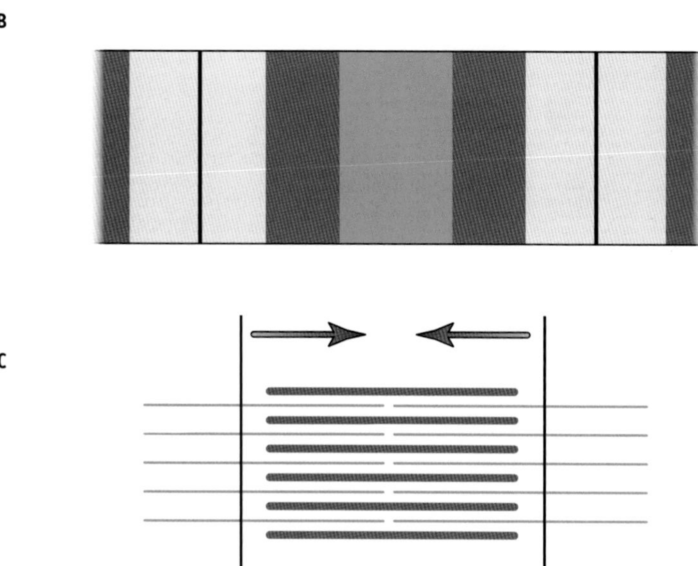

Figure 7.15 **A** The arrangement of actin and myosin filaments within a single muscle sarcomere when relaxed (extended). **B** The banding patterns created on an extended muscle myofibril. **C** The arrangement when the muscle contracts.

When the muscle contracts, the actin moves between the myosin as shown in Figure 7.15C. This shortens the length of the sarcomere, and hence the length of the muscle.

Q7.5 How many myofibrils are shown in Figure 7.14?

Q7.6 Look at the banding pattern on the contracted muscle fibre shown in Figure 7.16. Explain what has happened to the central, intermediate-coloured band, visible on an extended muscle in Figure 7.15B.

Figure 7.16 The banding pattern of a sarcomere when contracted.

How the sarcomere shortens

Actin molecules are associated with two other protein molecules called **troponin** and **tropomyosin**. Myosin molecules are shaped rather like golf clubs; the club shafts lie together as a bundle, with the heads protruding along their length (Figure 7.17). In a contraction, when the muscle shortens, the change in orientation of the myosin heads brings about the movement of actin. The myosin heads attach to the actin and dip forward, sliding the actin over the myosin. This is called the **sliding filament theory** of muscle action; the detailed sequence of events is described below and shown in Figure 7.19.

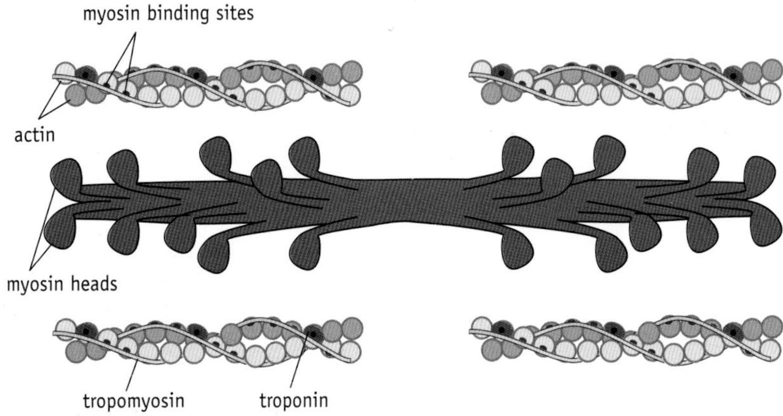

Figure 7.17 The myosin heads project on either side of the myosin molecule.

The sliding filament theory

When a nerve impulse arrives at a **neuromuscular junction** (Figure 7.18), calcium ions (Ca^{2+}) are released from the **sarcoplasmic reticulum**. This is a specialised type of endoplasmic reticulum: a system of membrane-bound sacs around the myofibrils (Figure 7.18A). The calcium ions diffuse through the sarcoplasm (the name given to cytoplasm in a muscle cell). This initiates the movement of the protein filaments as follows (Figure 7.19).

- Ca^{2+} attaches to the troponin molecule, causing it to move.
- As a result, the tropomyosin on the actin filament shifts its position, exposing myosin binding sites on the actin filaments.
- Myosin heads bind with myosin binding sites on the actin filament, forming cross-bridges.
- When the myosin head binds to the actin, ADP and P_i on the myosin head are released.
- The myosin changes shape, causing the myosin head to nod forward. This movement results in the relative movement of the filaments; the attached actin moves over the myosin.
- An ATP molecule binds to the myosin head. This causes the myosin head to detach.
- An ATPase on the myosin head hydrolyses the ATP, forming ADP and P_i.
- This hydrolysis causes a change in the shape of the myosin head. It returns to its upright position. This enables the cycle to start again.

The collective bending of many myosin heads combines to move the actin filaments relative to the myosin filament. This results in muscle contraction.

> **Activity**
>
> Building a model in **Activity 7.3** will let you check your understanding of the sliding filament theory. **A7.03S**

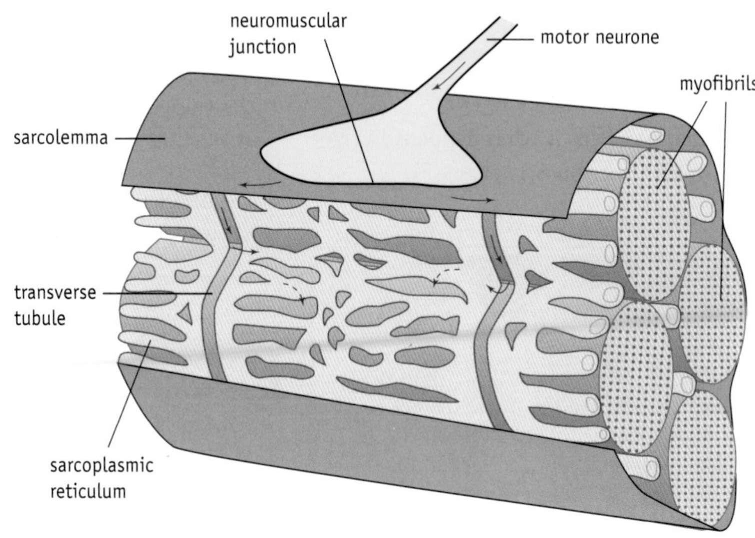
neuromuscular junction
motor neurone
myofibrils
sarcolemma
transverse tubule
sarcoplasmic reticulum
Key
route of nerve impulse
release of Ca^{2+} from sarcoplasmic reticulum
Nerve impulse reaches neuromuscular junction; spreads through muscle fibre via transverse tubules.
Ca^{2+} released from sarcoplasmic reticulum into sarcoplasm.

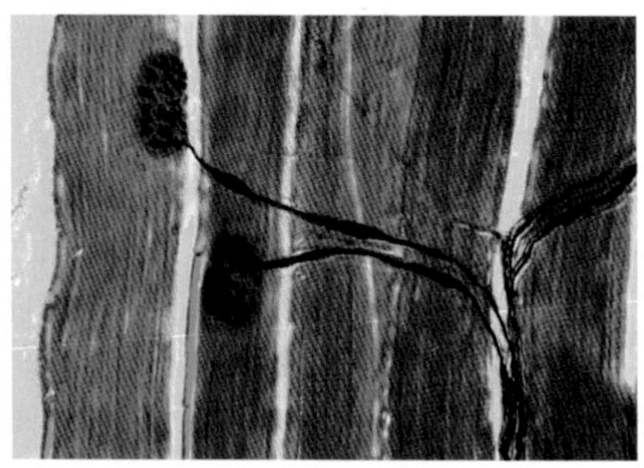

Figure 7.18 A Part of one muscle fibre (cell) showing how a nerve impulse initiates contraction of the fibre. The sarcolemma is the name given to the cell surface membrane of a muscle cell. **B** Coloured scanning electron micrograph of the neuromuscular junctions between a nerve cell (green) and a muscle fibre (red).

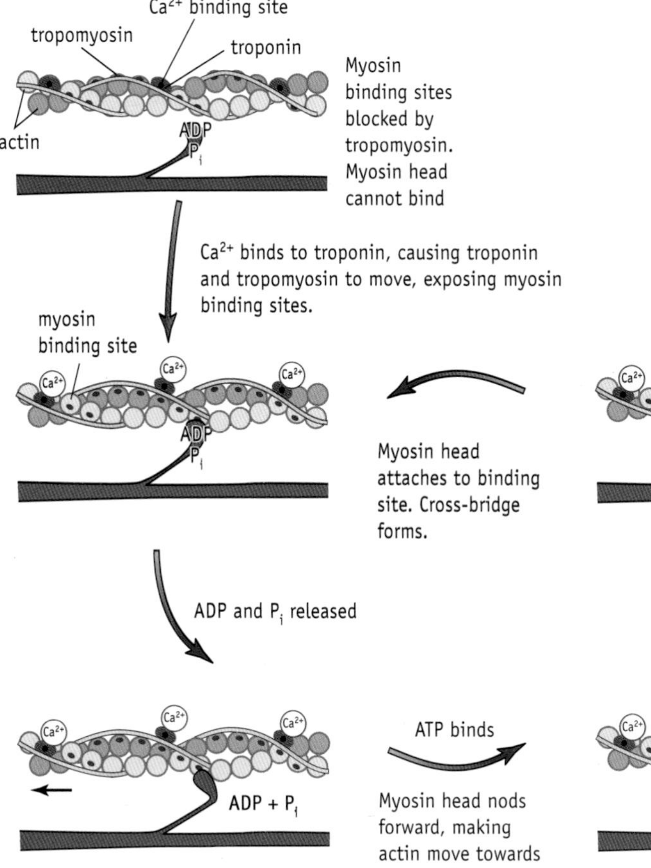
Ca^{2+} binding site
tropomyosin
troponin
actin
ADP
P_i
Myosin binding sites blocked by tropomyosin. Myosin head cannot bind
Ca^{2+} binds to troponin, causing troponin and tropomyosin to move, exposing myosin binding sites.
myosin binding site
Ca^{2+}
Myosin head attaches to binding site. Cross-bridge forms.
ADP and P_i released
ADP + P_i
ATP binds
Myosin head nods forward, making actin move towards centre of sarcomere.

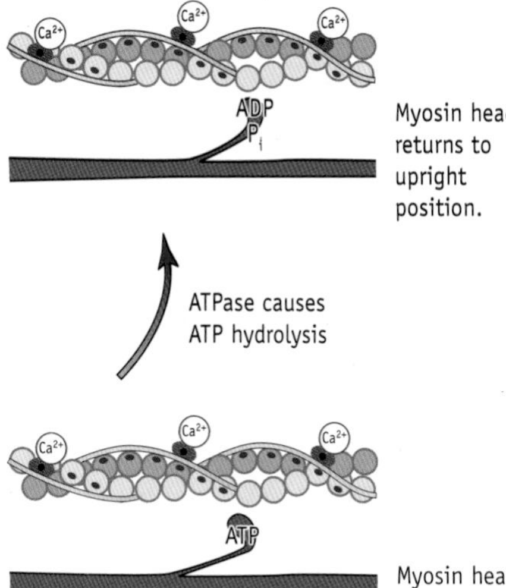
Ca^{2+}
ADP
P_i
Myosin head returns to upright position.
ATPase causes ATP hydrolysis
ATP
Myosin head detaches.

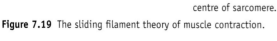

Figure 7.19 The sliding filament theory of muscle contraction.

When a muscle relaxes, it is no longer being stimulated by nerve impulses. Calcium ions are actively pumped out of the muscle sarcoplasm, using ATP. The troponin and tropomyosin move back, once again blocking the myosin binding sites on the actin.

In the absence of ATP, the cross-bridges remain attached. This is what happens in rigor mortis. Any contracted muscles become rigid (Topic 6, Section 6.1, page 85).

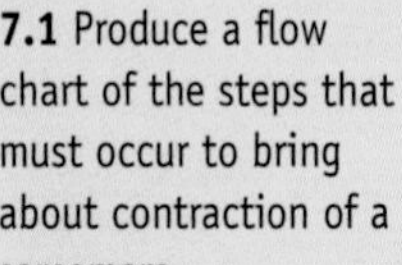

Checkpoint

7.1 Produce a flow chart of the steps that must occur to bring about contraction of a sarcomere.

Extension

Find out what a method of wrinkle control (Botox™) and a lethal form of food poisoning (botulism) have in common in **Extension 7.1. X7.01S**

Did you know?

Other types of muscles

Muscles found in such places as the gut wall, blood vessels, and the iris of the eye are known as smooth muscle: their fibres do not appear to be striped. These are small cells with a single nucleus. They have a similar mechanism of contraction to skeletal muscle, using myosin and actin protein filaments. But they are not arranged in the same way. Gap junctions – intercellular channels less than 2 nm in diameter – between the smooth muscle cells give cytoplasmic continuity between the cells. This allows chemical and electrical signals to pass between adjacent cells, and so allows synchronised contraction. Contractions in smooth muscle fibres are slower and longer lasting, and the fibres fatigue very slowly if at all.

The heart walls are made of specialised muscle fibres called cardiac muscle; these are striped and interconnected to ensure that a co-ordinated wave of contraction occurs in the heart (Figure 7.20). Cardiac muscle fibres do not fatigue. Neither smooth nor cardiac muscle are under conscious control – unless you became extremely good at certain forms of yoga. You will find out more about cardiac muscle contraction later in this topic.

Figure 7.20 Cardiac myofibrils. In the centre is an intercalated disc (strong, undulating dark blue line). Intercalated discs are double membranes that separate adjacent cells in the muscle fibres. They allow the nerve impulse to spread rapidly across the cells, synchronising muscle contraction. Numerous mitochondria (red) supply the muscle cells with energy.

7.2 Energy for action

Just staying alive, even if you're not doing anything active, uses a considerable amount of energy. This minimum energy requirement is called the **basal metabolic rate** (**BMR**) (see AS Topic 1). BMR is a measure of the minimum energy requirement of the body at rest to fuel basic metabolic processes; it is measured in kJ $g^{-1}h^{-1}$.

BMR is measured by recording oxygen consumption under strict conditions; no food is consumed for 12 hours before measurement, with the body totally at rest in a thermostatically controlled room. BMR is roughly proportional to the body's surface area. It also varies between individuals depending on their age and gender. Percentage body fat seems to be important in accounting for these differences.

Q7.7 Women generally have a higher percentage of body fat than men. Why might this account for the fact that women generally have lower BMRs?

Physical activity increases the body's total daily energy expenditure. Energy is needed for muscle contraction to move the body around, as you saw in Activity 1.17. An elite marathon runner uses energy at a rate of about 1.75 kJ s^{-1}, whereas a sprinter expends around 4 kJ s^{-1}. How does the muscle cell deal with these very different energy demands?

Releasing energy

Food is the source of energy for all animal activity. The main energy sources for most people are carbohydrates and fats, which have either just been absorbed from the gut or have been stored around the body. A series of enzyme-controlled reactions, known as **respiration**, is linked to ATP (adenosine triphosphate) synthesis. As you saw in Topic 5, cells use the molecule ATP (Figures 5.23 and 7.21A) as an energy carrier molecule. This is the cell's energy currency, coupling energy-yielding reactions and energy-requiring reactions, as in the muscle contraction described earlier in this topic.

ATP is created from ADP by the addition of inorganic phosphate (P_i). In solution, phosphate ions are hydrated, i.e. water molecules are bound to them. In order to make ATP, phosphate must be separated from these water molecules. This reaction requires energy. ATP in water is higher in energy than ADP and phosphate ions in water, so ATP in water is a way of storing chemical potential energy. ATP keeps the phosphate separated from the water, but they can be brought together in an energy-yielding reaction each time energy is needed for reactions within the cell.

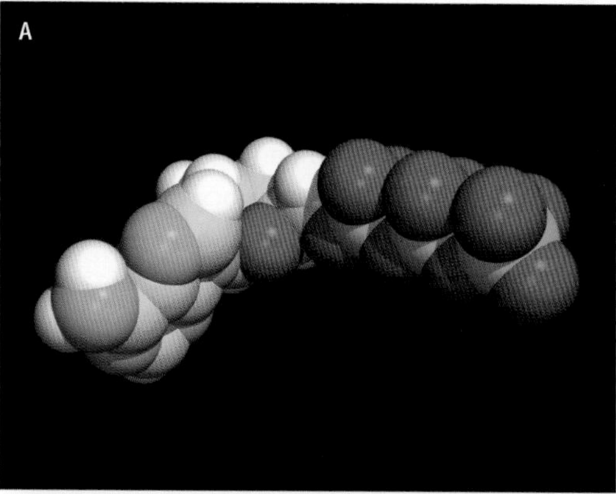

Figure 7.21 **A** Computer artwork of a molecule of ATP. The atoms are shown as spheres and colour-coded: carbon (yellow), oxygen (red), nitrogen (blue), hydrogen (white) and phosphorus (green). ATP is a universal molecule used by living organisms. **B** Here a firefly releases energy from ATP to create luminescence in its attempt to attract mates. Magnification × 3.

When one phosphate group is removed from ATP by hydrolysis, ADP (adenosine diphosphate) forms. A small amount of energy is required to break the bond holding the end phosphate in the ATP. Once removed, the phosphate group becomes hydrated. A lot of energy is released as bonds form between water and phosphate. This energy can be used to supply energy-requiring reactions in the cell. Some of the energy transferred during hydration of phosphate from ATP will raise the temperature of the cell; some is available to drive other metabolic reactions such as muscle contraction, protein synthesis or active transport. The hydrolysis of ATP is coupled to these other reactions:

$$\text{ATP in water} \rightarrow \text{ADP in water} + \text{hydrated } P_i + \text{energy transferred}$$

Support

To find out more about ATP and energy transfer in chemical reactions, look at the Biochemistry support on the website.

Carbohydrate oxidation

If exercise is low intensity, for example in long-distance walking and running, enough oxygen is supplied to cells to enable ATP to be regenerated through aerobic respiration of fuels. Fats and carbohydrates, like glucose (Figure 7.22), are oxidised to carbon dioxide and water; you are probably familiar with the summary equation for aerobic respiration:

$$C_6H_{12}O_6 + 6O_2 \rightarrow 6CO_2 + 6H_2O + \text{energy released}$$

Figure 7.22 Recall the structure of α-glucose studied in the AS course.

In Topic 5, photosynthesis was described as a process that separates hydrogen from oxygen by pulling apart water molecules. The hydrogen from water is stored by combining it with carbon dioxide to form carbohydrate.

In aerobic respiration, the hydrogen stored in glucose is brought together with oxygen to form water again. The bonds between hydrogen and carbon in glucose are not as strong as the bonds between hydrogen and oxygen atoms in water. So the input of energy needed to break the bonds in glucose and oxygen is not as great as the energy released when the bonds in carbon dioxide and water are formed. Overall, there is a release of energy, and this can be used to generate ATP.

Glucose and oxygen are not brought together directly, as this would release large amounts of energy quickly, which could damage the cell. Glucose is split apart in a series of small steps. Carbon dioxide is released as a waste product. Hydrogen from the glucose is eventually reunited with oxygen to release large amounts of energy as water is formed.

Having understood the significance of the overall reaction and energy changes in respiration, you need to look in more detail at the mechanisms taking place during this process.

Glycolysis first

The initial stages of carbohydrate breakdown, known as **glycolysis** (Figure 7.23), occur in the cytoplasm of cells, including the sarcoplasm of muscle cells. Glycolysis literally means 'splitting of sugar'.

Stores of glycogen (a polymer of glucose – see AS Topic 1) in muscle or liver cells must first be converted to glucose, a hexose sugar. Glucose is the main respiratory substrate. It is a good fuel – it can potentially yield 2880 kJ mol^{-1} – but it is quite stable and unreactive. Therefore, the first reactions of glycolysis need an *input* of energy from ATP to get things started. Two phosphate groups are added to the glucose from two ATP molecules, and this increases the reactivity of the glucose. It can now be split into two molecules of 3-carbon (3C) compounds. It is rather like lighting a candle; a match provides an initial input of energy before energy can be released from the fuel (candle wax).

Each intermediate 3C sugar is oxidised, producing a 3-carbon compound, **pyruvate**. Two hydrogen atoms are removed during the reaction and taken up by the **coenzyme NAD** (**nicotinamide adenine dinucleotide**), a non-protein organic molecule (Figure 7.24). The fate of these hydrogens and their role in ATP synthesis is covered later in this section.

Glucose is at a higher energy level than the pyruvate so, on conversion, some energy becomes available for the direct creation of ATP. Phosphate from the intermediate compounds is transferred to ADP, creating ATP. This is called **substrate-level phosphorylation**, because energy for the formation of ATP comes from the substrates; in this case the intermediate compounds are the substrates (Figure 7.23).

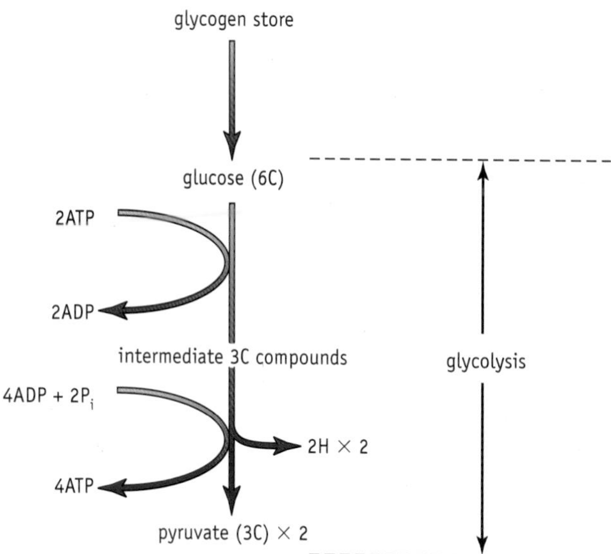

Figure 7.23 The glycolysis reactions in respiration take place under aerobic or anaerobic conditions.

In summary, glycolysis reactions yield a net gain of two ATPs, two pairs of hydrogen atoms, and two molecules of 3-carbon pyruvate as shown in Figure 7.23.

Q7.8 Why are the glycolysis reactions described as anaerobic?

What happens to the pyruvate depends on the availability of oxygen.

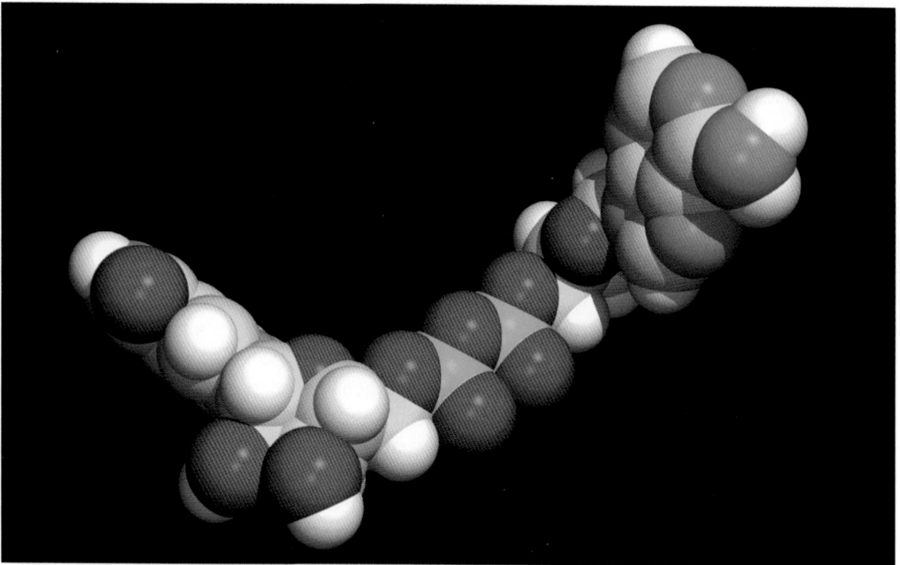

Figure 7.24 Coenzymes are small, organic, non-protein molecules that carry chemical groups between enzymes. NAD (shown here), FAD and coenzyme A are all examples of coenzymes. The atoms are shown as spheres and are colour-coded: carbon (yellow), oxygen (red), nitrogen (blue), hydrogen (white) and phosphorus (green).

The fate of pyruvate if oxygen is available

If oxygen is available, the 3C pyruvate created at the end of glycolysis passes into the mitochondria. There it is completely oxidised, forming carbon dioxide and water (see Figure 7.25).

The link reaction

In the first step, known as the **link reaction**, pyruvate is:

- **decarboxylated** (carbon dioxide is released as a waste product)
- **dehydrogenated** (two hydrogens are removed and taken up by the coenzyme NAD).

The resulting 2-carbon molecule combines with coenzyme A to form **acetyl coenzyme A** (**acetyl CoA** for short). As we shall see, the two hydrogen atoms released are involved in ATP formation. The coenzyme A carries the 2C acetyl groups to the **Krebs cycle**.

Activity

In **Activity 7.4** you can produce a summary diagram of glycolysis and the Krebs cycle. **A7.04S**

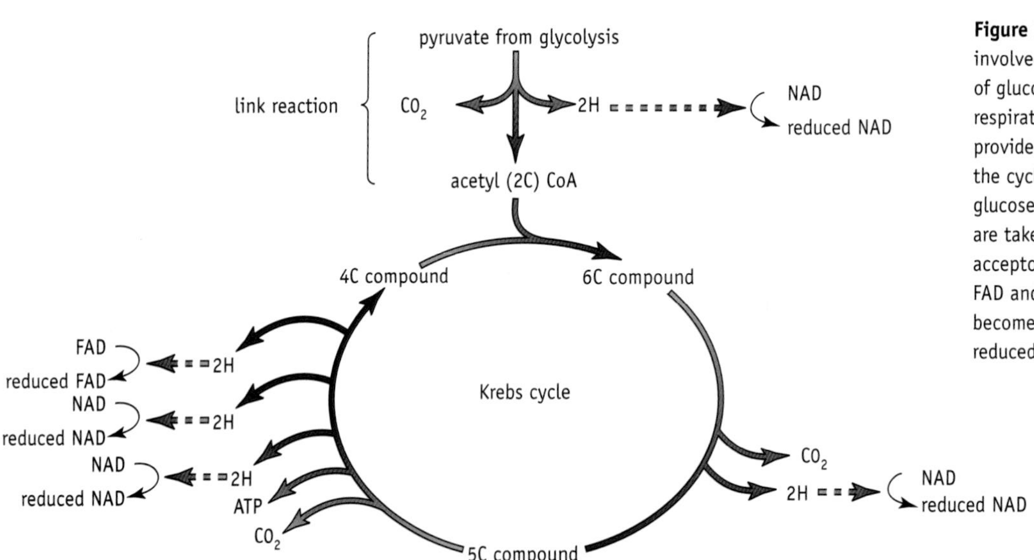

Figure 7.25 The reactions involved in the breakdown of glucose in aerobic respiration. Each glucose provides two pyruvates, so the cycle turns twice per glucose. The hydrogens are taken up by hydrogen acceptors: the coenzymes FAD and NAD, which become reduced FAD and reduced NAD.

Key biological principle: Understanding the chemistry of respiration

The chemical reactions inside cells are controlled by enzymes. There are four important types of reaction in the Krebs cycle:

- phosphorylation reactions, which add phosphate,
  e.g. ADP + P_i → ATP
- decarboxylation reactions, which break off carbon dioxide,
  e.g. pyruvate → acetyl CoA + CO_2
- redox reactions, where substrates are oxidised and reduced, e.g.
  (i) oxidised NAD + electrons → reduced NAD
  (ii) pyruvate → acetyl CoA + 2H (dehydrogenation)

When a molecule is oxidised, electrons, e^-, are lost from this substrate molecule, and the molecule that accepts the electrons is reduced (see Topic 5, Section 5.2). When a molecule loses hydrogen (dehydrogenation) it is also oxidised: the molecule that gains the hydrogen is reduced. In the previous example, pyruvate loses hydrogen to form oxidised acetyl CoA. Adding and removing oxygen is another type of redox reaction.

Support

To find out more about coenzymes and redox reactions, look at the Biochemistry support on the website.

Krebs cycle

Each 2-carbon acetyl CoA combines with a 4-carbon compound to create one with six carbons. In a circular pathway of reactions, the original 4-carbon compound is recreated. In these reactions, two steps involve decarboxylation with the formation of carbon dioxide. Four steps involve dehydrogenation, the removal of pairs of hydrogen atoms. In addition, one of the steps in the cycle involves substrate-level phosphorylation with direct synthesis of a single ATP (see the earlier section on 'Glycolysis' for details of substrate-level phosphorylation). This circular pathway of reactions (Figure 7.25) is known as the Krebs cycle, named after Sir Hans Krebs who worked out the cycle of reactions. The Krebs cycle takes place in the mitochondrial matrix (Figure 7.26) where the enzymes that catalyse the reactions are located.

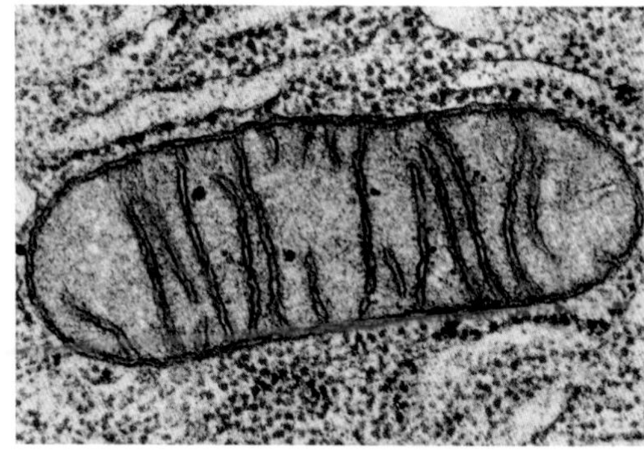

Figure 7.26 Coloured electron micrograph of a mitochondrion. Magnification × 41 000. Remind yourself of the structure of mitochondria by visiting the interactive cell in Activity 3.1.

In summary, each 2-carbon molecule entering the Krebs cycle results in the production of two carbon dioxide molecules, one molecule of ATP by substrate-level phosphorylation, and four pairs of hydrogen atoms, which are taken up by hydrogen acceptors – the coenzymes NAD and **FAD** (flavine adenine dinucleotide). The hydrogen atoms are subsequently involved in ATP production via the **electron transport chain**, described below.

Extension

Extension 7.2 explores the stages in the discovery of the Krebs cycle to illustrate the reactions involved. **X7.02S**

Fate of the hydrogens – the electron transport chain

Hydrogen atoms released during glycolysis, the link reaction, and the Krebs cycle are taken up by coenzymes. For most hydrogens produced, the coenzyme NAD is the hydrogen acceptor. But those released in one step of the Krebs cycle are accepted by the coenzyme FAD rather than NAD.

When a coenzyme accepts the hydrogen with its electron, the coenzyme is reduced, becoming reduced NAD or reduced FAD. The reduced coenzyme 'shuttles' the hydrogen atoms to the electron transport chain on the mitochondrial inner membrane. As shown in Figure 7.27, each hydrogen atom's electron and proton (H^+) then separate, with the electron passing along a chain of electron carriers in the inner mitochondrial membrane. This is known as the electron transport chain.

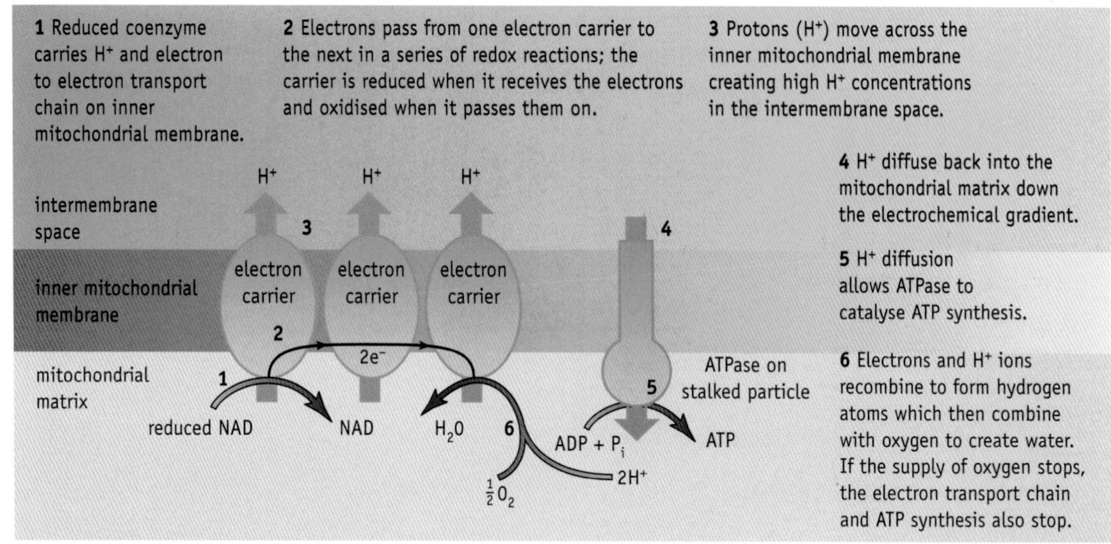

1 Reduced coenzyme carries H+ and electron to electron transport chain on inner mitochondrial membrane.
2 Electrons pass from one electron carrier to the next in a series of redox reactions; the carrier is reduced when it receives the electrons and oxidised when it passes them on.
3 Protons (H+) move across the inner mitochondrial membrane creating high H+ concentrations in the intermembrane space.
4 H+ diffuse back into the mitochondrial matrix down the electrochemical gradient.
5 H+ diffusion allows ATPase to catalyse ATP synthesis.
6 Electrons and H+ ions recombine to form hydrogen atoms which then combine with oxygen to create water. If the supply of oxygen stops, the electron transport chain and ATP synthesis also stop.
intermembrane space
inner mitochondrial membrane
mitochondrial matrix
H+
H+
H+
3
4
electron carrier
electron carrier
electron carrier
2
2e−
1
5
ATPase on stalked particle
reduced NAD
NAD
H2O
6
ADP + Pi
ATP
2H+
½O2

Figure 7.27 The electron transport chain and chemiosmosis result in ATP synthesis. The location of the electron transport chain within the mitochondrion can be seen in Figure 7.28.

ATP synthesis by chemiosmosis

How does the electron transport chain lead to ATP synthesis? This is explained by the **chemiosmotic theory** summarised in Figure 7.27. Peter Mitchell proposed the theory in 1961, publishing his ideas in the journal *Nature*. The theory was counter to the thinking at the time and did not gain immediate acceptance within the scientific community. Mitchell and his colleagues published further experimental evidence supporting the theory. Nowadays, it is very widely accepted. Mitchell was awarded the Nobel Prize for Chemistry in 1978 for his contribution to the understanding of biological energy transfer through the formulation of the chemiosmotic theory.

Energy is released as electrons pass along the electron transport chain. This energy is used to move hydrogen ions (originating from the hydrogen atoms released in glycolysis, the link reaction or the Krebs cycle) from the matrix, across the inner mitochondrial membrane, and into the intermembrane space. This creates a steep **electrochemical gradient** across the inner membrane. There is a large difference in concentration of H^+ across the membrane, and a large electrical difference, making the intermembrane space more positive than the matrix.

The hydrogen ions diffuse down this electrochemical gradient through hollow protein channels in stalked particles on the membrane. As the hydrogen ions pass through the channel, ATP synthesis is catalysed by ATPase located in each stalked particle. The hydrogen ions cause a conformational change (change in shape) in the enzyme's active site, so the ADP can bind.

Within the matrix, the H^+ and electrons recombine to form hydrogen atoms. These combine with oxygen to form water. The oxygen, acting as the final carrier in the electron transport chain, is thus reduced. This method of synthesising ATP is known as **oxidative phosphorylation**.

Q7.9 Why is the synthesis of ATP via the electron transport chain termed 'oxidative phosphorylation'?

Aerobic respiration is a many-stepped process, with each step controlled and catalysed by specific enzymes. It is summarised in Figure 7.28. Not all the steps are explained in this topic. You do not need to know all of them at A-level – you could spend a very long time just studying respiration if you looked at all the stages.

Activity

Using the context of mitochondrial diseases, **Activity 7.5** has an interactive animation looking at the electron transport chain and chemiosmosis. **A7.05S**

Support

To find out more about coenzymes, redox reactions and electron transport chains, visit the Biochemistry support on the website.

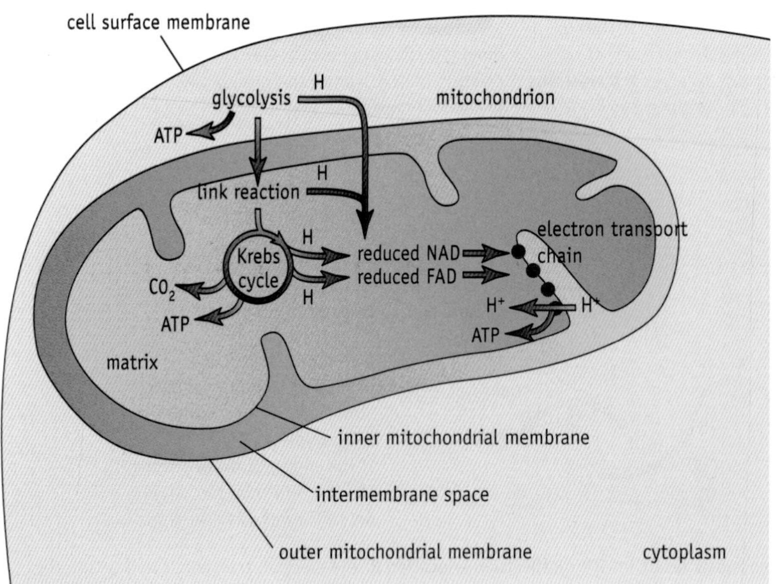

Figure 7.28 A summary of where the respiration reactions occur.

How much ATP is produced?

Depending on which textbook you read, you are likely to find different numbers for the number of ATP molecules produced from the complete aerobic respiration of each glucose molecule. This doesn't mean that one book is right and another wrong; it means that the total number of ATPs produced by one glucose molecule can vary according to the efficiency of the cell. Also, scientists are continually revising their estimates in the light of new knowledge.

A simple explanation would give a maximum number of 38 ATP molecules made per glucose molecule. This is based on the assumption that each reduced NAD that is reoxidised results in the formation of three ATP molecules and each reduced FAD results in production of two ATP molecules. Figure 7.29 shows where the 38 ATP come from.

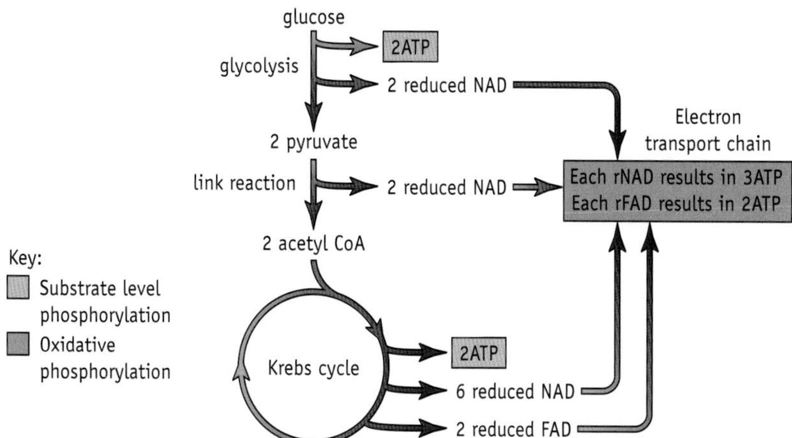

Figure 7.29 ATP are generated by substrate-level phosphorylation, and oxidative phosphorylation via the electron transport chain.

It is now widely accepted that this simple calculation is not accurate. This is because the electrochemical gradient across the inner membrane is also involved in transport of other ions and molecules across the membrane. For example, the exchange of ADP and ATP between the mitochondrial matrix and the cytoplasm uses up some H^+. Thus not all the H^+ ions are available for generation of ATP molecules. The actual yield of ATP from one glucose is probably around 30 ATP.

Q7.10 **a** Unlike numbers of ATP, there is no argument about the number of molecules of reduced NAD produced per molecule of glucose during aerobic respiration. How many reduced NAD are made?

b Some textbooks state that each reduced NAD gives rise to 2.5 ATP molecules, and each reduced FAD results in 1.5 ATP. If this is the case, how many ATP molecules would be produced by:

i each turn of the Krebs cycle

ii in total by aerobic respiration, as summarised in Figure 7.29.

Complete oxidation, for example by combustion in oxygen, of one mole of glucose molecules (180 g) releases 2880 kJ. Only 1163 kJ of energy is released from the ATP made from one mole of glucose; this is just 40% of the total potential chemical energy stored in the glucose. This figure assumes that 38 molecules of ATP are produced from aerobic respiration of one molecule of glucose. If only 30 molecules of ATP are

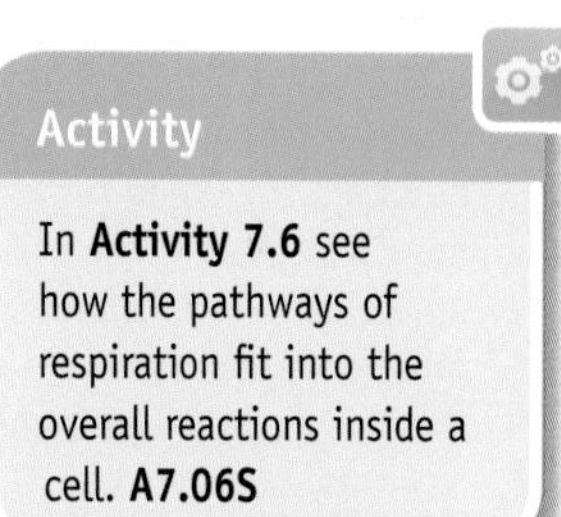

produced from aerobic respiration of glucose, this figure falls to 918 kJ, close to 32%. The remaining energy raises the temperature of the cell. This helps to increase the rate of metabolic reactions and, in mammals and birds, to maintain core body temperature.

Did you know?

Fatty acid oxidation

Fats can also be respired to release energy and are a richer energy store than carbohydrates – glycogen releases 17 kJ g^{-1}, whereas triglycerides release 37 kJ g^{-1}. In fatty acid oxidation, the glycerol and fatty acids that make up triglycerides are separated. The fatty acids are broken down in a series of reactions, each generating the same 2-carbon compound, and these can be fed into the Krebs cycle for oxidation (Figure 7.30).

Because fatty acids can only be respired through the Krebs cycle, fats can only be a fuel for aerobic respiration, and cannot be used when oxygen is not available. Glucose can be respired aerobically or anaerobically.

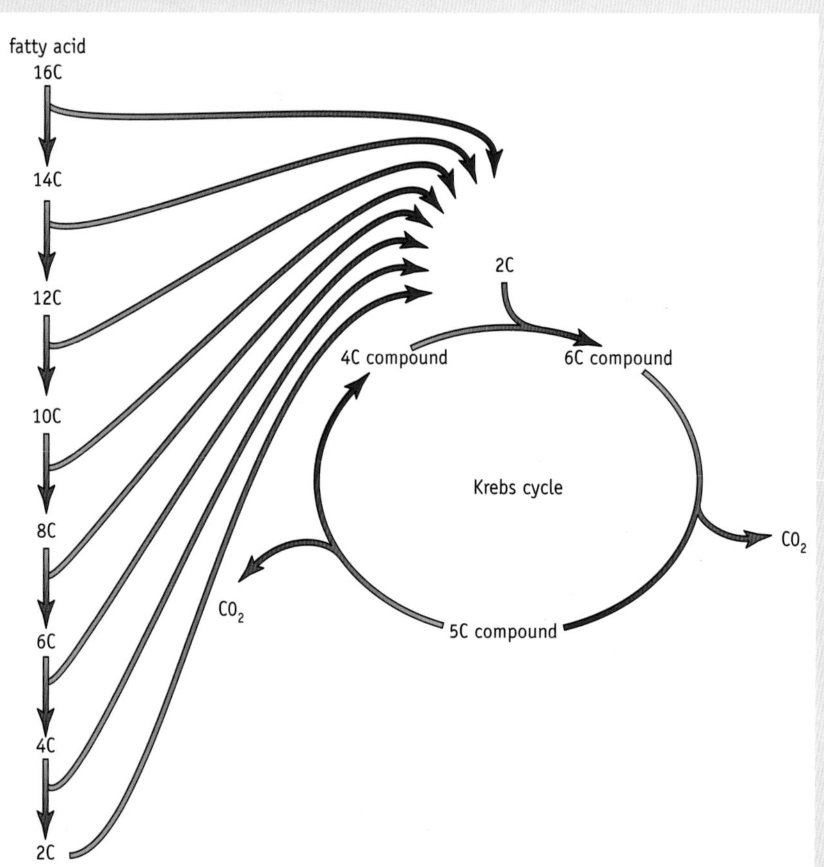

Figure 7.30 Oxidation of fatty acids occurs via the Krebs cycle.

Did you know?

Using proteins

Like their ancient Olympian counterparts, some modern athletes believe that a high-protein diet is the key to successful competition. However, excess intake of any fuel, including protein, is converted into fat. Amino acids will only be built into muscle protein if there is appropriate training. The body may use the excess protein as fuel through respiration, but to do this the amino acids must first be deaminated, producing urea. Water is needed to excrete the extra urea. As urine output increases, the body's fluid requirements also increase.

Rate of respiration

In small organisms, the rate of aerobic respiration can conveniently be determined by measuring the uptake of oxygen using a respirometer (Figure 7.31). Respiration is a series of enzyme-catalysed reactions, so its rate will be influenced by any factor affecting the rate of these enzyme-controlled reactions. For example, enzyme concentration, substrate concentration, temperature and pH will all affect the rate of respiration.

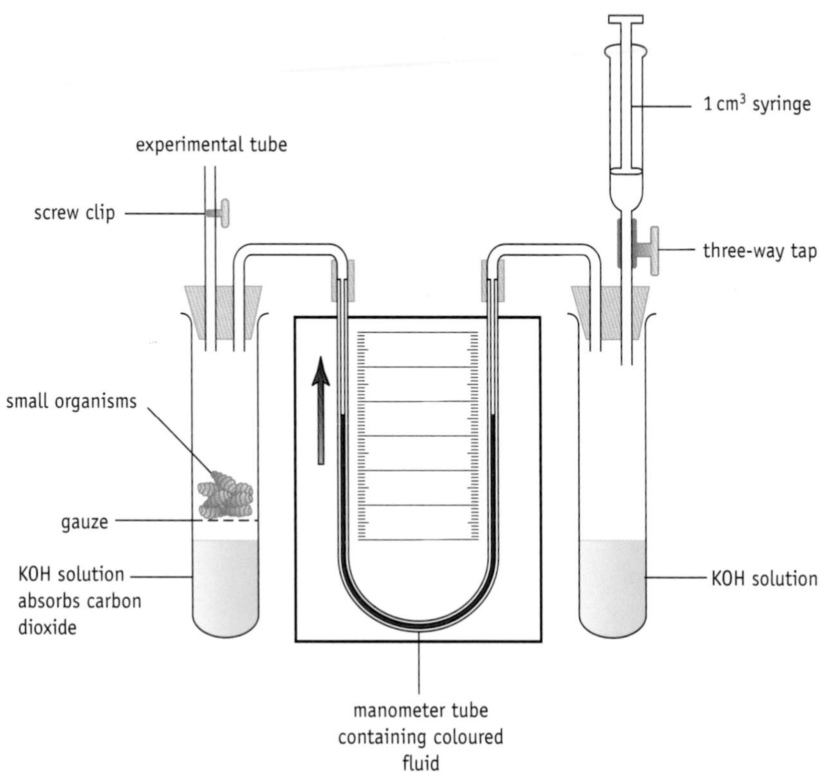

Figure 7.31 A respirometer. As the living organisms in the experimental tube take up oxygen, the fluid in the manometer will move in the direction of the arrow. The potassium hydroxide (KOH) solution absorbs any carbon dioxide produced by the organisms through respiration. The second tube compensates for any changes in volumes due to variations in gas pressure or temperature inside the apparatus.

The concentration of ATP in the cell also has a role in the control of respiration. ATP inhibits the enzyme in the first step of glycolysis, the phosphorylation of glucose. The enzyme responsible for glucose phosphorylation can exist in two different forms. In the presence of ATP, the enzyme has a shape that makes it inactive; it cannot catalyse the reaction. As the ATP is broken down, the enzyme is converted back to the active form and catalyses the phosphorylation of glucose again. This is known as end point inhibition: the end product inhibits an early step in the metabolic pathway, so controlling the whole process.

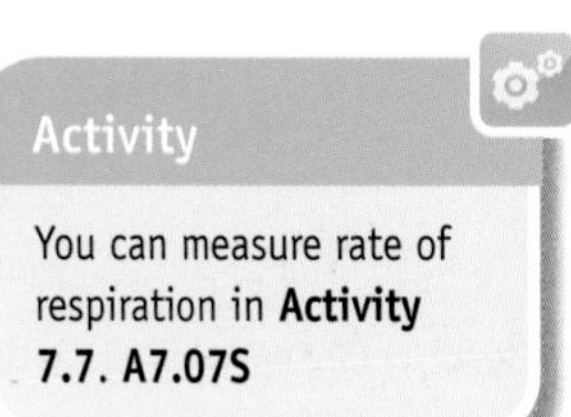
Activity

You can measure rate of respiration in **Activity 7.7. A7.07S**

Q7.11 What is the advantage of having a system of enzyme-controlled reactions to transfer energy from food fuels?

The fate of pyruvate without oxygen

Anaerobic respiration

At the start of any exercise and during intense exercise, for example a 400 m race, oxygen demand in the cells exceeds supply (Figure 7.32). Without oxygen to accept the hydrogen ions and electrons, the electron transport chain ceases; the reduced NAD created during glycolysis, the link reaction and the Krebs cycle is not oxidised. Without a supply of oxidised NAD, most respiration reactions cannot continue.

However, it is possible to oxidise the reduced NAD created during glycolysis in the absence of oxygen. The pyruvate produced at the end of glycolysis is reduced to **lactate** and the oxidised form of NAD is regenerated. In this way, anaerobic respiration allows the athlete to continue by partially breaking down glucose to make a small amount of ATP (Figure 7.33). The net yield is just two ATP molecules per glucose molecule – in other words, only 61 kJ of energy made available in the form of ATP. This process has only 2% efficiency.

Figure 7.32 The women's world record for running 400 m is 47.6 seconds. The intensity of the activity is such that the muscle cells do not get enough oxygen for aerobic respiration. How do such athletes fuel their performance?

The end product of anaerobic respiration is lactate, which builds up in the muscles and must be disposed of later. Animal tissues can tolerate quite high levels of lactate, but lactate forms lactic acid in solution. This means that, as lactate accumulates, the pH of the cell falls, inhibiting the enzymes that catalyse the glycolysis reactions. The glycolysis reactions and the physical activity depending on them cannot continue.

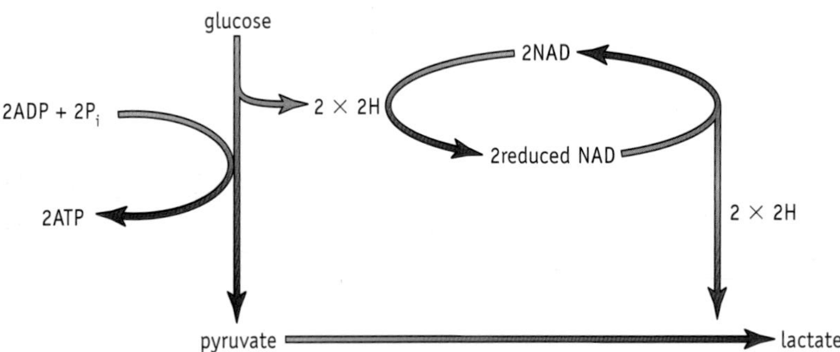

Figure 7.33 Anaerobic respiration.

The effect of lactate build-up

Enzymes function most efficiently over a narrow pH range. Many of the amino acids that make up an enzyme have positively or negatively charged groups. As hydrogen ions from the lactic acid accumulate in the cytoplasm, they neutralise the negatively charged groups in the active site of the enzyme. The attraction between charged groups on the substrate and in the active site will be affected (Figure 7.34). The substrate may no longer bind to the enzyme's active site.

Activity

Complete the worksheet on anaerobic respiration in **Activity 7.8. A7.08S**

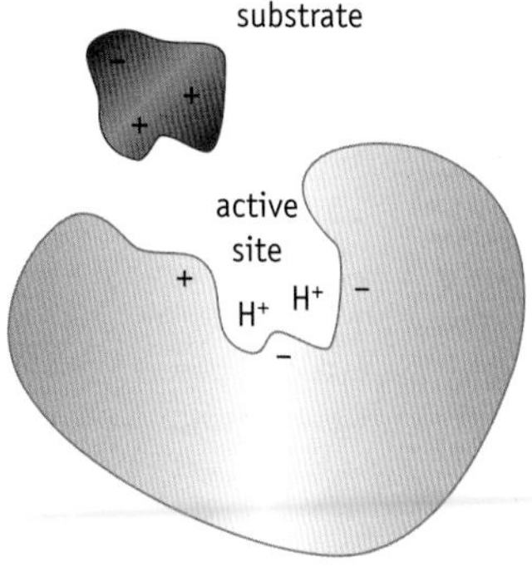

As pH falls, hydrogen ions prevent the substrate from being attracted to the charged groups in the active site.

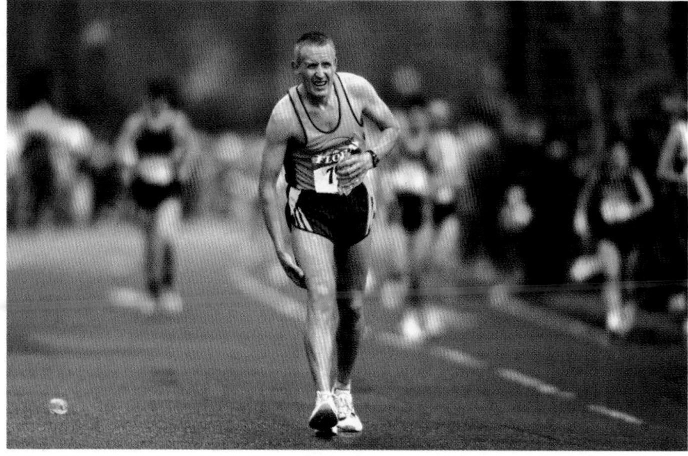

Figure 7.34 As lactate builds up within cells, the pH eventually falls, inhibiting the glycolysis reactions. The athlete (or anyone sprinting for a bus!) or any other exhausted animal has to slow down or stop so that oxygen supply can meet demand. Lactate build-up may lead to muscular cramp.

Getting rid of lactate

After a period of anaerobic respiration, most of the lactate is converted back into pyruvate. It is oxidised directly to carbon dioxide and water via the Krebs cycle, thus releasing energy to synthesise ATP. As a result, oxygen uptake is greater than normal in the recovery period after exercise (see Figure 7.35). This excess oxygen requirement is called the **oxygen debt**, or **post-exercise oxygen consumption**. It is needed to fuel the oxidation of lactate. Some lactate may also be converted into glycogen and stored in the muscle or liver.

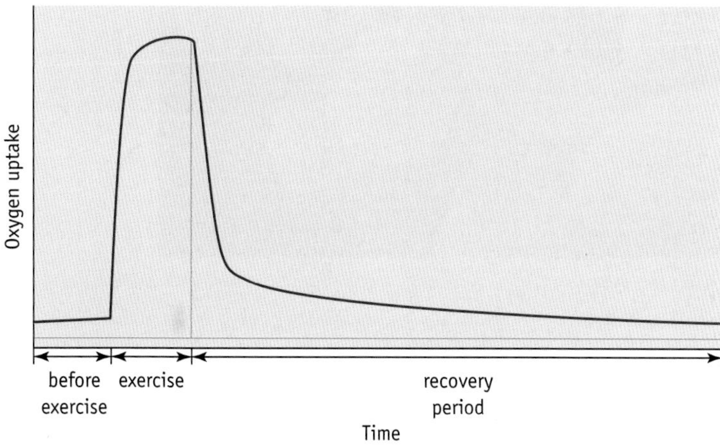

Figure 7.35 Oxygen uptake in the recovery period.

Q7.12 Why are athletes in training advised not to simply stop or lie down after strenuous exercise, but rather to aim for active recovery through gentle exercise?

Did you know?

Alcoholic fermentation

Yeast cells adopt a different tactic for dealing with anaerobic conditions. They reduce pyruvate to ethanol and CO_2 using the hydrogen from reduced NAD, thus recreating oxidised NAD and allowing glycolysis to continue (Figure 7.36). This anaerobic respiration, also known as fermentation, is much exploited in the brewing industry. Yeast cells are facultative anaerobes. Aerobes are organisms that respire using oxygen; anaerobes respire without using oxygen. Facultative anaerobes use either aerobic or anaerobic respiration depending on the conditions.

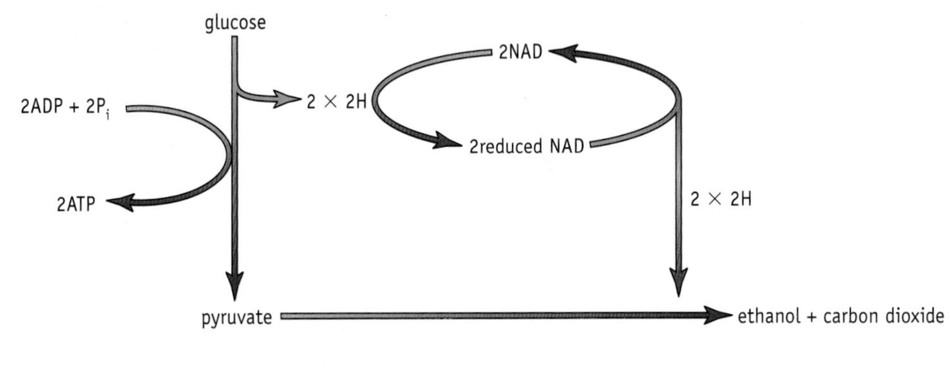

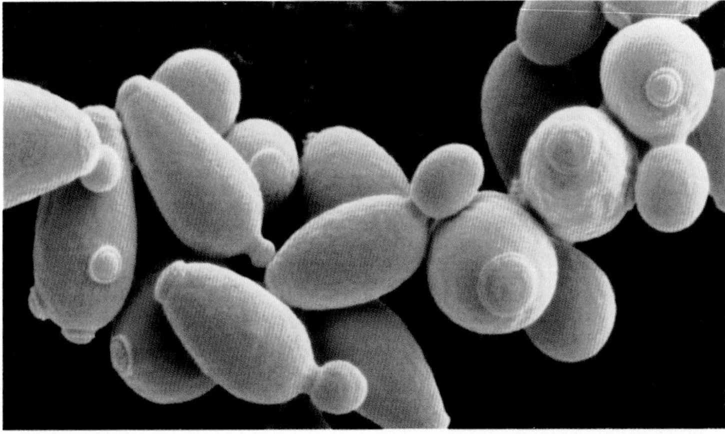

Figure 7.36 Yeast cells (bottom) can respire both aerobically and anaerobically. In the absence of oxygen they use alcohol fermentation. Magnification × 1600.

Supplying instant energy

Cells store only a tiny amount of ATP, enough in humans to allow a couple of seconds of explosive, all-out exercise. Therefore the recharging of the ATP store has to be very rapid.

At the start of any type of exercise, the immediate regeneration of the ATP is achieved using **creatine phosphate** (sometimes also called phosphocreatine, PC). This is a substance stored in muscles that can be hydrolysed to release energy. This energy can be used to regenerate ATP from ADP and phosphate, the phosphate being provided by the creatine phosphate itself. Creatine phosphate breakdown begins as soon as exercise starts (triggered by the formation of ADP). The reactions can be represented as:

Checkpoint

7.2 Draw up a table of comparisons between aerobic and anaerobic respiration.

$$\text{creatine phosphate} \rightarrow \text{creatine} + P_i$$

$$ADP + P_i \rightarrow ATP$$

These can be summarised as:

$$\text{creatine phosphate} + ADP \rightarrow \text{creatine} + ATP$$

The reactions do not require oxygen, and provide energy for about 6–10 seconds of intense exercise. This is known as the ATP/PC system. It is relied upon for regeneration of ATP during bursts of intense activity, for example when throwing or sprinting (Figure 7.37). Later, creatine phosphate stores can be regenerated from ATP when the body is at rest.

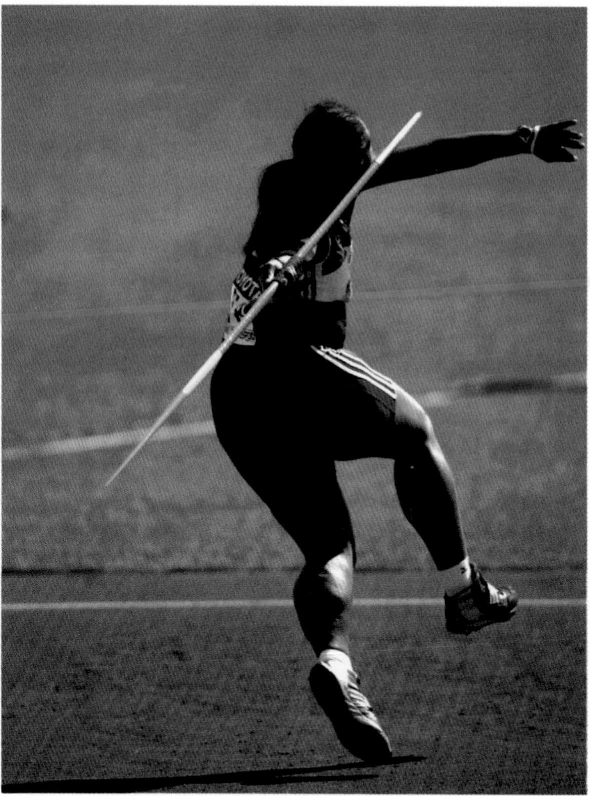

Figure 7.37 This javelin thrower relies almost entirely on the ATP/PC system to supply the energy burst for her powerful throw.

Three energy systems

At the start of any exercise, aerobic respiration cannot meet the demands for energy because the supply of oxygen to the muscles is insufficient (Figure 7.38). The lungs and circulation are not delivering oxygen quickly enough, and ATP will be regenerated without using oxygen. First the ATP/PC system and then the anaerobic respiration system allow ATP regeneration.

In endurance-type exercise, an increase in blood supply to the muscles ensures higher oxygen supply to the muscle cells. Aerobic respiration can regenerate ATP as quickly as it is broken down. This allows the exercise to be sustained for long periods.

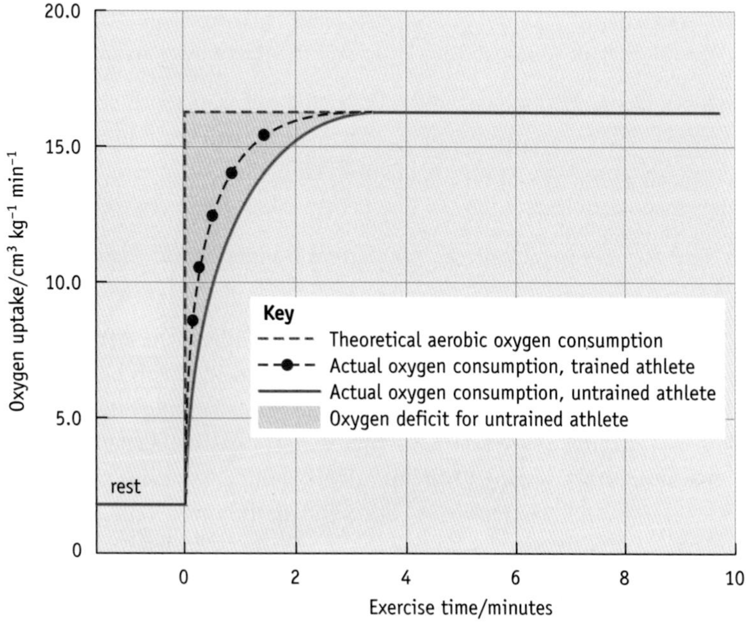

Figure 7.38 The oxygen deficit is the difference between the actual oxygen consumption and the theoretical oxygen consumption had the exercise been completed entirely aerobically.

Q7.13 State which energy system:

a a cheetah will use in its sprint to catch prey

b a wildebeest will use during the majority of its migration.

Q7.14 The table below shows the percentage contributions of the three energy systems during various sports. Which represents:

a volleyball

b hockey

c long-distance running?

Sport	ATP/PC	Anaerobic glycolysis to lactate	Aerobic respiration
A	60	20	20
B	90	10	0
C	10	20	70

Q7.15 Will the trained or untrained athlete create a larger oxygen deficit during the exercise period shown on the graph in Figure 7.38? Give a reason for your answer.

Q7.16 A cheetah is a carnivore, so its diet is largely protein and fat. How might it gain carbohydrate stores for use in anaerobic respiration?

7.3 Peak performance

The ability to undertake prolonged periods of strenuous but submaximal activity (e.g. running but not flat-out sprinting) is dependent on maintaining a continuous supply of ATP for muscle contraction. This, in turn, depends on **aerobic capacity**, that is, the ability to take in, transport and use oxygen.

At rest we consume about 0.2–0.3 litres of oxygen per minute. This is known as $\dot{V}O_2$. ($\dot{V}$ means volume per minute.) This increases to 3–6 litres a minute during maximal aerobic exercise, known as **$\dot{V}O_2$(max)**. $\dot{V}O_2$(max) is often expressed in units of ml min^{-1} kg^{-1} of body mass. Successful endurance athletes have a higher $\dot{V}O_2$(max). $\dot{V}O_2$(max) is dependent on the efficiency of uptake and delivery of oxygen by the lungs and cardiovascular system, and the efficient use of oxygen in the muscle fibres. A fit person can work for longer and at a higher intensity, using aerobic respiration without accumulating lactate, than can someone who does not undertake regular aerobic exercise.

Activity

In **Activity 7.9** you can measure your $\dot{V}O_2$(max). **A7.09S**

Most of us have run at some time and ended up breathing heavily with a pounding heart, or have seen a pet dog collapse panting at the end of a run (Figure 7.39). Without any conscious thought, the cardiovascular and ventilation systems adjust to meet the demands of the exercise, ensuring that enough oxygen and fuel reaches the muscles, and removing excess carbon dioxide and lactate. These systems are also important in redistribution of energy in temperature control. The major changes are to **cardiac output** (volume of blood pumped by the heart in a minute), breathing (also called ventilation) rate, and the depth of breathing. When running, adequate oxygen supply is maintained by:

- increasing cardiac output
- faster rate of breathing
- deeper breathing.

The more efficient the cardiovascular and ventilation systems, the better suited an individual will be to aerobic-endurance type exercise.

Look back at AS Topic 1 to refresh your memory of the structure and function of the heart, and Topic 2 to remind yourself about the lungs and gas exchange.

Figure 7.39 Catching your breath after running is the one time you may be conscious of changes occurring in your cardiovascular and ventilation systems.

Cardiac output

The volume of blood pumped by the heart in one minute is called the cardiac output. This increases during exercise. At rest, cardiac output is approximately 5 dm^3 per minute in both trained and untrained individuals. But it can rise to about 30 dm^3 min^{-1} in a trained athlete making maximum effort.

Cardiac output depends on the volume of blood ejected from the left ventricle (the **stroke volume**) and the **heart rate**:

$$\text{Cardiac output (CO)} = \text{stroke volume (SV)} \times \text{heart rate (HR)}$$

Activity

Activity 7.10 lets you look at the effect of exercise on cardiac output. **A7.10S**

Stroke volume

The stroke volume is the volume of blood pumped out of the left ventricle each time the ventricle contracts, measured in cm^3. (The volume pumped from both the left and right ventricles is virtually identical. Think about it!) For most adults at rest, about 50–90 cm^3 is pumped into each of the pulmonary artery and the aorta when the ventricles contract.

The heart draws blood into the atria as it relaxes during diastole. How much blood the heart pumps out with each contraction is determined by how much blood is filling the heart, that is, the volume of blood returning to the heart from the body. During exercise there is greater muscle action, so more blood returns to the heart in what is known as the **venous return**. In diastole, during exercise, the heart fills with a larger volume of blood. The heart muscle is stretched to a greater extent, causing it to contract with a greater force, and so more blood is expelled. This increases stroke volume and cardiac output.

When the body is at rest, the ventricles do not completely empty with each beat; approximately 40% of the blood volume remains in the ventricles after contraction. During exercise stronger contractions occur, ejecting more of the residual blood from the heart. In Figure 7.40 you can see the effect of exercise on stroke volume.

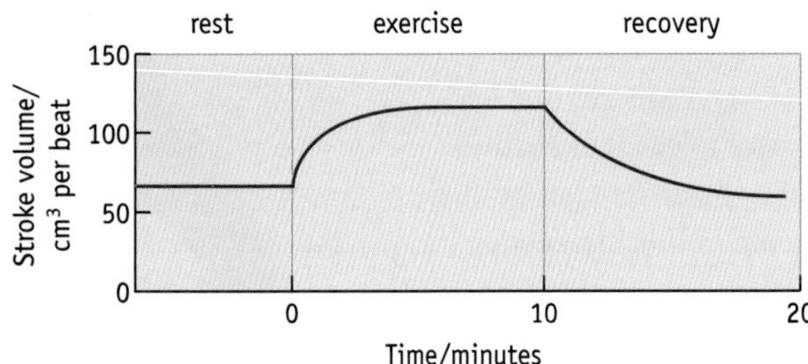

Figure 7.40 The effect of exercise on stroke volume.

Heart rate

Each of us has a slightly different resting heart rate. Measure your heart rate by counting your pulse while sitting at your desk. With each beat of your heart, a pulse of blood is ejected. This can be felt passing along the arteries; you can feel it fairly easily at the wrist (radial artery) or at your neck (carotid artery). The average heart rate for males is 70 beats per minute (bpm) while for females it is 72 bpm. The average fit person's heart rate would be around 65 bpm.

Differences in resting heart rate are caused by many factors. For example, our hearts differ in size, owing to differences in body size and genetic factors. A larger heart will usually have a lower resting heart rate. It will expel more blood with each beat and so, other things being equal, does not have to beat as frequently to circulate the same volume of blood around the body. Endurance training produces a lower resting heart rate, largely due to an increase in the size of the heart, resulting from thickening of the heart muscle walls. The cyclist Miguel Indurain, five times winner of the Tour de France, had a resting pulse rate of 28 beats per minute.

Look at the graphs in Figure 7.41 to see the effect of exercise on heart rate and cardiac output.

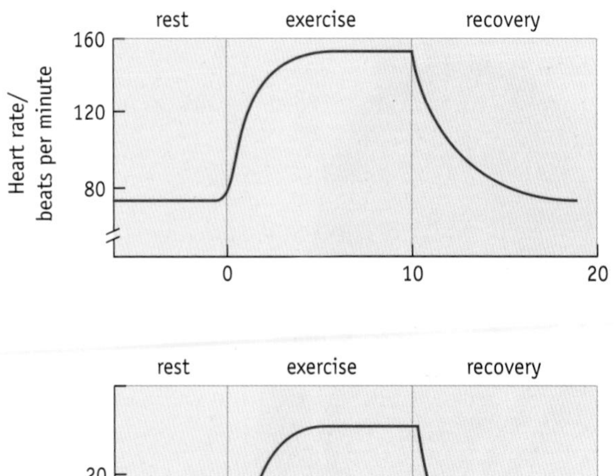

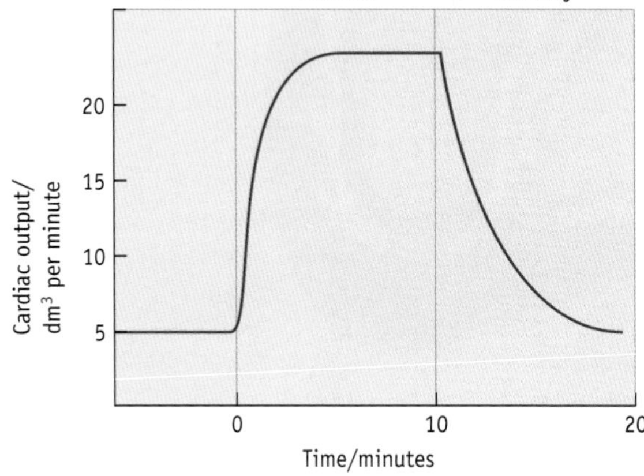

Figure 7.41 The effect of exercise on heart rate and cardiac output.

Q7.17 **a** A person has a resting stroke volume of 75 cm^3. They take their pulse rate and find that it is 70 beats per minute. What is their cardiac output?

b An endurance athlete has the same cardiac output at rest as the person in part a, but has a resting heart rate of 50 bpm. What is their stroke volume?

c The cyclist Lance Armstrong, seven times winner of the Tour de France, has a resting heart rate of 33 bpm. What will his resting stroke volume be, assuming his resting cardiac output is much the same as everyone else's?

Control of heart rate

Remind yourself about the cardiac cycle by reviewing AS Topic 1, Section 1.1.

How does the heart beat?

The heart can beat without any input from the nervous system, even when it is removed from the body and placed in glucose and salt solution. This shows that the heart muscle is **myogenic**; it can contract without external nervous stimulation. Contraction of cardiac muscle is initiated by small changes in the electrical charge of cardiac muscle cells. When these cells have a slight positive charge on the outside, they are said to be polarised. When this charge is reversed, they are depolarised. A change in polarity spreads like a wave from cell to cell, and causes the cells to contract.

Depolarisation starts at the **sinoatrial node** (**SAN**). This is a small area of specialised muscle fibres located in the wall of the right atrium, beneath the opening to the superior vena cava (Figure 7.42). The sinoatrial node is also known as the **pacemaker**.

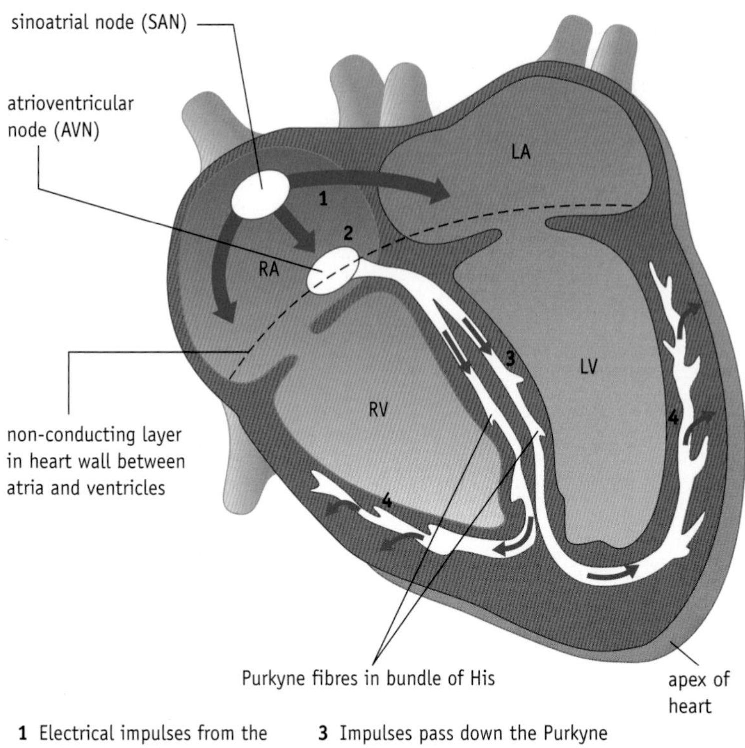

1 Electrical impulses from the SAN spread across the atria walls, causing contraction.

2 Impulses pass to the ventricles via the AVN.

3 Impulses pass down the Purkyne fibres to the heart apex.

4 The impulses spread up through the ventricle walls causing contraction from the apex upwards. Blood is squeezed into the arteries.

Figure 7.42 Follow the route taken by the electrical impulses passing through the heart, initiating contraction of the atria and then the ventricles.

The SAN generates an electrical **impulse**; this spreads across the right and left atria, causing them to contract at the same time. The impulse also travels to some specialised cells called the **atrioventricular node** (**AVN**). From here, the impulse is conducted to the ventricles after a delay of about 0.13 seconds.

Why is it important that the impulse is delayed? The delay ensures that the atria have finished contracting, and that the ventricles have filled with blood before they contract.

After this delay, the signal reaches the **Purkyne fibres** (sometimes called Purkinje fibres). These are large, specialised muscle fibres that conduct impulses rapidly to the apex (tip) of the ventricles. There are right and left bundles of fibres, and they are collectively called the **bundle of His**.

The Purkyne fibres continue around each ventricle, and divide into smaller branches that penetrate the ventricular muscle. These branches carry the impulse to the inner cells of the ventricles, and from here it spreads through the entire ventricles.

The first ventricular cells to be depolarised are at the apex of the heart, so that contraction begins at this point and travels *upwards* towards the atria. This produces a wave of contraction moving up the ventricles, pushing the blood into the aorta and the pulmonary artery.

Activity

Activity 7.11 includes an animation of the heart's conductive pathways, and lets you investigate its working in more detail. **A7.11S**

Checkpoint

7.3 Draw a flow chart to summarise the information you have just read about the electrical activity of the heart.

Measuring electrical activity

The electrical activity of the heart can be detected and displayed on an **electrocardiogram** (**ECG**), a graphic record of the electrical activity during the cardiac cycle. This is the most common test to check for problems with the heart. The test is easy to carry out, and any patient with suspected cardiac problems will probably have one.

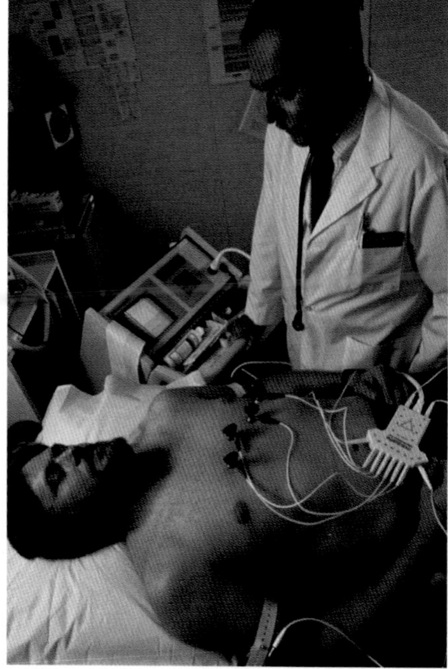

Figure 7.43 Twelve electrodes are used to give twelve views of the heart. By using different combinations of electrodes, the ECG can detect electrical currents as they spread in different directions across different regions of the heart.

How is an ECG carried out?

In an ECG, electrodes are attached to the person's chest and limbs to record the electrical currents produced during the cardiac cycle (see Figure 7.43). When there is a change in **polarisation** of the cardiac muscle, a small electrical current can be detected at the skin's surface. This is what an ECG measures.

An ECG is usually performed while the patient is at rest, lying down and relaxed, but it may be used in a stress test. This involves doing an ECG before, during and after a period of exercise, and is used to detect heart problems that emerge only when the heart is working hard. Breathing rate and blood pressure may also be measured and recorded.

What does the ECG trace show us?

Look at Figure 7.44. Note which wave on the ECG represents each stage in the electrical activity of the heart.

- **P wave** – depolarisation of the atria, leading to atrial contraction (atrial systole).
- **PR interval** – the time taken for impulses to be conducted from the SAN across the atria to the ventricles, through the AVN.
- **QRS complex** – the wave of depolarisation resulting in contraction of the ventricles (ventricular systole).
- **T wave** – repolarisation (recovery) of the ventricles during the heart's relaxation phase (diastole).

The ECG does not show atrial repolarisation because the signals generated are small, and are hidden by the QRS complex.

The ECG can be used to measure heart rate. Each square represents 0.2 seconds; five squares represent 1 second (Figure 7.44), and 300 squares pass through the ECG machine in 1 minute. You can work out the time for one complete cardiac cycle by multiplying the number of squares between QRS complexes by 0.2 (the length in seconds of a square), and then dividing this into 60. Alternatively you can divide 300 by the number of squares per beat. The two methods give the same answer.

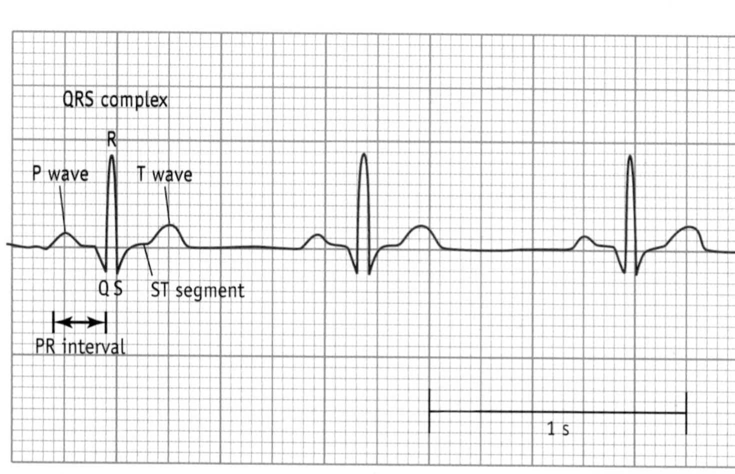

Figure 7.44 A normal ECG trace. The vertical axis shows electrical activity; the horizontal axis shows time.

Q7.18 Work out the time for one cardiac cycle and the heart rate for the patient whose ECG trace is shown in Figure 7.44.

A heart rate of less than 60 beats per minute (60 bpm) is known as bradycardia. It is common in fit athletes at rest, but can also be a symptom of heart problems. Possible causes include hypothermia, heart disease, or use of medicines or drugs.

Tachycardia, a heart rate greater than 100 bpm, is normally the result of anxiety, fear, fever or exercise. It can also be a symptom of coronary heart disease, heart failure, use of medicines or drugs, fluid loss or anaemia.

During a period of ischaemia, heart muscle does not receive blood due to atherosclerosis causing a blockage of coronary arteries. The normal electrical activity and rhythm of the heart are disrupted, and arrhythmias (irregular beatings caused by electrical disturbances) can affect a larger area of heart muscle than that affected by the initial ischaemia (Figure 7.45A).

Activity

Can you work out the patients' heart problems from the ECGs in **Activity 7.12? A7.12S**

A

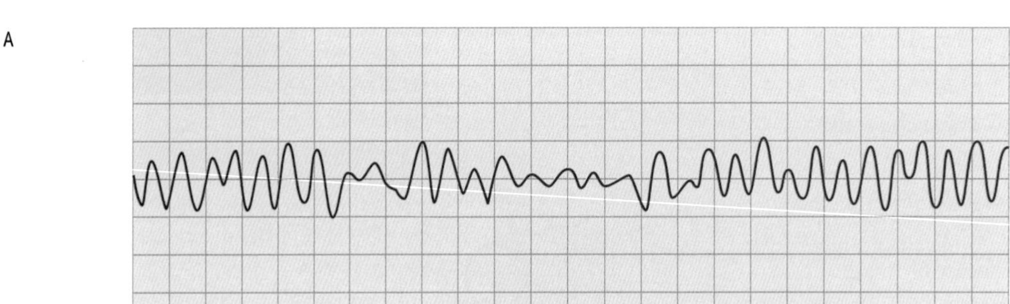

B

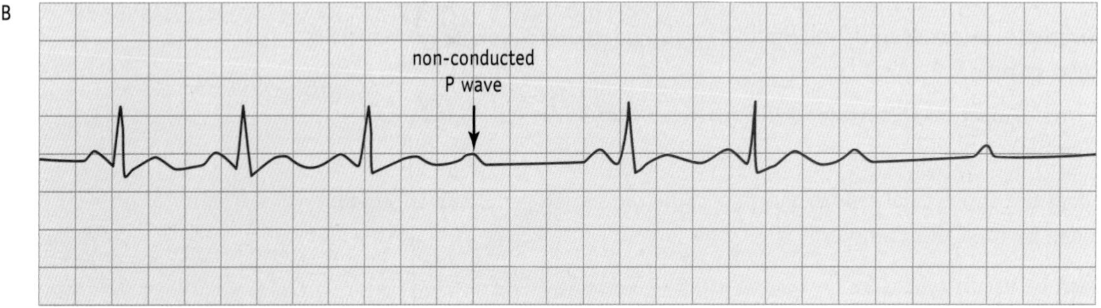

C

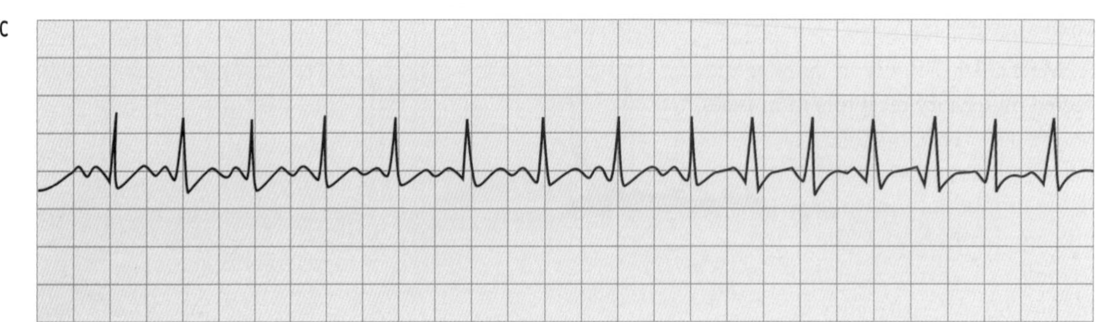

Figure 7.45 **A** In ventricular fibrillation, irregular stimulation of the ventricles makes them contract in a weak and uncoordinated manner. This leads to a fall in blood pressure, and results in sudden death unless treated immediately. **B** Notice that the P wave is not followed by a QRS complex, showing there is a break in the conduction system of the heart. **C** This ECG trace shows abnormal rhythms that could lead to cardiac arrest.

An ECG trace can provide information about abnormal heartbeats, areas of damage, and inadequate blood flow. Abnormalities on an ECG can also indicate an elevated risk of sudden death, for example as a result of cardiac-conduction abnormalities, or due to hypertrophic cardiomyopathy – an inherited condition in which gene mutations cause abnormally thick walls in the left ventricle.

Look at Figure 7.45 and then try answering the questions that follow.

Q7.19 How does the trace in **a** Figure 7.45A and **b** Figure 7.45C differ from that of a normal ECG?

Q7.20 Look at Figure 7.45B. Where do you think that the break in the conduction system occurs?

Q7.21 The ECG in Figure 7.45C is from a young woman who had collapsed at a club. Suggest what might have happened to produce these rapid rhythms.

Heart rate is under the control of the nervous system, and is also affected by hormonal action.

Nervous control of heart rate

Heart rate is under the control of the **cardiovascular control centre** located in the medulla of the brain.

Nerves forming part of the **autonomic nervous system** (the nervous system that you have no control over) lead from the cardiovascular control centre to the heart. There are two such nerves going from the cardiovascular control centre to the heart – a **sympathetic nerve** (accelerator) and the **vagus nerve**, which is a **parasympathetic nerve** (decelerator). See Table 7.1 for a comparison of the functions of these different nerve types. Stimulation of the sinoatrial node (SAN) by the sympathetic nerve causes an increase in the heart rate, whereas impulses from the vagus nerve slow down the rate (Figure 7.46).

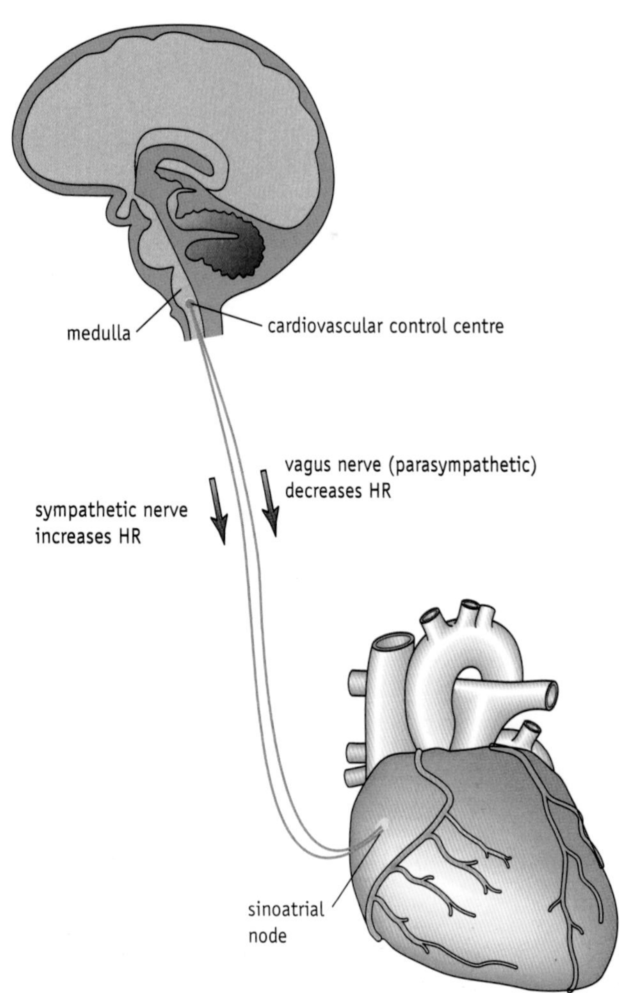

Figure 7.46 Control of the heart rate by the cardiovascular control centre.

The cardiovascular control centre detects accumulation of carbon dioxide and lactate in the blood, reduction of oxygen, and increased temperature. Mechanical activity in muscles and joints is detected by sensory receptors in muscles, and impulses are sent to the cardiovascular control centre. These changes result in a higher heart rate.

At the sound of the starting pistol, or the sight of prey (an animal's next meal), skeletal muscles contract, and stretch receptors in the muscles and tendons are stimulated. They send impulses to the cardiovascular control centre. This in turn raises the heart rate via the sympathetic (accelerator) nerve. There is an increase in venous return, which leads to a rise in the stroke volume. The elevated heart rate and stroke volume together result in higher cardiac output, thus transporting oxygen and fuel to muscles more quickly.

Blood pressure rises with higher cardiac output. To prevent it rising too far, pressure receptors in the aorta and in the carotid artery send nerve impulses back to the cardiovascular control centre. Inhibitory nerve impulses are then sent from here to the sinoatrial node. In this way, an excessive rise in blood pressure is avoided through negative feedback, which prevents a further rise in heart rate.

The autonomic (unconscious) part of your nervous system is made up of two sets of nerves:

- Sympathetic – stimulation of the sympathetic nerves prepares the body's systems for action (for the fight or flight response).
- Parasympathetic – stimulation of the parasympathetic nerves controls the body's systems when resting and digesting.

Examples of the opposing effects of the two types of nerve:

Organ or tissue	Effect of sympathetic stimulation	Effect of parasympathetic stimulation
intercostal muscles	increases breathing rate	decreases breathing rate
heart	increases heart rate and stroke volume	decreases heart rate and stroke volume
gut	inhibits peristalsis*	stimulates peristalsis

Table 7.1 Sympathetic and parasympathetic nerves.
*Muscle contractions in the gut wall that move food through the gut.

Hormonal effects on heart rate

Fear, excitement and shock cause the release of the hormone **adrenaline** into the bloodstream from the adrenal glands located above the kidneys. Adrenaline has an effect on the heart rate similar to stimulation by the sympathetic nerve. It has a direct effect on the sinoatrial node, increasing the heart rate to prepare the body for any likely physical demands.

Adrenaline also causes dilation of the arterioles supplying skeletal muscles, and constriction of arterioles going to the digestive system and other non-essential organs; this maximises blood flow to the active muscles. Adrenaline causes an anticipatory increase in heart rate before the start of a race.

Q7.22 The carotid artery at the side of the neck is sometimes used to measure heart rate (pulse rate).

a Suggest why pressing on the carotid artery might reduce pulse rate, thereby giving a false reading.

b Where could you take a pulse more reliably?

Q7.23 Look at Figure 7.41. Explain why the heart rate rises before the start of exercise, and why this may be an advantage.

Q7.24 Describe and explain the changes that occur in blood distribution after a person has started to exercise.

> **Activity**
> In **Activity 7.13** you can measure lung volumes and breathing rate.
> **A7.13S**

Breathing

Lung volumes

The volume of air we breathe in and out at each breath is our **tidal volume**. At rest this is usually about 0.5 dm^3. When exercise begins, we increase our breathing rate and depth of breathing. The maximum volume of air we can inhale and exhale is our **vital capacity**. In most people this is 3–4 dm^3, but in large or very fit people it can be 5 dm^3 or more. Singers and those playing wind instruments may also have a large vital capacity. Lung volumes, including tidal volume and vital capacity, can be measured using a spirometer (see Figure 7.47).

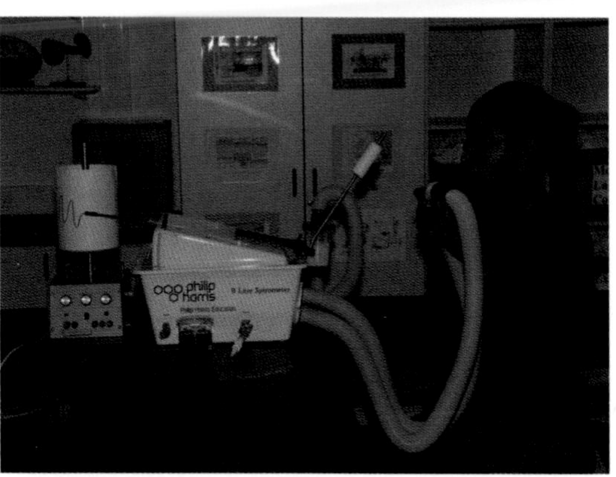

Figure 7.47 A spirometer can be used to measure lung volumes; changes in volume can be recorded as a trace. A nose-clip is worn to ensure accurate results.

The volume of air taken into the lungs in one minute is the **minute ventilation**. This is calculated by multiplying the tidal volume (the average volume of one breath, in dm^3) by the breathing rate (number of breaths per minute).

Q7.25 At rest, the average person takes 12 breaths per minute. Assuming a tidal volume of 0.5 dm^3, calculate the volume of air breathed in each minute.

Q7.26 From the spirometer trace in Figure 7.48:

a determine the tidal volume

b determine the vital capacity

c work out the minute ventilation

d work out the volume of oxygen consumed between points A and B on the trace, and calculate the rate of oxygen consumption.

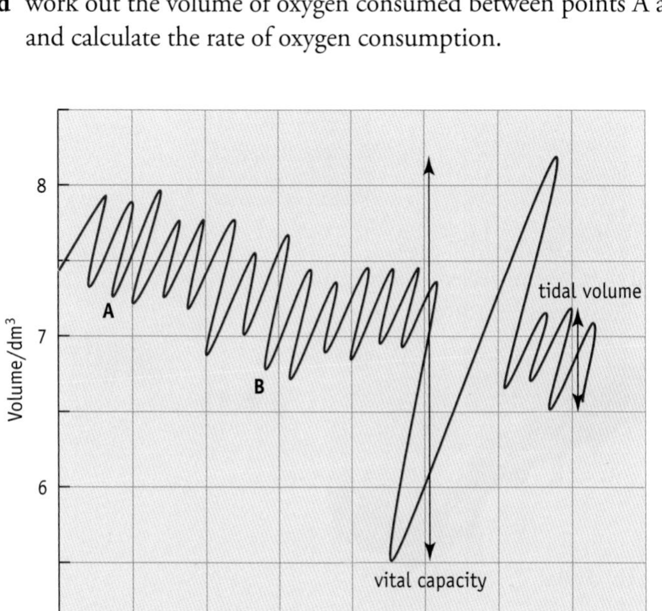

Figure 7.48 A spirometer trace showing quiet breathing, with one maximum breath in and out. The trace can be used to measure depth and frequency of breathing. The fall in the trace is due to the consumption of oxygen by the subject. The rate of oxygen consumption can be calculated by dividing the decrease in volume by time for the fall. Try analysing this trace by answering Question 7.26.

The control of breathing

The **ventilation centre** in the medulla oblongata of the brain controls breathing; this is summarised in Figure 7.49.

Inhalation

The ventilation centre sends nerve impulses every 2–3 seconds to the external intercostal muscles and diaphragm muscles. Both these sets of muscles contract causing inhalation.

During deep inhalation, not only are the external intercostals and diaphragm muscles stimulated, but the neck and upper chest muscles are also brought into play.

Exhalation

As the lungs inflate, stretch receptors in the bronchioles are stimulated. The stretch receptors send inhibitory impulses back to the ventilation centre. As a consequence, impulses to the muscles stop and the muscles relax, stopping inhalation and allowing exhalation.

Exhalation is caused by the elastic recoil of the lungs (like a deflating balloon), and by gravity helping to lower the ribs. Not all of the air in the lungs is exhaled with each breath. The air remaining in the lungs, residual air, mixes with the air inhaled with each breath.

The internal intercostal muscles only contract during deep exhalation. For example, during vigorous exercise a larger volume of air is exhaled, leaving less residual air in the lungs.

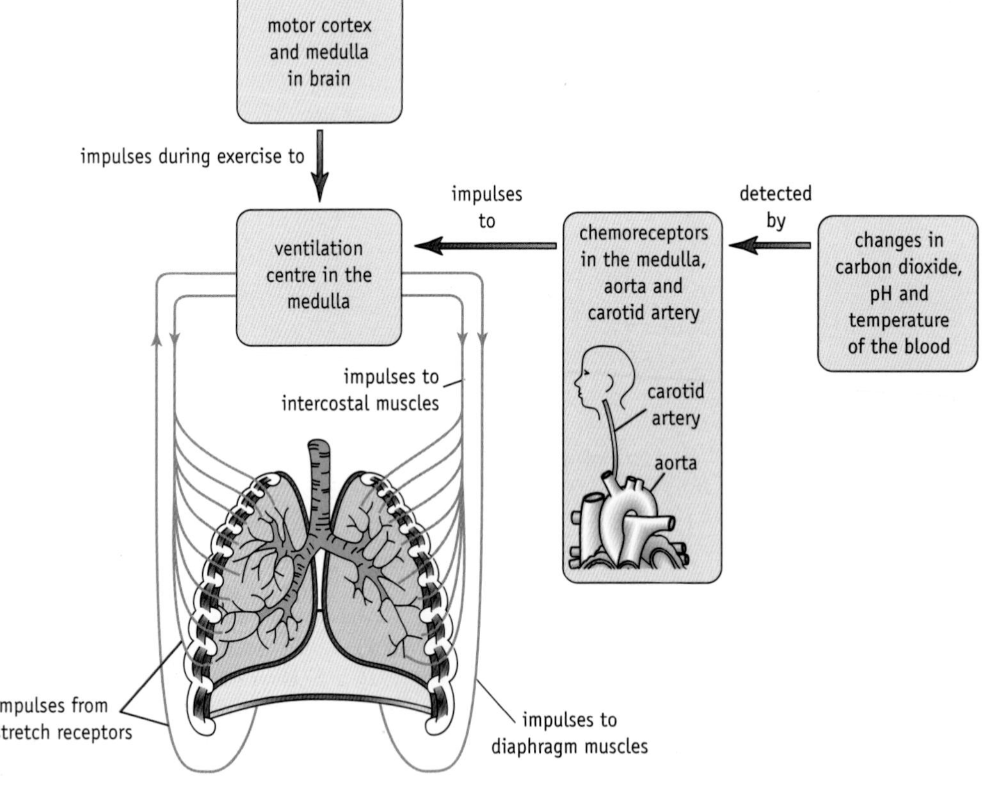

Figure 7.49 Control of breathing.

Controlling breathing rate and depth

At rest, the most important stimulus controlling the breathing rate and depth of breathing is the concentration of dissolved carbon dioxide in arterial blood, via its effect on pH. A small increase in blood carbon dioxide concentration causes a large increase in ventilation. This is achieved as follows:

- Carbon dioxide dissolves in the blood plasma, making carbonic acid.
- Carbonic acid dissociates into hydrogen ions and hydrogencarbonate ions, thereby lowering the pH of the blood.

$$CO_2 + H_2O \rightleftharpoons H_2CO_3 \rightleftharpoons H^+ + HCO_3^-$$

- Chemoreceptors sensitive to hydrogen ions are located in the ventilation centre of the medulla oblongata. They detect the rise in hydrogen ion concentration.
- Impulses are sent to other parts of the ventilation centre.
- Impulses are sent from the ventilation centre to stimulate the muscles involved in breathing.

There are also chemoreceptors in the walls of the carotid artery (Figure 7.50) and aorta. These are stimulated by changes in pH resulting from changes in carbon dioxide concentration. These chemoreceptors monitor the blood before it reaches the brain, and send impulses to the ventilation centre.

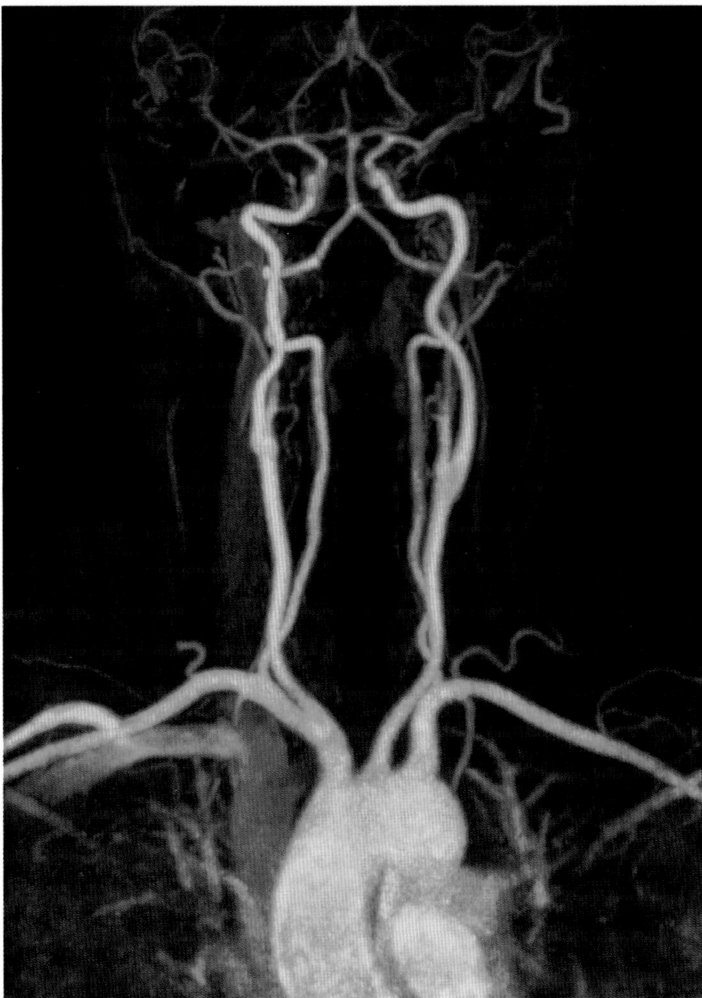

Figure 7.50 Coloured magnetic resonance angiograph of the blood vessels between the heart and the base of the brain. Bottom centre is the aortic arch, which curves over the heart. The carotid arteries are clearly visible in the neck. The carotid arteries carry blood to the head. Their walls contain pressure receptors sensitive to blood pressure, and chemoreceptors sensitive to changes in pH. The subclavian arteries supplying the arms are also visible.

Increasing carbon dioxide and the associated fall in pH leads to an increase in rate and depth of breathing, through more frequent and stronger contraction of the appropriate muscles. The more frequent and deeper breaths maintain a steep concentration gradient of carbon dioxide between the alveolar air and the blood (Figure 7.51). This in turn ensures efficient removal of carbon dioxide and uptake of oxygen. The opposite response occurs with a decrease in carbon dioxide. The control of carbon dioxide levels in the blood is an example of homeostasis operating via negative feedback. You will look at this in detail later in this topic.

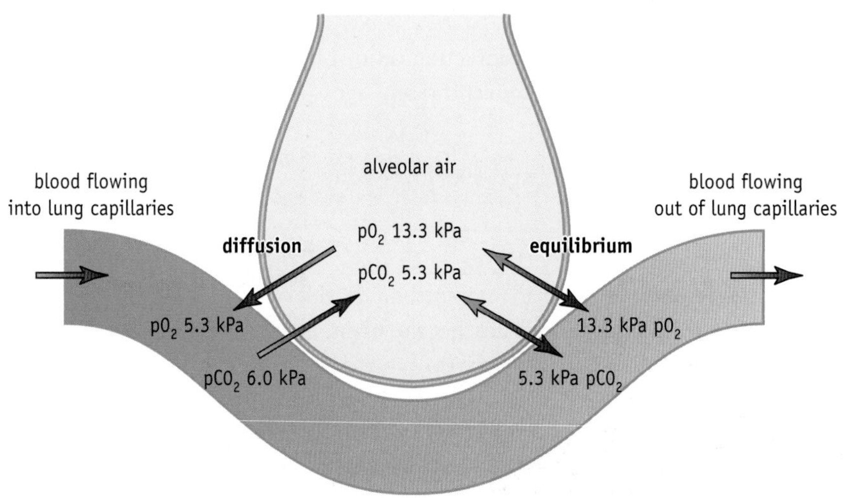

The gas partial pressure values give a measure of concentration.
At sea level the total atmospheric pressure is approximately 101 kPa. 21% of the atmosphere is oxygen, so 21% of the total pressure, about 21 kPa, is due to oxygen. This is the partial pressure of oxygen in the atmosphere at sea level.

Figure 7.51 The exchange of carbon dioxide and oxygen between a lung capillary and an alveolus.

Activity

In **Activity 7.14** you can investigate the control of ventilation rate. **A7.14S**

Extension

To find out how the blood carries oxygen and carbon dioxide, read **Extension 7.3. X7.03S**

Controlling breathing during exercise

The motor cortex of the brain is the region that controls movement. As soon as exercise begins, impulses from the motor cortex have a direct effect on the ventilation centre in the medulla, increasing ventilation sharply. Ventilation is also increased in response to impulses reaching the ventilation centre from stretch receptors in tendons and muscles involved in movement. The various chemoreceptors sensitive to carbon dioxide levels and to changes in blood temperature increase the depth and rate of breathing via the ventilation centre. The control mechanisms are summarised in Figure 7.49. There are receptors sensitive to changing oxygen concentrations in the blood; however, they are rarely stimulated under normal circumstances.

Q7.27 During vigorous exercise, the concentration of oxygen in the lungs is higher than when at rest. Suggest the reasons for, and the advantage of, this elevated oxygen level.

Q7.28 Suggest why it is beneficial that stimulation of stretch receptors in the muscles increases ventilation.

Q7.29 When a person breathes air containing 80% oxygen, the minute ventilation is reduced by 20%. Explain how this occurs.

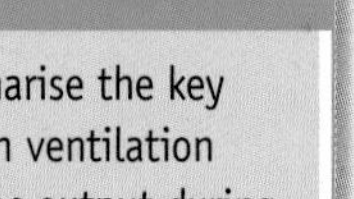

Checkpoint

7.4 Summarise the key changes in ventilation and cardiac output during exercise. Describe the sequence of events that changes the breathing rate when there is a decline in carbon dioxide in the blood.

Did you know?

Altitude sickness

The control of breathing can go wrong if we climb too quickly to high altitude. As oxygen becomes scarce (at about 4000 m), the blood oxygen concentration can decrease to a very low level (Figure 7.52). This triggers the ventilation centre in the medulla oblongata to make us breathe deeply and rapidly, especially if we are doing a hard climb. The hard panting flushes out too much carbon dioxide, resulting in a rise of blood pH. The ventilation centre may respond by stopping breathing altogether for a few seconds, so we alternate between heavy panting and not breathing at all. If this happens it is essential to descend quickly to a lower altitude.

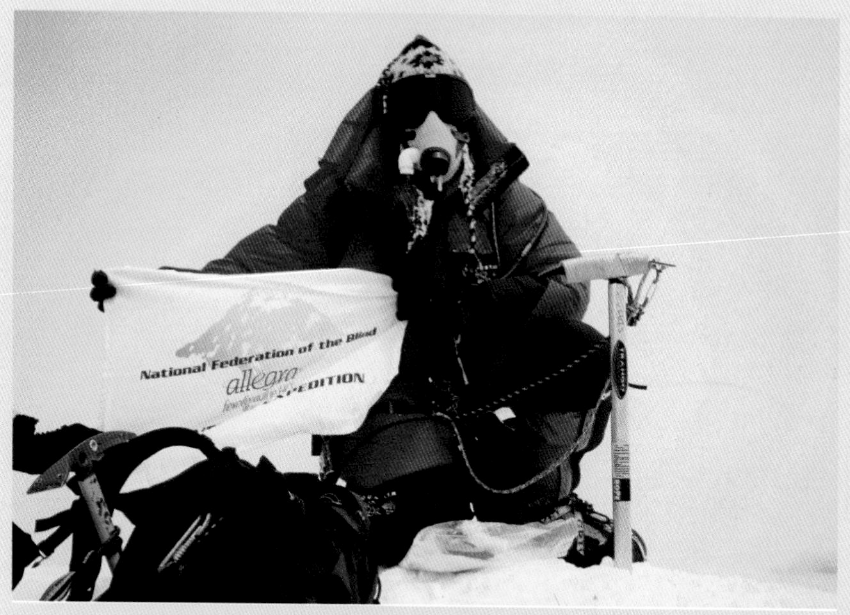

Figure 7.52 At 4000 m there is only about 62% of the oxygen available at sea level; this falls to about 48% at 5000 m. At the highest camp on Mount Everest at the South Col (7950 m), there is only 38% of the oxygen available at sea level; at the summit (8848 m), there is only 33%. To avoid altitude (mountain) sickness, climbers use extra oxygen.

All muscle fibres are not the same

Aerobic capacity is dependent not only on uptake and transport of oxygen to the muscles, but also on efficiency of use once it reaches the muscle fibres. (Remember that a muscle fibre is another term for a muscle cell.) Although the skeletal muscles of all mammals are almost identical in their macroscopic and microscopic structure, it is possible to identify two distinct types of fibre.

If you eat chicken (which has the same types of muscle fibres as humans and other mammals) you may have noticed that some parts of the flesh are darker and others lighter. The breast meat, where the flight muscles are, is pale (the 'white meat') whereas the leg muscles are darker. They are different because the two regions contain two different types of muscle fibre, reflecting the different functions of the muscles.

Slow twitch fibres

Chickens spend most of their time on the ground, standing still or walking. The darker muscle in the legs and body is made up of fibres called **slow twitch fibres**. These are specialised for slower, sustained contraction, and can cope with long periods of exercise. To do this they need to carry out a large amount of aerobic respiration.

Key features of slow twitch muscle fibres

The slow twitch muscle fibres have many mitochondria and high concentrations of respiratory enzymes to carry out the aerobic reactions. They also contain large amounts of the dark red pigment **myoglobin**, which gives them their distinctive colour. Myoglobin is a protein similar to haemoglobin (the oxygen-carrying pigment found in red blood cells). It has a high affinity for oxygen, and only releases it when the concentration of oxygen in the cell falls very low; it therefore acts as an oxygen store within muscle cells. Slow twitch fibres are associated with numerous capillaries to ensure a good oxygen supply.

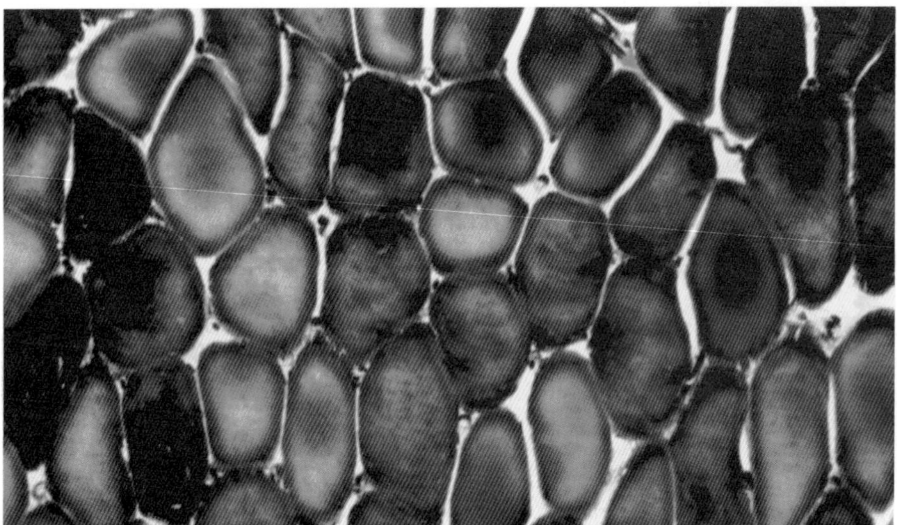

Figure 7.53 Light microscope cross-section of skeletal muscle fibres. The fibres are seen end on. Notice how some of the fibres are very dark. These contain more myoglobin and are slow twitch.

Fast twitch fibres

Chickens can perform near vertical take-offs in moments of panic or to reach higher perches, though they cannot fly for long. The paler flight muscle is largely made up of a different type of muscle fibre called **fast twitch fibres**. These fibres are specialised to produce rapid, intense contractions. The ATP used in these contractions is produced almost entirely from anaerobic glycolysis.

Key features of fast twitch muscle fibres

The fast twitch fibres have few mitochondria. (Remember that glycolysis does not occur in mitochondria.) They also have very little myoglobin, so have few reserves of oxygen and few associated capillaries. Their reliance on anaerobic respiration means there is a rapid build-up of lactate, so the fast twitch muscle fibres fatigue easily. With aerobic training, fast twitch fibres can take on some of the characteristics of slow twitch fibres. For example, they may have more mitochondria, allowing them to use aerobic respiration reactions when contracting.

Q7.30 Study Table 7.2 and then answer the following questions.

Slow twitch fibres	Fast twitch fibres
red (a lot of myoglobin)	white (little myoglobin)
many mitochondria	few mitochondria
little sarcoplasmic reticulum	extensive sarcoplasmic reticulum
low glycogen content	high glycogen content
numerous capillaries	few capillaries
fatigue resistant	fatigue quickly

Table 7.2 Characteristics of the two muscle fibre types.

a What is the significance of the large number of mitochondria in slow twitch fibres?

b Why do fast twitch fibres need more sarcoplasmic reticulum?

c Why will a large amount of myoglobin be advantageous to the slow twitch fibres?

d How will each type of fibre regenerate ATP?

e Which type of fibre will build up an oxygen debt more quickly? Give a reason for your answer.

f Should 'high creatine phosphate level' be added to the slow or fast twitch column?

Activity

In **Activity 7.15** you can compare the fast and slow twitch muscles in a mackerel. **A7.15S**

What makes a sprinter?

In mammals, these two types of muscle fibre are not separated, but are found together in all the skeletal muscles (Figure 7.53). The proportion of each type appears to be genetically determined, and varies between different people. Research has shown that successful endurance athletes such as marathon runners, rowers and cross-country skiers, with high aerobic capacity, have a higher proportion of slow twitch muscle fibres – up to 80%. In contrast, sprinters may have as few as 35% slow twitch fibres. Throwers and jumpers have a more or less equal proportion of the two types.

Individuals may be better suited to a particular type of sport if they naturally have a higher proportion of fibres used in that activity. This, of course, will not be the only factor that contributes to an individual's success in a sport. For example, an individual with a highly efficient cardiovascular system will be well suited to aerobic exercise.

Q7.31 Which type of fibre might dominate the leg muscles of **a** the cheetah and **b** the wildebeest?

Q7.32 Researchers analysed muscle samples from the leg muscle of wild African cheetahs and found that 83% of the muscle was made up of fast twitch fibres. The total mitochondrial volume density of the muscle ranged from 2.0 to 3.9%. Maximum activities for pyruvate kinase and lactate dehydrogenase in the muscle were high, 1519.00 ± 203.60 and 1929.25 ± 482.35 μmol min^{-1} g wet wt^{-1}, respectively. What conclusion can be drawn from these findings?

7.4 Breaking out in a sweat

The maximum distance a cheetah can sprint is approximately 500 m. After this point, not only will the fast twitch muscle fibres have fatigued due to the build-up of lactate, but the body temperature will have risen. Cheetahs must stop to recover.

The demands of physical activity can increase metabolic activity by up to ten times, releasing energy. This could potentially increase core temperature by around 1 °C every five to ten minutes. This energy must be dispersed to maintain thermal balance.

The marathon, more than any other sport, has a history of heat-related deaths. In 490 BC, Pheidippides, an intercity messenger of Ancient Greece, ran 26 miles from Marathon to Athens with instructions that the Athenians should not surrender to the Persian fleet. Legend has it that at the end of his journey, Pheidippides dropped dead from exhaustion. At the 1912 Olympic Games in Stockholm, the Portuguese runner Lazaro collapsed from heat stroke after running 19 miles. He died the following day. Nowadays marathons tend to be scheduled in the early morning to reduce the chance of heat stroke. However, this is unpopular with TV schedulers, and runners are sometimes required to race in more dangerous conditions to maximise advertising revenues. At the Athens Olympics in 2004, the marathons were run at peak viewing time in temperatures of 33 °C (men) and 35 °C (women).

Activity

In **Activity 7.16** you can investigate body temperature. **A7.16S**

Q7.33 **a** What is the optimal temperature range for human cells?

b What will happen to metabolic reactions if body temperature falls below or rises above the normal range?

Figure 7.54 During strenuous exercise, enough heat is produced to raise our body temperature by 1 °C every five to ten minutes were it not dissipated.

Key biological principle: Homeostasis

The need for homeostasis

In humans, if cells are to function properly, the body's internal conditions must be maintained within a narrow range of the cells' optimum conditions. The maintenance of a stable internal environment is called **homeostasis**.

This is partly achieved by maintaining stable conditions within the blood, which in turn gives rise to the tissue fluid that bathes the body's cells. In the blood the concentration of glucose, ions and carbon dioxide must be kept within narrow limits. In addition, the water potential (determined by the concentration of solutes in the blood), pH and temperature of the blood are tightly regulated.

The role of negative feedback

Each condition that is controlled has a **norm value** or **set point** that the homeostatic mechanisms are 'trying' to maintain. **Receptors** are used to detect deviations from the norm. These receptors are connected to a control mechanism, which turns on or off **effectors** (muscles and glands) to bring the condition back to the norm value.

For example, as blood glucose concentration rises above the norm, this information is fed back to the control mechanism. The same thing happens if it falls below the norm. Effectors then act to shift the blood glucose concentration back towards the norm. Because a deviation from the norm results in a change in the opposite direction, back to the norm, the process is known as **negative feedback.** These corrections mean that the actual value fluctuates in a narrow range around the norm. This is summarised in Figure 7.55.

Q7.34 Look back at the control of ventilation in Section 7.3. Explain how this is an example of negative feedback.

A Conditions controlled by homeostasis fluctuate around the norm value.

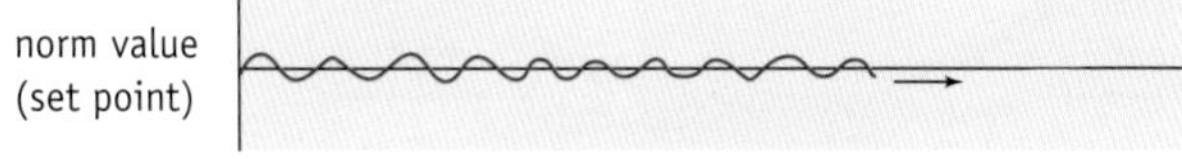

B The condition is controlled by negative feedback.

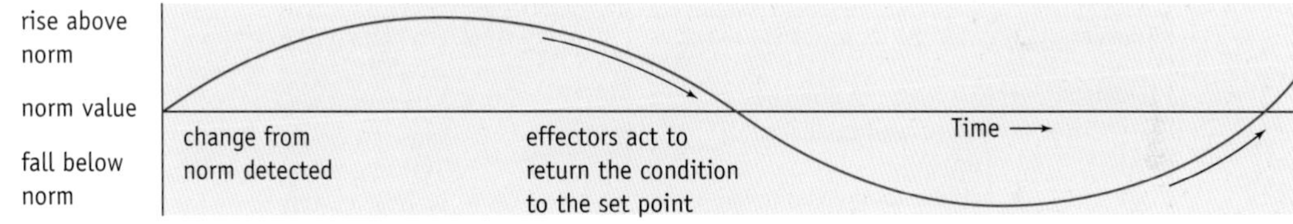

Figure 7.55 A summary of the principles of negative feedback in homeostatic control.

Temperature control

Thermoregulation is the control of body temperature. Our core body temperature is very stable at about 37.5 °C; this is about 0.5 °C higher than oral temperature, measured with a thermometer under the tongue (Figure 7.56). This body temperature allows enzyme-controlled reactions to occur at a reasonable rate. At lower temperatures, the reactions would occur too slowly for the body to remain active; at higher temperatures the enzymes would denature. A rise of only 5 °C in core body temperature can be fatal.

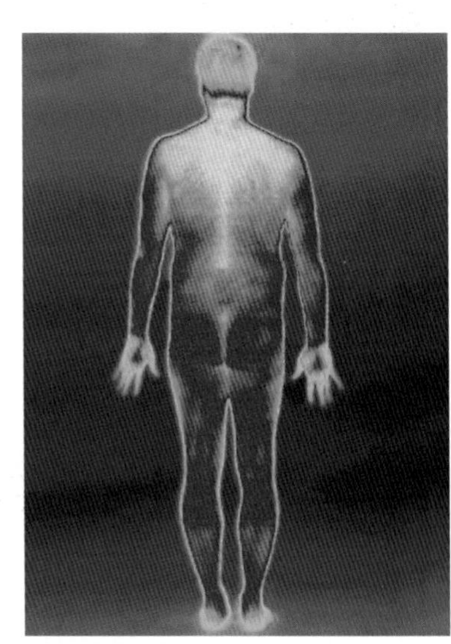

Figure 7.56 Thermograms show surface temperature by recording emission of infrared radiation. The temperature range goes from hot (white) to cold (blue). In humans it is the core body temperature that is maintained by homeostasis.

Temperature control receptors and effectors

In humans, temperature is maintained by a negative feedback system. This is summarised in Figure 7.57. This system involves receptors that detect changes in the blood temperature. These receptors are located in a structure in the brain called the hypothalamus. The hypothalamus is the control mechanism and acts as a thermostat, turning on the effectors necessary to return the temperature to the norm.

There are also thermoreceptors in the skin that detect temperature changes. If the skin is warm, then impulses are sent to the hypothalamus initiating the heat-loss responses and inhibiting heat-gain responses. If the skin is cold, the opposite happens. These changes help keep the core body temperature near its optimum.

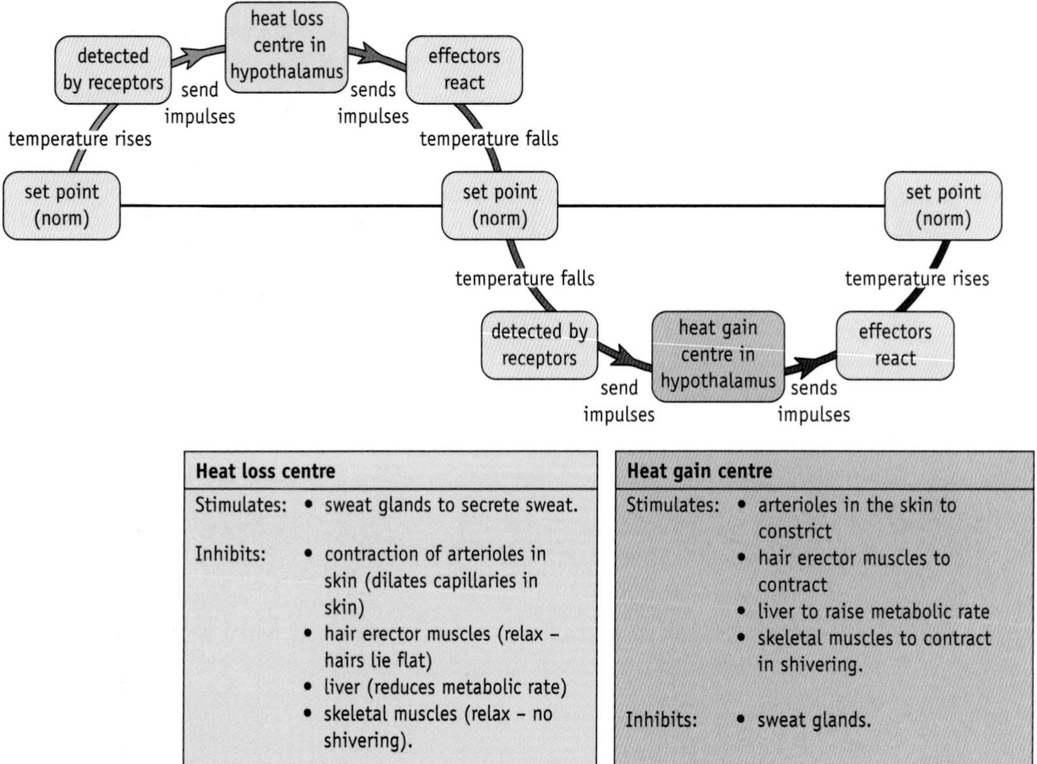

Heat loss centre	
Stimulates:	• sweat glands to secrete sweat.
Inhibits:	• contraction of arterioles in skin (dilates capillaries in skin) • hair erector muscles (relax – hairs lie flat) • liver (reduces metabolic rate) • skeletal muscles (relax – no shivering).

Heat gain centre	
Stimulates:	• arterioles in the skin to constrict • hair erector muscles to contract • liver to raise metabolic rate • skeletal muscles to contract in shivering.
Inhibits:	• sweat glands.

Figure 7.57 Negative feedback in thermoregulation.

There are several ways that the skin can help in the control of temperature, including sweating, hair attitude (flat or raised), route of blood flow, and shivering. The structures in the skin and the methods involved are summarised in Figure 7.58. Shivering is the uncontrolled contraction of normally voluntary muscles, and can increase heat production six-fold. Shivering transfers energy to muscle tissue, which helps maintain body temperature.

Q7.35 What effectors are activated when the body temperature falls below the norm?

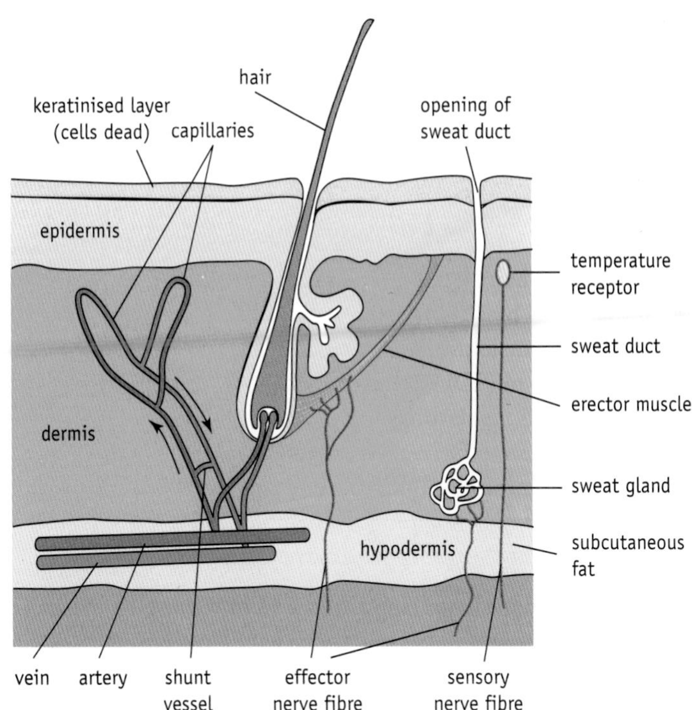

Sweat released on the skin via the sweat duct evaporates, taking heat energy from the skin. Sweat glands are stimulated by nerves from the hypothalamus. In temperate climates, such as the UK, sweat glands produce approximately 1 litre of watery fluid per day; in hot dry conditions this can increase to as much as 12 litres.

Hairs are raised in cold weather by contractions of the erector muscles. This is a reflex that we have no control over. It produces the characteristic goose pimples on the flesh. The aim is to trap a layer of air that insulates the body. Due to our shortage of hair this is not very effective in humans compared to other mammals and birds. Therefore most of us wear clothes as insulation.

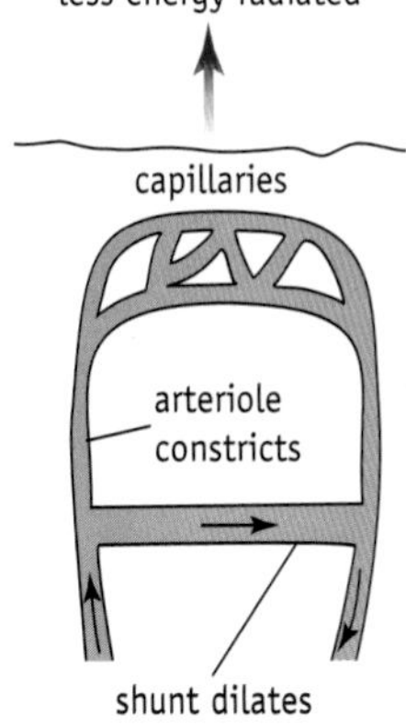

Energy is lost from the blood flowing through the surface capillaries by radiation.

In colder conditions the muscles in the arteriole walls contract, causing the arterioles to constrict (get narrower), reducing the blood supply to the surface capillaries. Blood is diverted through the shunt vessel which dilates (gets wider) as more blood flows through it. Blood flows further from the skin surface so less energy is lost. This is known as **vasoconstriction**.

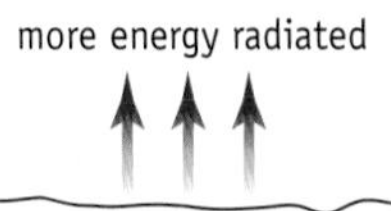

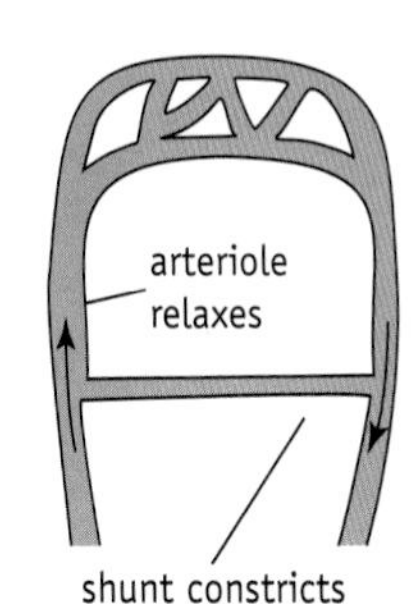

Constriction of the arterioles and shunts is controlled by the hypothalamus.

In warm conditions the **shunt vessel** constricts and muscles in the walls of the arterioles relax. Blood flows through the arterioles, making them dilate (become wider). Blood flows closer to the surface so more energy is lost. This is known as **vasodilation**.

Figure 7.58 The skin is the major organ involved in thermoregulation. It is important in preventing and in promoting heat loss.

Activity

Complete the summary diagram on homeostasis and thermoregulation in **Activity 7.17. A7.17S**

Extension

Find out about homeostasis and control of blood concentration in **Extension 7.4. X7.04S.**

Temperature regulation during exercise

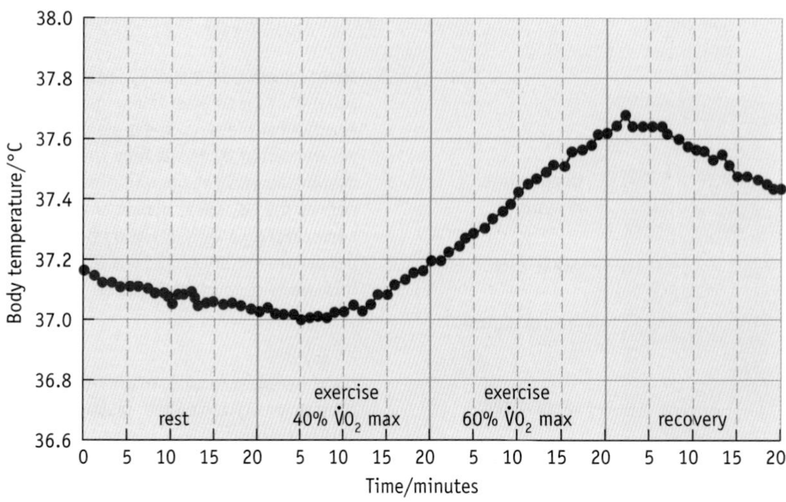

Figure 7.59 The effect of a 40-minute period of exercise on temperature (rectal measurements). *Source*: NASA.

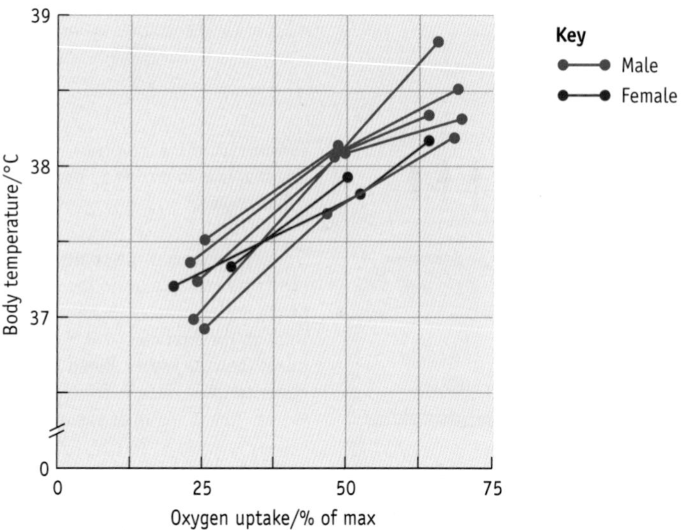

Figure 7.60 Changes in core body temperature with increasing exercise intensity, expressed as oxygen uptake.

During exercise, core body temperature rises (Figure 7.59); this rise is related to the intensity of exercise (Figure 7.60). The comparatively slight rise in temperature does not indicate a failure of the body to regulate the temperature, but without mechanisms to redistribute energy, this increase would quickly reach dangerous levels.

Once the hypothalamus detects a deviation from the norm in core temperature, it starts a chain of actions to counteract the deviation and bring body temperature back to the norm. Figure 7.61 summarises how energy is transferred to and from the body in temperature regulation. Table 7.3 will remind you about the different methods of energy transfer.

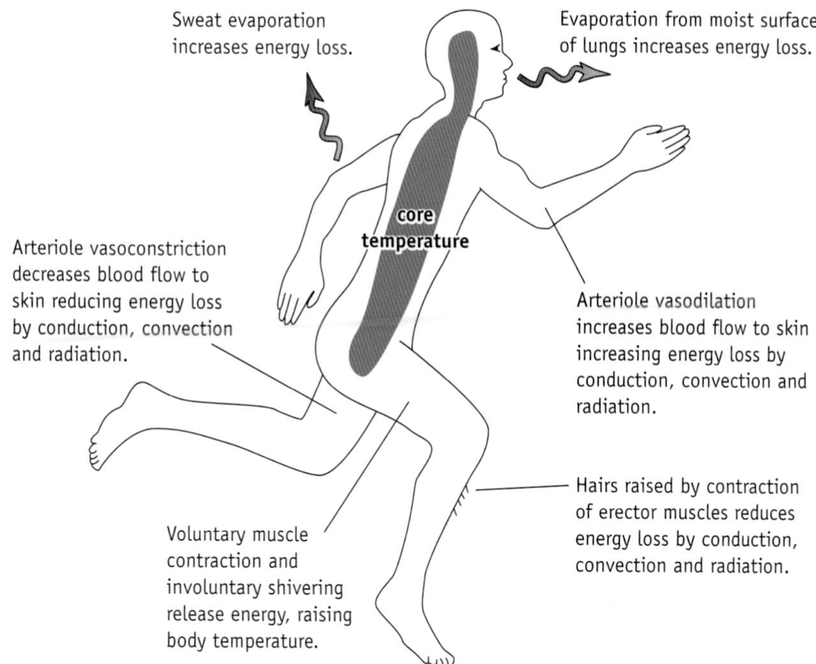

Figure 7.61 How energy is transferred to and from the body.

Radiation	Energy can be radiated from one object to another through air, or even through a vacuum, as electromagnetic radiation. Our bodies are usually warmer than the surrounding environment, so we radiate energy. Of course, this operates both ways – a person is warmed by energy radiated directly from the Sun even in sub-freezing conditions such as a cold but sunny winter's day.
Conduction	Energy loss by conduction involves direct contact between objects, and energy transfer from one to another. Sitting on a cold rock on a hot day will cause body cooling, as energy is transferred from the body to the rock by conduction.
Convection	Air lying next to the skin will be warmed by the body (unless air temperature exceeds body temperature). As the air expands and rises, or is moved away by air currents, it will be replaced by cooler air, which is then warmed by the body. This energy loss by bulk movement of air is called convection. Trapping a layer of still air next to the skin using fur or thermal underwear is an effective method of thermal insulation, because it reduces convection.
Evaporation	Energy is needed (latent heat of evaporation) to convert water from liquid to vapour. It takes 2400 kJ of energy to evaporate 1 dm^3 of water. The energy required to evaporate sweat is drawn from the body, cooling it. However, the process of sweating only has this cooling effect when the water evaporates. Mammals, birds and reptiles may also pant to keep cool, by evaporation of water from the gas exchange surfaces. In conditions of high humidity, evaporation becomes virtually impossible. Controlling body temperature becomes much more difficult when the air is saturated with water vapour.

Table 7.3 Methods of energy transfer.

Q7.36 **a** The **marathon runner** in Figure 7.54 feels cooler after dousing himself in water. How does this help to cool him?

b Why does high humidity make marathon running more dangerous?

Q7.37 Cats (including the big cats) can only sweat from the skin surface of their paws and nose. What other method might a cheetah rely on for transfer of energy to the environment?

At 37.5 °C, human core body temperature is normally higher than the surroundings, so energy will be transferred to the environment. In very cold environments, excessive cooling may occur and the body can lose thermal balance – the core body temperature starts to fall. The hypothalamus detects this and immediately does its best to regulate the internal temperature by increasing metabolic rate and slowing energy loss. Although not common during exercise, there are occasions when the body faces this challenge (see Figure 7.62).

Q7.38 **a** What will be the major route of energy loss for the cross-Channel swimmer in Figure 7.62?

b Swimming the English Channel in summer means spending between 10 and 20 hours in water that is usually between 13 and 16 °C. What mechanisms are used to maintain body temperature and enable survival without wearing a wetsuit?

Figure 7.62 Channel swimming exposes the body to cold stress.

7.5 Overdoing it

Although we will never run as fast as a sprinting cheetah, over the years top athletes have been getting faster. At the end of the 1920s the 100 m world records stood at 10.4 and 12.0 seconds for men and women respectively. Since then, the times have steadily fallen, and currently (2008) stand at 9.69 and 10.49 seconds. This improvement in performance has been achieved through more frequent and targeted training, improved nutrition, and advances in the design and materials used for athletes' clothing, footwear and tracks.

But some athletes attempt to do more than their body can physically tolerate, and reach the point where they have inadequate rest to allow for recovery (Figure 7.63). This is known as overtraining. 'Burnout' symptoms due to overtraining can persist for weeks or months. The symptoms are varied and, in addition to poor athletic performance and chronic fatigue, can include immune suppression leading to more frequent infections, and increased wear and tear on joints, which may need surgical repair.

Figure 7.63 Top sportspeople can suffer from overtraining. At the end of the season, some professional footballers have been found to have very low white blood cell counts.

Excessive exercise and immune suppression

Athletes engaged in heavy training programmes seem more prone to infection than normal. Sore throats and flu-like symptoms (upper respiratory tract infections, URTIs) are more common. Some scientists have suggested that there is a U-shaped relationship between risk of infection and amount of exercise (Figure 7.64).

Q7.39 **a** What do these data suggest about the risk of upper respiratory tract infections related to exercise?

b Look at Figure 7.64. Explain the difference between correlation and cause with reference to this example.

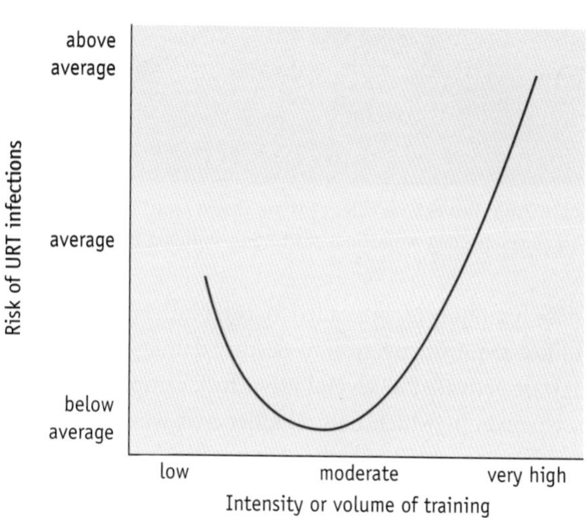

Figure 7.64 Moderate exercise seems to offer some protection against upper respiratory tract infections.

In a study of participants in the Los Angeles marathon in 1984, it was found that 13% of the runners reported upper respiratory tract infections in the week after the race. A control group of runners of comparable fitness who had not taken part had an infection rate of only 2% in the week after the race. There is much discussion about whether this is a true cause and effect relationship.

Two main factors have been suggested as contributing to higher infection rates: increased exposure to pathogens, and suppressed immunity with hard exercise. The location of the competition and any necessary travel may expose the athlete to a greater range of infected people and unfamiliar microorganisms. This alone could increase the occurrence of infection, even if overtraining did not suppress their immunity. Participation in team sports will also bring players into close contact with others, and increase the chances of transmission of infection.

Effects of exercise on immunity

Several research studies have shown that components of the non-specific and specific immune systems are affected by both moderate and excessive exercise.

Moderate exercise

Moderate exercise increases the number and activity of a type of lymphocyte called **natural killer cells** (Figure 7.65), which are found in blood and lymph. Unlike B and T cells, they do not use specific antigen recognition, but provide non-specific immunity against cells invaded by viruses and cancerous cells.

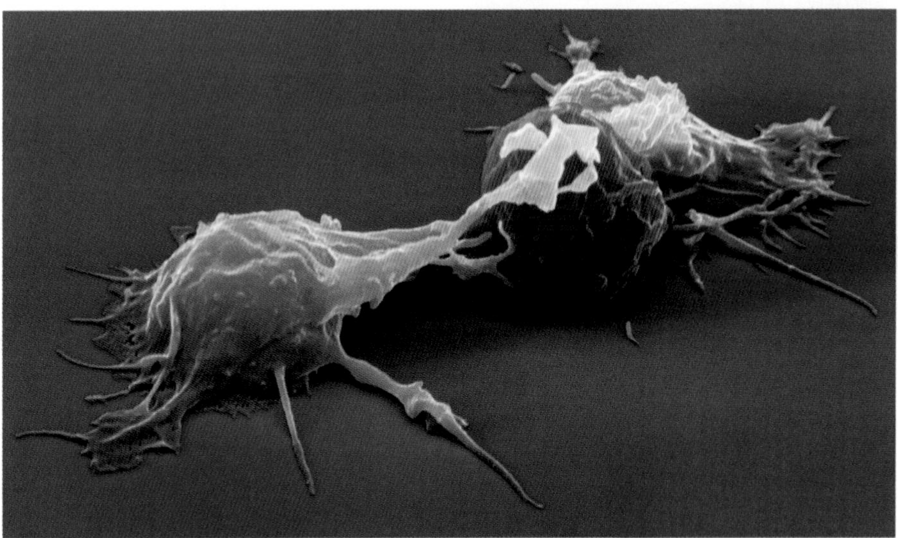

Figure 7.65 Two natural killer cells are shown attacking a cancer cell. The natural killer cells are yellow, with numerous long projections starting to surround the red cancer cell. Magnification × 5 600.

These lymphocytes are called 'natural killers' because it was once thought that they did not require activation to destroy damaged cells. However, it is now known that they are activated in several ways, for example by cytokines and interferons. The exact mechanism by which natural killer cells work is not yet thoroughly understood, but they seem to target cells that are not displaying 'self' markers. The killer cells release the protein perforin, which makes pores in the targeted cell membrane. These pores allow other molecules, such as proteases, to enter and cause apoptosis (death of the cell and all of its contents). Natural killer cells thus offer non-specific protection against URTIs and other infections.

Q7.40 Why is cell apoptosis likely to have a better outcome than cell lysis in the case of cells infected by a virus, such as one that causes URTI?

Vigorous exercise

Research shows that, during recovery after vigorous exercise, the number and activity of some cells in the immune system falls. These include:

- natural killer cells
- phagocytes
- B cells
- T helper cells.

The specific immune system is depressed as a consequence of these changes.

The decrease in helper T cells reduces the amount of cytokines available to activate lymphocytes. This in turn reduces the quantity of antibody produced. It has also been suggested that an inflammatory response occurs in muscles due to damage to muscle fibres caused by heavy exercise; this may reduce the available non-specific immune response against upper respiratory tract infections.

Q7.41 **a** Which of the white blood cells mentioned above secretes antibodies?

b What are the main features of the non-specific immune response to infection?

c How will the action of killer T cells be affected by the decrease in T helper cell numbers?

There is also much debate about whether the effects of intense exercise are caused by the activity itself, or by related psychological stress due to heavy training schedules and competition. Both physical exercise and psychological stress cause secretion of hormones such as adrenaline and cortisol (this is a hormone also secreted by the adrenal glands – Figure 7.72). Both of these hormones are known to suppress the immune system.

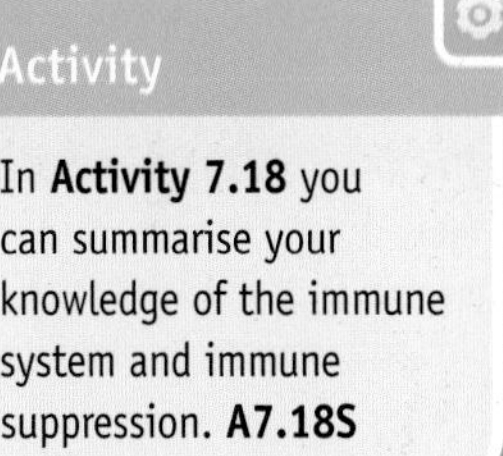

Activity

In **Activity 7.18** you can summarise your knowledge of the immune system and immune suppression. **A7.18S**

Q7.42 How does the above evidence support the idea that moderate exercise enhances immunity, while excessive exercise suppresses it?

How are joints damaged by exercise?

Professional athletes, such as football, hockey and rugby players, risk developing joint injuries due to the high forces the sport generates on their joints. Repeated forces on joints such as the knee can lead to wear and tear of one or more parts of the joint. Many joint disorders are associated with such overuse, some of which can also result from ageing. These disorders are typically associated with pain, inflammation and restricted movement of the joint. Treatment usually involves rest, ice, compression and elevation (RICE), anti-inflammatory painkillers and, if necessary, surgical repair.

Knees are particularly susceptible to wear and tear injuries. The problems include the following:

- The articular cartilage covering the surfaces of the bones wears away so that the bones may actually grind on each other, causing damage that can lead to inflammation and a form of arthritis.

- Patellar tendonitis (jumper's knee) occurs when the kneecap (patella) does not glide smoothly across the femur due to damage of the articular cartilage on the femur.
- The bursae (fluid sacs) that cushion the points of contact between bones, tendons and ligaments can swell up with extra fluid. As a result, they may push against other tissues in the joint, causing inflammation and tenderness. Bursitis of the knee is also known as 'housemaid's knee'; it was common in housemaids due to the repetitive kneeling associated with their work.
- Sudden twisting or abrupt movements of the knee joint often result in damage to the ligaments.

Activity

Use **Activity 7.19** to analyse and interpret data on possible disadvantages of exercising too much and too little. **A7.19S**

How can medical technology help?

Improvements in medical technology over recent years, including the development of prosthetic limbs and keyhole surgery procedures, have enabled the disabled and those with injuries to participate more fully in sport.

Keyhole surgery

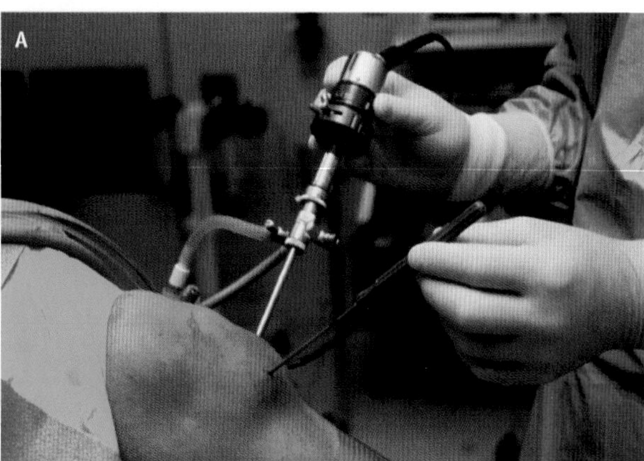

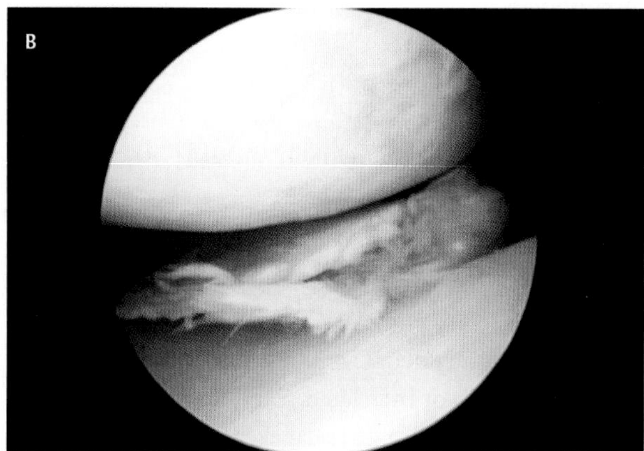

Figure 7.66 Arthroscopy (**A**) allows surgeons to see within joints and to repair damage, such as the degenerative tear in the pad of cartilage (meniscus) of this knee joint (**B**). There are two menisci in each knee. These crescent-shaped pieces of cartilage reduce friction in the moving joint and increase its stability.

Injuries to joints can limit the amount of exercise a person can take, and have often shortened the careers of professional athletes. Surgical operations to repair damage used to be painful and recovery took a long time. The main reason for this was the large incision needed to remove or repair even very small structures. A large hole had to be made to allow access for the surgeon's hands and instruments, and also to let in enough light to allow the surgeon to see what they were doing. These large incisions caused a good deal of bleeding, a lot of pain, increased risk of infection, and prolonged recovery after the operation.

With the advent of **keyhole surgery**, using fibre optics or minute video cameras, all this has changed. It is now possible to repair damaged joints or to remove diseased organs through small holes. Keyhole surgery on joints is known as **arthroscopy** (Figure 7.66A). This literally means 'to look within the joint'.

To carry out an arthroscopic procedure, the surgeon makes one or two small incisions, often only about 4 mm long. A small camera and light source are inserted, allowing the inside of the joint to be seen (Figure 7.66B) and a diagnosis to be made or confirmed. If surgery is required, miniature instruments are inserted through the incisions.

The inside of most joints can be viewed with an arthroscope. Damage to the **cruciate ligaments** in the knee can be tackled particularly effectively by keyhole surgery. The knee is a hinge joint held together by four ligaments. You should remember, from earlier on in this topic, that ligaments attach bone to bone. They also help to control joint movement and to prevent the joint becoming overstretched. Two of the four knee ligaments, the cruciate ligaments, are found deep inside the joint and are attached to the end of the femur and to the top of the tibia (see Figure 7.9, page 137).

The posterior cruciate ligament normally prevents the knee from being bent too far back; most athletic injuries to it occur during a fall onto a bent knee. The anterior cruciate ligament normally prevents the knee from being bent too far forwards; it can be damaged by sudden turning, pivoting or cutting manoeuvres. This happens particularly during sports such as football and basketball, although skiers, gymnasts, and rugby and hockey players are also at risk.

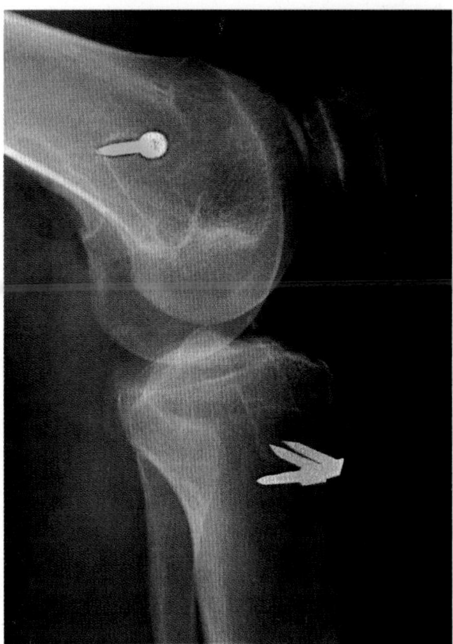

Figure 7.67 Coloured X-ray of a knee joint showing metal pins (pink) inserted to mend a torn cruciate ligament (blue).

A popping or snapping sound can sometimes be heard when a cruciate ligament is torn. If surgery is carried out, the ligament can be repaired (Figure 7.67), and the knee joint can be stabilised so that further injury is less likely. This is important, as repeated damage to a joint can affect the cartilage and shorten an athlete's career.

Because only a small incision is made, recovery after keyhole surgery can be rapid, and only a short stay in hospital is needed. It is not uncommon for top athletes to return to normal athletic activities within a few weeks of cruciate ligament surgery.

Q7.43 If a cruciate ligament sustains so much damage that it is not possible to repair it, the damaged ligament may be replaced with a tendon from around the same knee.

a Describe the normal role of a tendon.

b Suggest two reasons why the tendon used to repair a ligament is taken from the same injured joint.

c Ligaments are more elastic than tendons. Suggest what advice a footballer who has had a torn ligament in her knee replaced with one of her tendons might be given to help her get fit again.

d You might think removing a tendon to use as a ligament would just cause a different problem for the athlete. Suggest why it doesn't.

Prostheses

A **prosthesis** (plural prostheses) is an artificial body part used by someone with a disability to enable him or her to regain some degree of normal function or appearance. By using specialised prostheses, disabled athletes can be more physically active and perform at higher levels (Figure 7.68).

There have been significant developments in prosthetic limbs over recent years, with the introduction of variations in design for different activities. For example, athletes with prosthetic legs may use a dynamic response prosthetic foot. Such a foot changes its shape under body weight, but returns to the original shape on lifting off the ground. This puts spring into the step and provides sure footing.

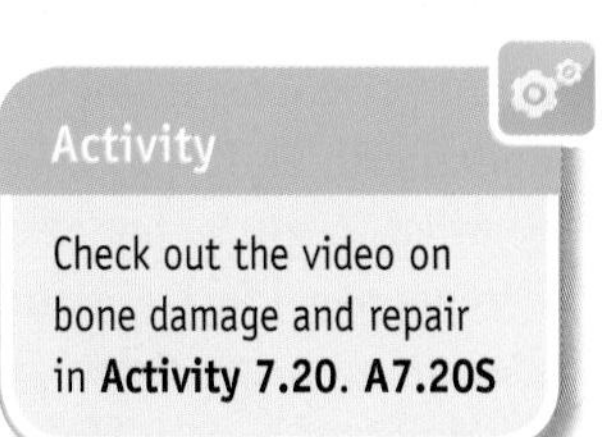

Activity

Check out the video on bone damage and repair in **Activity 7.20. A7.20S**

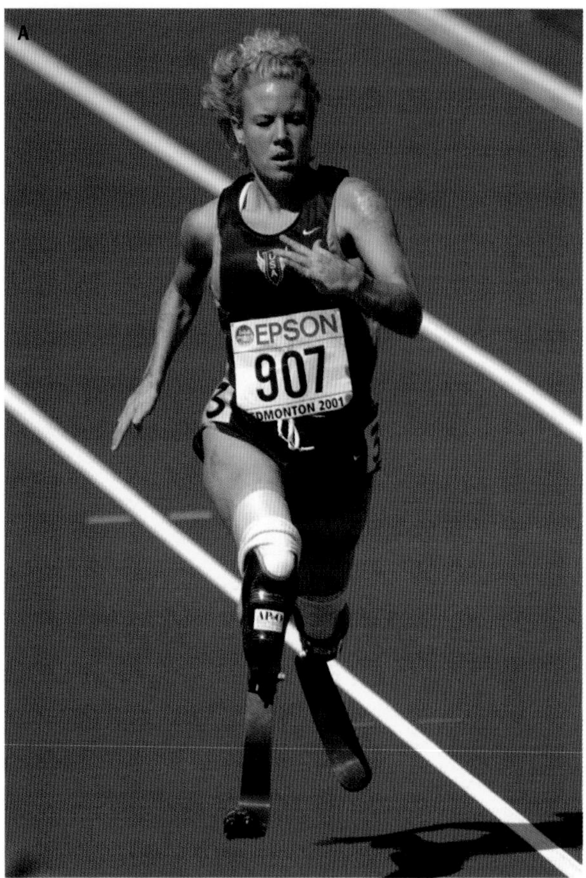

Figure 7.68 Two different types of leg prostheses in use by athletes. The female runner (A) is wearing two prostheses that are specially designed for sprinting and used only for this purpose. The female athlete (B) is wearing her normal prosthesis – can you tell which leg it is?

Prosthetic feet may also be articulated (have joints) or not. Articulated feet are better on uneven surfaces, and so useful in sports such as golf. In some sports, a foot may not be required at all; a flipper could be used for swimming, or a pedal-binding for cycling. High friction surfaces can be added to provide better grip for prostheses used for rock climbing.

Prosthetics may also be used to replace damaged or diseased joints that have not responded to medical or other therapy. The damage can be due to wear and tear in sport, but may also result from diseases such as arthritis. In extreme cases of arthritis, so much cartilage has been lost that bone scrapes on bone.

Replacing hip and knee joints with artificial joints is a particularly successful use of prosthetics (Figure 7.69). To replace a knee joint, the surgeon makes an incision in the affected knee. Once the patella is moved out of the way, the ends of the femur and tibia are trimmed to fit the prosthesis, taking care not to damage any of the ligaments. The underneath of the patella is also trimmed so that an artificial piece may be fitted. A special bone cement is used to attach the new surfaces. Often a metal such as stainless steel or titanium is used on the end of the femur; metal, ceramic or plastic such as polyethylene is attached to the tibia and patella. However, new materials are being developed and tested all the time, to improve surface-to-surface contact and durability.

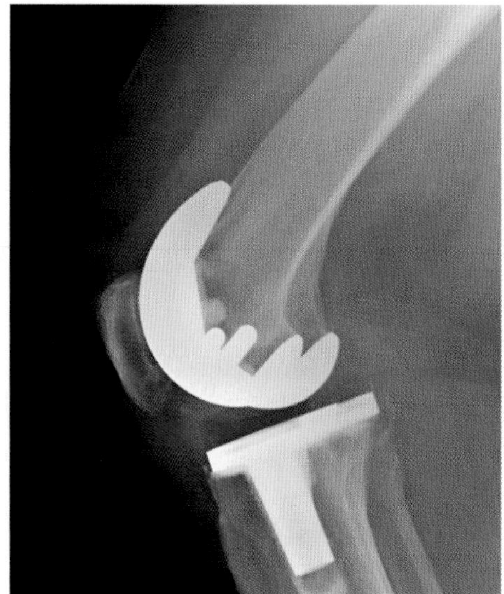

Figure 7.69 X-ray showing side view of a prosthetic knee joint. The artificial parts, made from a metal alloy, show up as white. One half of the prosthesis is attached to the femur above the knee, and the other half is attached to the tibia below the knee.

A fit person will recover within a few months and should then be able to walk unaided and without pain using their 'new knee'. It is usually recommended that contact sports are avoided after a knee joint replacement, but low impact sports such as swimming and golf should be fine.

Taking enough exercise

Although overuse can result in damage to joints and bones, physical activity has many advantages. These include the following:

- Increasing arterial vasodilation lowers blood pressure; this reduces the risk of coronary heart disease and stroke.
- Physical activity increases the level of blood HDLs, which transport cholesterol to the liver where it is broken down, and reduces LDLs, which are associated with development of atherosclerosis in coronary heart disease and stroke (see AS Topic 1, Section 1.3).
- A balance between energy input and output helps to maintain a healthy weight.
- Increased sensitivity of muscle cells to insulin improves blood glucose regulation, and reduces the likelihood of developing type II diabetes.
- Physical activity increases bone density and reduces its loss during old age; this delays the onset and slows the progress of the bone-wasting disease osteoporosis.
- Exercise reduces the risks of some cancers.
- It improves mental well-being.

The Department of Health's advice on physical activity recommends that adults take 30 minutes of at least moderate activity on at least five days a week. Physical activity includes walking, cycling, gardening, and active hobbies or sports. Taking too little exercise means that you do not gain the health benefits described above, and you are at a higher risk of coronary heart disease, stroke, cancer, obesity, diabetes and osteoporosis.

Weblink

Find out more about osteoporosis from the National Osteoporosis Society website, and diabetes from the Diabetes UK website.

A sedentary lifestyle combined with overeating and drinking can lead to weight gain and potentially to obesity. A person is considered obese if their body mass index is over 30 (BMI = body mass in kg/height^2 in m^2, see AS Topic 1). Obesity leads to high blood pressure and high blood LDL levels, which increase the risk of coronary heart disease and stroke.

Obesity also increases the risk of developing type II diabetes (often known as non-insulin dependent diabetes or late-onset diabetes). High blood glucose levels due to eating sugar-rich foods can reduce sensitivity of cells to insulin and result in type II diabetes. The body does not produce enough insulin and body cells do not respond to insulin that is produced, so blood sugar levels cannot be controlled. There is decreased absorption of glucose from the blood; cells break down fatty acids and proteins instead, which leads to weight loss.

Checkpoint

7.5 Produce annotated lists of the disadvantages of exercising **a** too much and **b** too little.

7.6 Improving on nature

Performance-enhancing substances

The use of drugs to enhance performance in sport is known as doping. It is thought that this term originates from the South African word 'doop', which referred to an alcoholic stimulant drink used in certain tribal ceremonies.

Doping is not just a recent problem; throughout history, some athletes have sought a competitive edge by the use of chemicals. As long ago as the third century BC, some of the Ancient Greeks were known to ingest hallucinogenic mushrooms to improve their athletic performance (Figure 7.70). Roman gladiators used stimulants in the Circus Maximus to overcome fatigue and injury. The important social status of sport, and the high economic value of victory, both then and now, has placed great pressure on athletes to be the best. This pressure has increased the abuse of performance-enhancing drugs.

In the earliest times a simple foot-race was the only event

Figure 7.70 Ancient drug cheats? At the Beijing Olympics in 2008, 40 cheats were caught before the Games, and four more were discovered during them.

The World Anti-Doping Agency (WADA) is funded equally by sports movements and governments of the world. It coordinated the development and implementation of the World Anti-Doping Code, which is a collection of anti-doping policies for all sports and all countries. The Code prohibits a wide range of substances that can be taken for their performance-enhancing effects. It also prohibits such practices as blood doping (taking extra blood or red blood cells), and artificially enhancing the uptake, transport or delivery of oxygen using drugs or haemoglobin products. Gene doping or the non-medicinal use of cells, genes or gene expression that might affect athletic performance are also prohibited.

Weblink

Look at the World Anti-Doping Agency website for the full list of banned substances and practices in sport.

Competition athletes with medical conditions requiring prescription of any of the prohibited drugs need to obtain permission to use them. It is the responsibility of the athlete to check whether a drug is banned or not, and to ensure that he or she does not inadvertently take the banned substance (Figure 7.71). They can get a medical certificate if they have had an illness or an operation that required the use of banned drugs.

Human growth hormone, insulin and erythropoietin (EPO) are on the list of prohibited substances for athletes, yet these are all peptide hormones produced naturally in the body. Another group of substances, also banned by the World Anti-Doping Agency, includes testosterone. Testosterone also occurs naturally in the human body, in both males and females, and belongs to the group of hormones known as steroid hormones. What are hormones, and why are they used as performance-enhancing drugs?

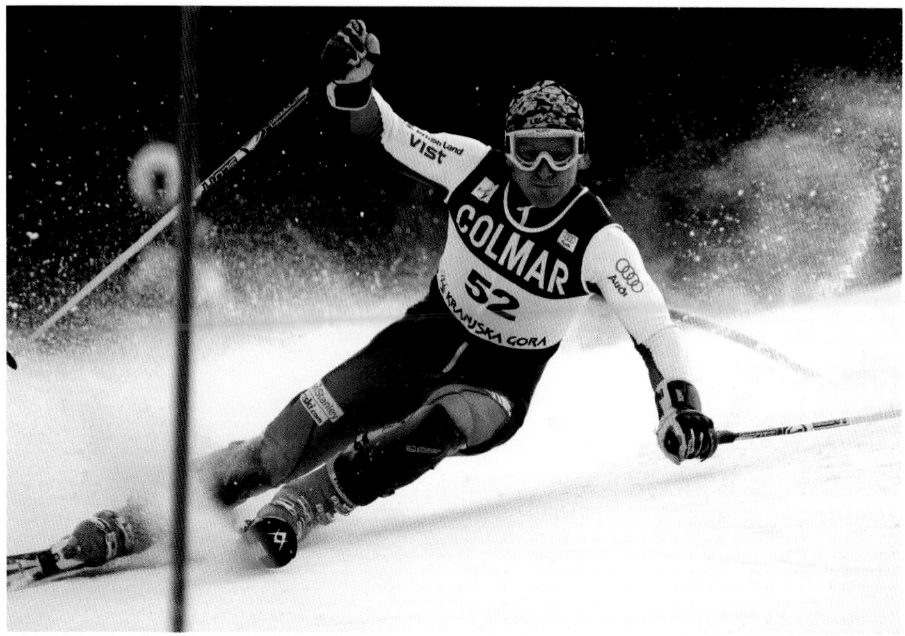

Figure 7.71 Alain Baxter, the British skier, was stripped of his bronze medal at the 2002 Winter Olympics for testing positive for a banned substance after using a nasal inhaler.

Hormones

Hormones are chemical messengers, released directly into the blood from endocrine glands (Figure 7.72 and Table 7.4). Unlike exocrine glands, such as sweat glands and salivary glands, endocrine glands do not have ducts. It is important that the cells in the endocrine glands that make hormones are not themselves affected by their products. For this reason, most hormones are produced either in an inactive form or packaged within secretory vesicles by the Golgi apparatus. The vesicles fuse with the cell surface membrane, releasing their contents by exocytosis.

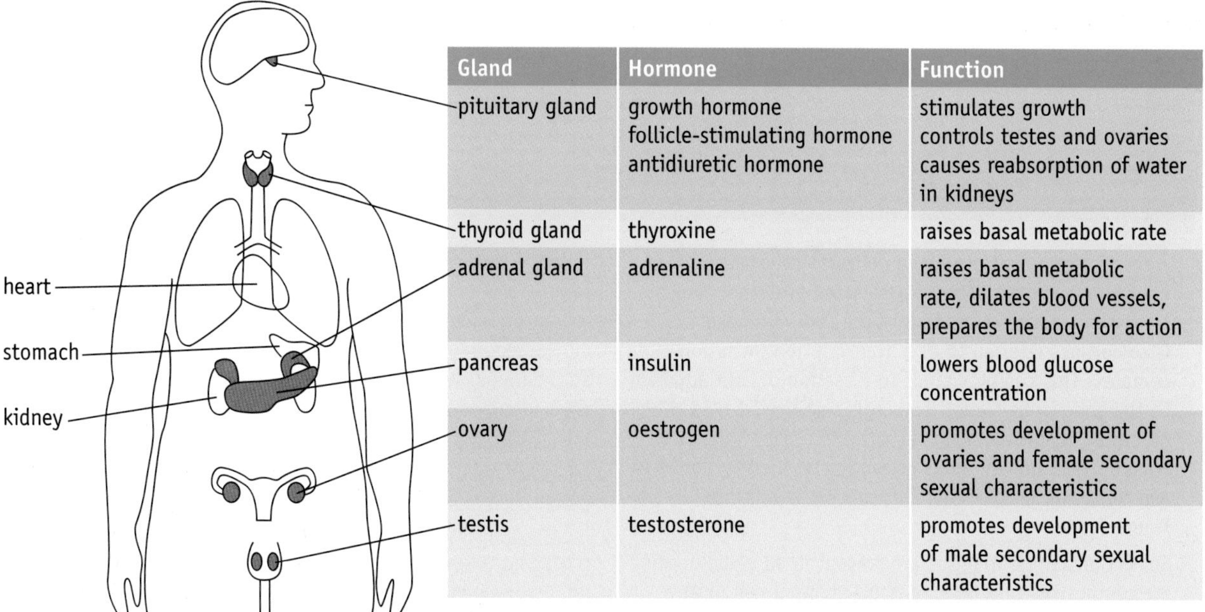

Gland	Hormone	Function
pituitary gland	growth hormone follicle-stimulating hormone antidiuretic hormone	stimulates growth controls testes and ovaries causes reabsorption of water in kidneys
thyroid gland	thyroxine	raises basal metabolic rate
adrenal gland	adrenaline	raises basal metabolic rate, dilates blood vessels, prepares the body for action
pancreas	insulin	lowers blood glucose concentration
ovary	oestrogen	promotes development of ovaries and female secondary sexual characteristics
testis	testosterone	promotes development of male secondary sexual characteristics

Figure 7.72 and Table 7.4 The main endocrine glands, with some examples of hormones they produce.

Each hormone affects only specific target cells, modifying their activity. Hormones are carried around the body in the bloodstream. They either enter the target cells or bind to complementary receptor molecules on the outside of the cell membranes. Each hormone brings about characteristic responses, resulting from its effect on enzymes. Some hormones bind to receptors on the cell surface, producing a second messenger that can activate enzymes within the cell. Others act on the cell by direct or indirect control of transcription.

A Action of peptide hormones

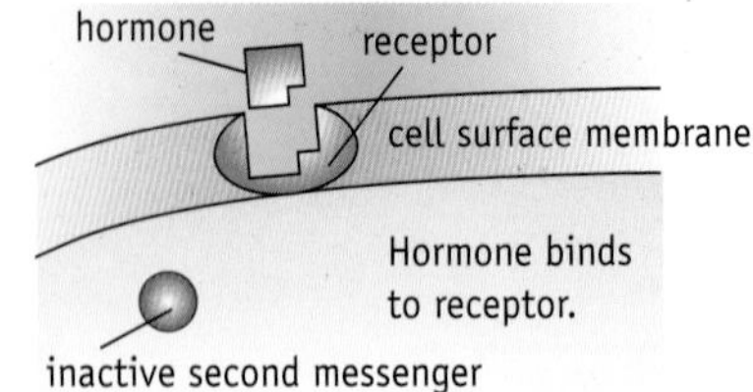
hormone
receptor
cell surface membrane
Hormone binds to receptor.
inactive second messenger

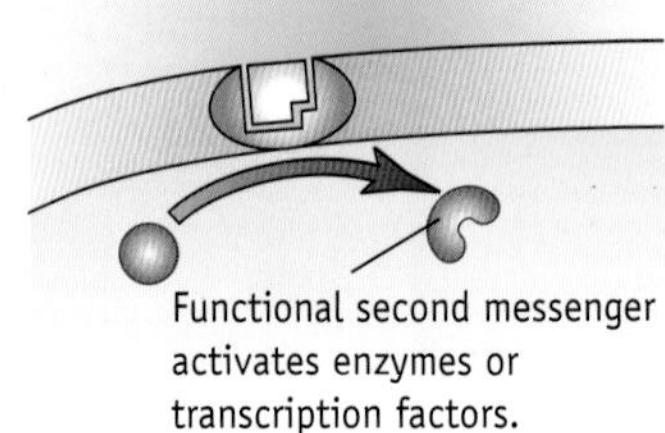
Functional second messenger activates enzymes or transcription factors.

B Action of steroid hormones

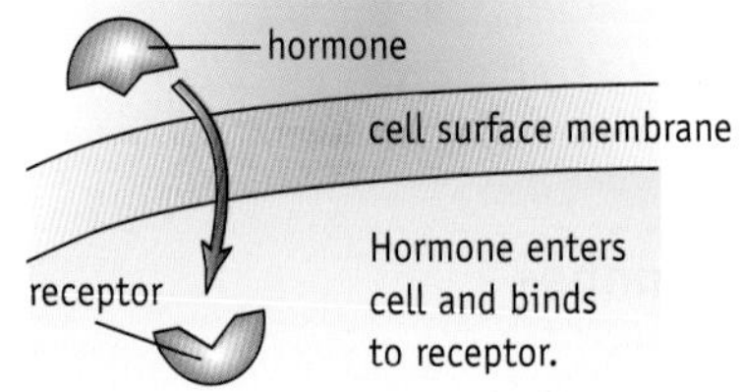
hormone
cell surface membrane
Hormone enters cell and binds to receptor.
receptor

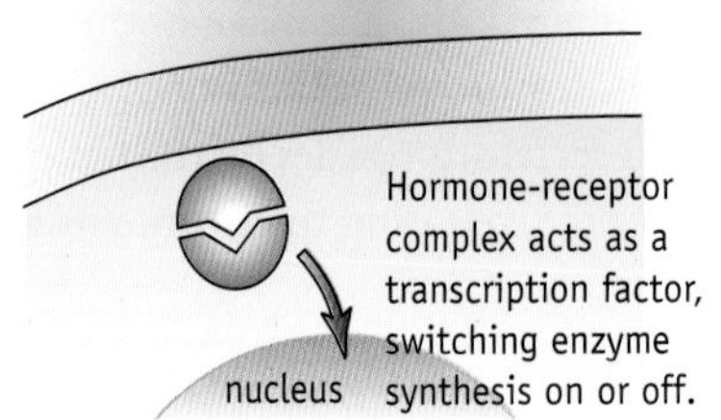
Hormone-receptor complex acts as a transcription factor, switching enzyme synthesis on or off.
nucleus

Figure 7.73 How hormones affect target cells.

How hormones affect cells

Peptide hormones are protein chains, varying from about 10 to 300 amino acids in length. Insulin, for example, has 51 amino acids. Despite being relatively small molecules, peptide hormones are not able to pass through cell membranes easily, because they are charged. Instead they bind to a receptor on the cell membrane; this then activates another molecule in the cytoplasm, called a second messenger. The functional second messenger brings about chemical changes within the cell (Figure 7.73) directly or indirectly by affecting gene transcription. Peptide hormones include EPO, human growth hormone and insulin.

Steroid hormones are formed from lipids and have complex ring structures. Testosterone, like all steroid hormones, passes through the cell membrane and binds directly to a receptor molecule within the cytoplasm. Once activated, the hormone-receptor complex brings about characteristic responses, resulting from its effect on transcription. The hormone-receptor complex functions as a **transcription factor**, switching enzyme synthesis on or off.

Q7.44 Why can steroid hormones pass through the cell membrane?

Q7.45 Most hormones travel all round the body in the blood, and will come into contact with many cells. Why do hormones only produce a response within cells of their target organs?

Key biological principle: How transcription factors work

Transcription is initiated by an enzyme called RNA polymerase and a cluster of associated protein transcription factors binding to the DNA. The result is sometimes referred to as a 'transcription initiation complex'. This complex binds to a section of DNA adjacent to the gene to be transcribed. This section is called the promoter region. Only when the transcription initiation complex has formed and is correctly attached to the DNA will transcription proceed. This process is outlined in Figure 7.74.

Some transcription factors are present in all cells; others are synthesised only in a particular type of cell or at a particular stage of development. Most are created in an inactive form, and are then converted into the active form by the action of hormones, growth factors or other regulatory molecules. The gene remains switched off until all the required transcription factors are present in their active forms. The transcription initiation complex can then form and attach to the promoter region successfully.

Transcription of a gene can be prevented by protein repressor molecules attaching to the DNA of the promoter region. This blocks the attachment sites for transcription factors, preventing formation of the transcription initiation complex. Alternatively, protein repressor molecules can attach to the transcription factors themselves, preventing them from forming

CONTINUED ▶

the transcription initiation complex. Or the repressor molecules may actually be inactive transcription factors. Whatever the mechanism, the gene is switched off; it is not transcribed within this cell. Activator molecules in the cell, on the other hand, stimulate the binding of the transcription initiation complex.

As you saw in AS Topic 3, the structure and function of each type of cell is determined by the proteins it contains, which is in turn a result of the particular genes expressed. The control of transcription initiation is one of the important mechanisms that determines whether a gene is expressed.

Checkpoint

7.6 Write a short paragraph that summarises how transcription factors: **a** switch on and **b** switch off the transcription of a gene.

Activity

In **Activity 7.21** you look at the role of transcription factors in more detail. **A7.21S**

Genes are switched on by successful formation and attachment of the transcription initiation complex to the promoter region.

Genes remain switched off by failure of the transcription initiation complex to form and attach to the promoter region. This is due to the absence of protein transcription factor(s) or the action of repressor molecules.

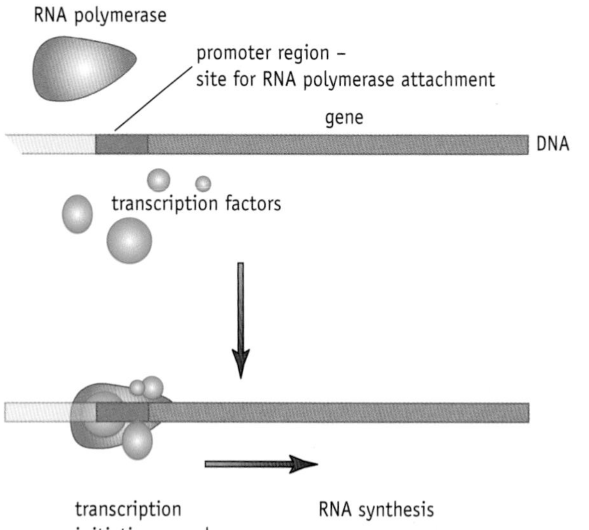

Figure 7.74 DNA transcription will start only when the transcription initiation complex is attached to the promoter region of the gene to be transcribed.

Hormones used to enhance performance

Erythropoietin

Erythropoietin (EPO) is a peptide hormone produced naturally by the kidneys. It stimulates the formation of new red blood cells in bone marrow. EPO can be produced using DNA technology, and is used to treat anaemia. As it is a natural substance, it has been difficult to test to see whether raised EPO levels are natural or not.

Q7.46 Explain how taking EPO would increase the performance of an endurance athlete.

Q7.47 Suggest how EPO might stimulate formation of new blood cells.

There are health risks associated with this substance. If EPO levels are too high, the body will produce too many red blood cells, which can increase the risk of thrombosis, possibly leading to heart attack and stroke. Injections of EPO have been implicated in the deaths of several athletes.

Scientists have developed a technique capable of distinguishing between synthetic and natural EPO.

Q7.48 What is meant by thrombosis?

Q7.49 Explain why EPO is not taken by sprint athletes.

Q7.50 Why would it be important to distinguish between natural and synthetic EPO?

Testosterone

Testosterone is a steroid hormone (made from cholesterol) produced in the testes by males and in small amounts by the adrenal glands in both males and females. Testosterone is one of a group of male hormones known as androgens, from the Greek *andros*, meaning male or man.

Testosterone causes the development of the male sexual organs. During adolescence it is responsible for development of the male secondary sexual characteristics, for example the deepening of the voice, growth of facial and body hair, and skeletal and muscular changes. Character changes such as increased aggressiveness have been attributed to testosterone.

Testosterone binds to androgen receptors, which are numerous on cells in target tissues. They modify gene expression to alter the development of the cell; for example they will increase anabolic reactions such as protein synthesis in muscle cells, increasing the size and strength of the muscle.

Figure 7.75 Anabolic steroids are used to increase muscle development. Heavy weight or resistance training is necessary for anabolic steroids to exert any beneficial effect on performance.

Athletes and body builders (Figure 7.75) may use injections of testosterone to increase muscle development, but this is not very effective as testosterone is quickly broken down. To overcome this problem, synthetic **anabolic steroids** such as nandrolone have been manufactured by chemical modification of testosterone.

Medical experts see significant dangers in the use – and particularly the gross over-use – of anabolic steroids. For example, anabolic steroids can cause high blood pressure, liver damage, changes in the menstrual cycle in women, decreased sperm production and impotence in men, kidney failure and heart disease. They can increase aggression in both men and women. In women the androgenic (masculinising) side effects are not generally thought to be desirable.

Originally developed for the treatment of muscle-wasting diseases, anabolic steroids are also used in the treatment of osteoporosis. In the UK they are prescription-only drugs. They are classified as Class C drugs under the Misuse of Drugs Act, with the maximum penalty for the illegal *possession* of steroids currently (2008) standing at two years' imprisonment and/or a fine. *Supplying* a Class C drug, such as an anabolic steroid, can lead to heavier penalties, even if no money has changed hands. The International Olympic Committee, in accordance with the World Anti-Doping Code, has banned the use of anabolic steroids. The illegal use of steroids not only occurs in human sport but also in animal sports such as horse- and dog-racing.

Anabolic steroids and their by-products can be detected relatively easily in urine samples by the technique of mass spectrometry. However, as these substances occur naturally it is difficult to set a level above which an athlete can confidently be said to be doping. Testosterone and a related compound, epitestosterone, are both found in urine. When an athlete takes an anabolic steroid, the ratio of testosterone to epitestosterone (the T:E ratio) increases. The World Anti-Doping Code states that an athlete with a T:E ratio above 4:1 is guilty of doping.

Activity

In **Activity 7.22** you interpret data on the effects of testosterone. **A7.22S**

Performance-enhancing substances not banned

Creatine

One example of a substance that is taken for its performance-enhancing effect but is not banned is creatine; it is considered to be a nutrition supplement.

Many athletes take dietary supplements containing the amino acid derived compound known as creatine. Creatine is naturally found in meat and fish. Once ingested it is absorbed unchanged and carried in the blood to tissues such as skeletal muscle. Creatine is also synthesised in the body from the amino acids glycine and arginine. Creatine supplements have been reported to increase the amounts of creatine phosphate (CP) in muscles. The theoretical benefit of increased CP storage is an improvement in performance during repeated, short-duration, high-intensity exercise. Research has shown improvements in activities such as sprinting, swimming and rowing. The use of creatine supplements combined with heavy weight training has been associated with increases in muscle mass and maximal strength, and a decrease in recovery time.

Figure 7.76 Creatine supplements can be bought in high street stores.

Creatine is considered to be a nutritional supplement (Figure 7.76) and is therefore not on the list of prohibited substances, so its use is not banned. Some adverse effects of taking creatine supplements have been reported. These include diarrhoea, nausea, vomiting, high blood pressure, kidney damage and muscle cramps.

Checkpoint

7.7 Outline the uses and misuses of the drugs creatine, testosterone and erythropoietin. Suggest ethical arguments for and against the use of drugs to improve sporting performance.

Should performance-enhancing substance use be banned?

The pressure to succeed in competitive sport is ever-increasing, not only due to the desire to be the best but also for the financial rewards and greater media interest. The desire to win combined with pressure and expectations from coaches, sponsors and the general public can be such that some athletes are prepared to take drugs that will enhance their performance, even if there are associated risks.

The International Olympic Committee (IOC) and other sporting bodies consider that the use of performance-enhancing substances is unhealthy and against the ethics of sport. The IOC started drug testing in 1968 after a Tour de France cyclist died from an amphetamine overdose; random testing began in 1989. The ban aims to protect the health of athletes and ensure that there is fair competition.

There are some people who consider that the use of substances is ethically acceptable, arguing that athletes have a right to decide whether they take the drug or not, deciding for themselves if the potential benefit is worth the risk to their health. Those who oppose this view may say that frequently the athletes do not make a properly informed decision, lacking information about the possible health consequences, and coming under pressure from others to take illegal drugs.

The idea that drug-free sport is fair is disputed by those who maintain that drug use is acceptable on the grounds that there is already inequality of competition due to the differences in time available for training and in resources. Some individuals may know that drug use is against the rules in terms of both the governing bodies of sport and the idea of fair play, but they are unwilling to be at a competitive disadvantage so choose not to adhere to the rules.

Activity

Think about the ethical arguments about doping in **Activity 7.23. A7.23S**

Activity

Use **Activity 7.24** to check your notes using the topic summary provided. **A7.24S**

Review

Now that you have finished Topic 7, complete the end-of-topic test before starting Topic 8.

Grey matter

Why a topic called Grey matter?

The brain, with over 10^{12} neurones (nerve cells), is the most complicated organ in the body and makes even our most sophisticated computers seem simple. The brain influences our every sensation, emotion, thought, memory and action. At each moment of every day it is bombarded with sensory information from the world around us, and interprets this information to create a meaningful view of the world. Looking at the world is not merely inspecting a simple picture, like observing a slide show projected on a screen inside our heads. The information is processed to provide us with our experience of the world.

But sometimes things may not be as they seem, as the anthropologist Colin Turnbull found in the 1950s when he was working in what is now called the Democratic Republic of Congo. He and Kenge, one of the Bambuti people who are used to living in dense forest with only small clearings, went out onto the grassland plains of the former Congo. Looking across the wide open plain, Kenge turned to Turnbull and asked 'What insects are those?' (Figure 8.1). When Turnbull told Kenge that the 'insects' were buffalo, Kenge laughed and simply could not believe it.

How does the nervous system function to let any of us look across a plain? Why did Kenge get the wrong impression? Was his visual development faulty or was he misinterpreting what he was seeing?

Figure 8.1 Kenge thought the buffalo in the distance were insects.

It is not just Kenge who is sometimes mistaken by what he sees. Have a look at Figure 8.2 and decide which of the lines is longer. Now measure them to find out if you were correct.

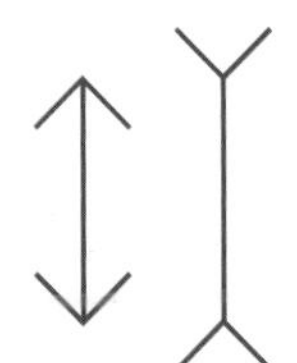

Figure 8.2 Which line is longer?

All in the synapses

With upwards of 10^{14} interconnections between its neurones, the working of the brain is dependent on its synapses and their neurotransmitters. How do these function and how can they go wrong? Imbalances in naturally-occurring brain chemicals and drugs that can cross the blood-brain barrier can affect synapses and be bad for the health. How are synapses affected by conditions such as depression and Parkinson's disease and by the use of, for example, MDMA (ecstasy)? How can we treat conditions such as depression and Parkinson's disease? How can gene technology be used to improve treatments for these types of conditions?

Plants have no nervous system so how do they detect and respond to their environment?

Overview of the biological principles covered in this topic

In this topic, you will revisit ideas about receptors and effectors introduced in Topics 6 and 7 by examining the pupil light reflex. Building on these ideas you will consider the detection of stimuli in greater detail, as exemplified by light detection by receptor cells in the retina, and how the brain and eyes combine to enable visual perception. This leads to discussion of the transmission of nerve impulses along axons and across synapses, before contrasting nervous and hormonal coordination. You go on to consider how plants, which lack a nervous system, can detect stimuli and coordinate responses.

You will investigate the structures of the different regions of the brain and the evidence that links structure and function, including the use of imaging techniques you met in Topic 1. You will look at visual development and in particular the need for stimulation of synapses and the role of synapses in learning. Throughout this topic the contribution of nature and nurture to brain development is highlighted.

You will look at how some diseases and drugs affect the brain to illustrate how chemicals can affect synaptic transmission. You will consider how some of these diseases are being treated and how gene technology is helping improve treatments.

You will have the opportunity to discuss the ethical issues related to the use of animals in research and the risks and benefits of using genetically modified organisms.

8.1 The nervous system and nerve impulses

As Kenge and Colin Turnbull emerged from the forest, how did their eyes and brains work to let them look across the plain? Seeing is possible because the cells of the nervous system (Figure 8.3) are able to conduct **nerve impulses** and pass them to one another. In fact all our senses, emotions, memories and thoughts are dependent on nerve impulses.

The nervous system is highly organised (Figure 8.4), receiving, processing and sending out information, as we saw with temperature control and control of heart rate in Topic 7.

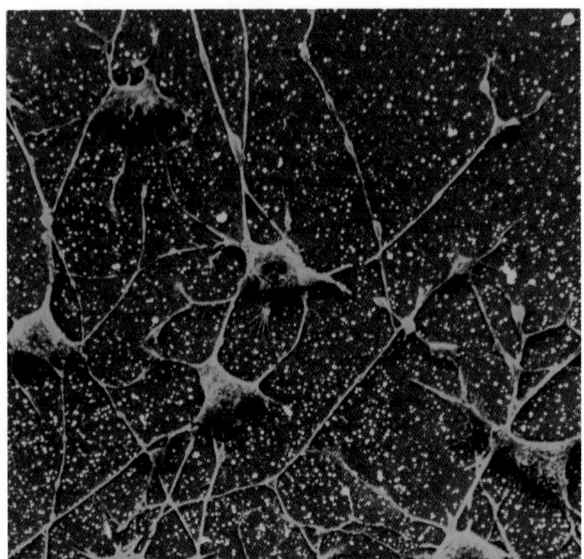

Figure 8.3 Nerve cells such as these form the basis of the nervous system.

What are nerve cells like?

It is important to distinguish between a **neurone**, which is a single cell, and a nerve. A nerve is a more complex structure containing a bundle of the axons of many neurones surrounded by a protective covering.

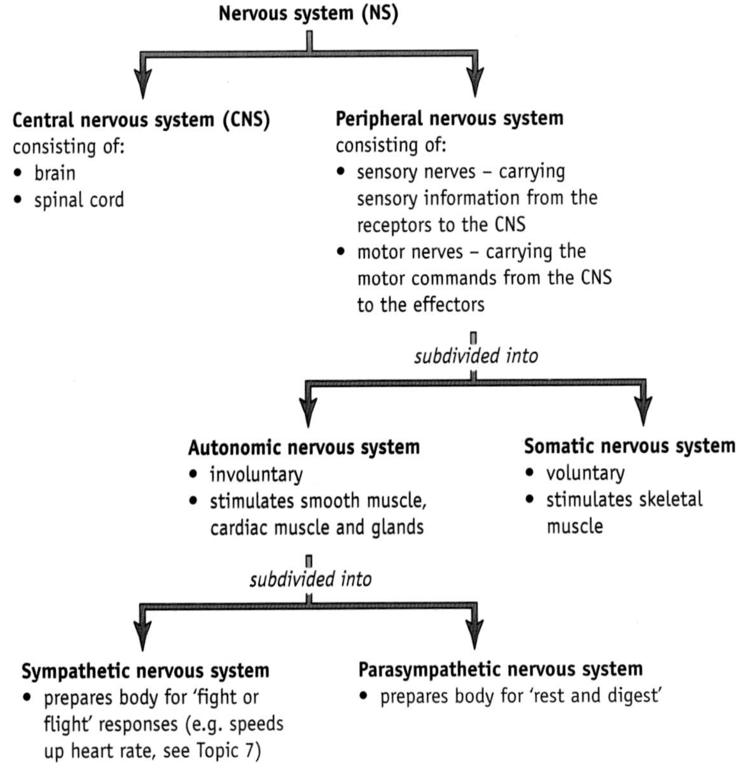

Figure 8.4 The organisation of the nervous system.

Although there are different types of neurone, they all have the same basic characteristics. The **cell body** contains the nucleus and cell organelles within the cytoplasm. There are two types of thin extensions from the cell body:

- very fine **dendrites** conduct impulses *towards* the cell body
- a single long process, the **axon**, transmits impulses *away from* the cell body.

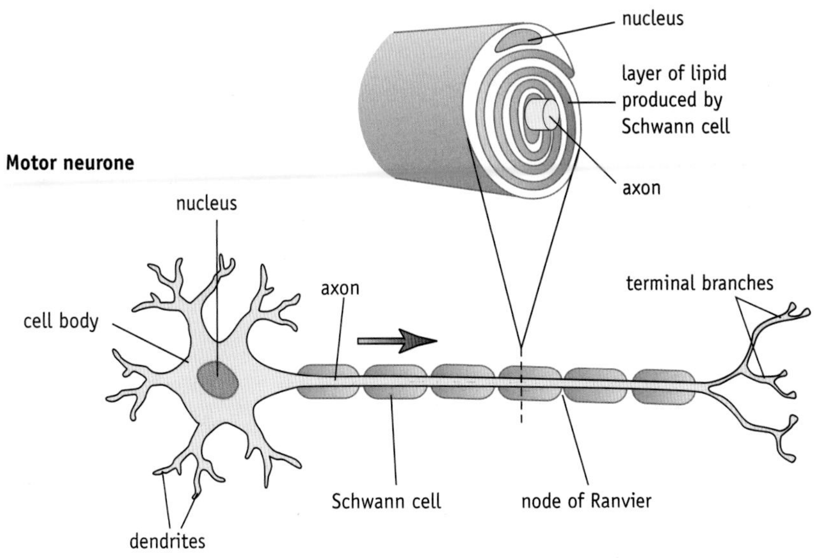

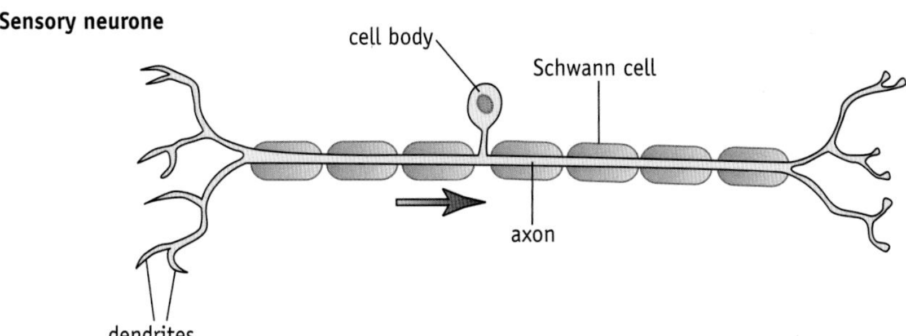

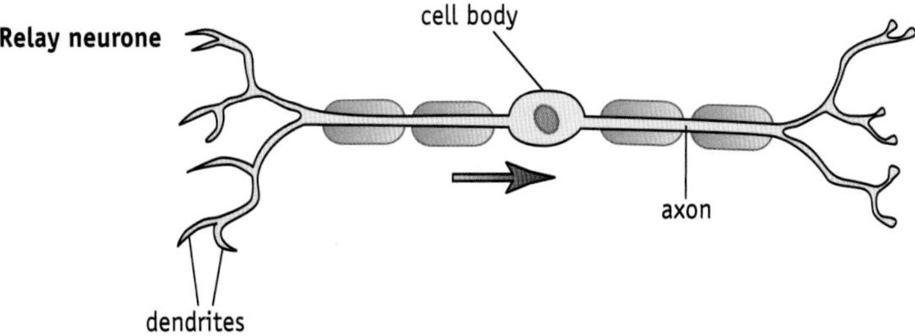

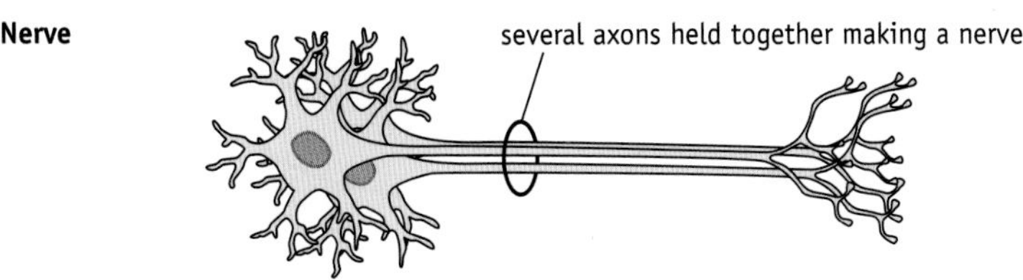

Figure 8.5 The structure of neurones.

There are three main types of neurone (Figure 8.5):

- **Motor neurones** – the cell body is always situated within the central nervous system (CNS) and the axon extends out, conducting impulses from the CNS to effectors (muscles or glands). The axons of some motor neurones can be extremely long, such as those that run the full length of the leg. Motor neurones are also known as effector neurones.
- **Sensory neurones** – these carry impulses from sensory cells to the CNS.
- **Relay neurones** – these are found mostly within the CNS. They can have a large number of connections with other nerve cells (Figure 8.3). Relay neurones are also known as connector neurones and as interneurones.

There is usually a fatty insulating layer called the **myelin sheath** around the axon. This is made up of **Schwann cells** wrapped around the axon (Figure 8.5). As we shall see later, the sheath affects how quickly nerve impulses pass along the axon (page 211). Not all animals have myelinated axons – they are not found in invertebrates and some vertebrate axons are also unmyelinated.

Checkpoint

8.1 Draw up a table comparing the structure and location of motor, relay and sensory neurones.

Reflex arcs

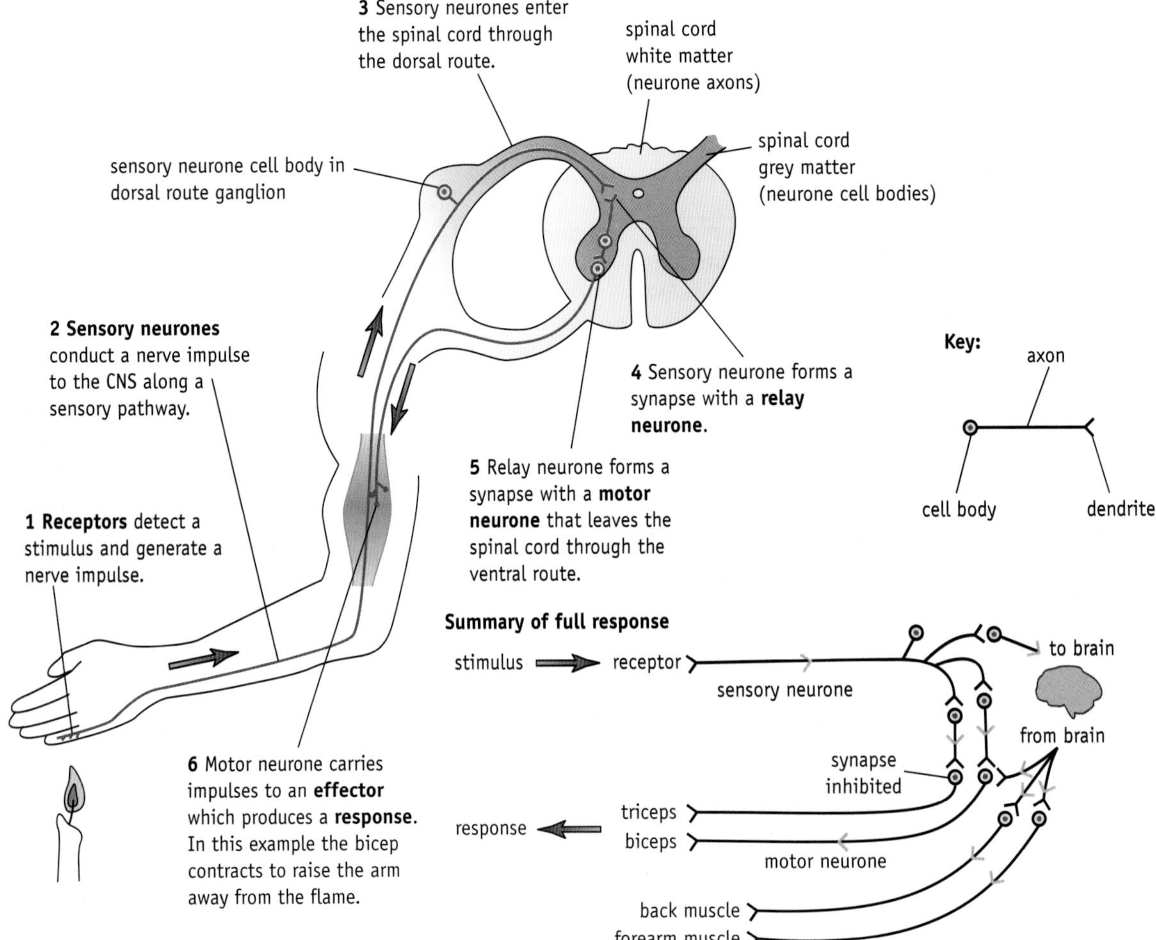

Figure 8.6 A reflex arc allowing withdrawal of the hand.

Nerve impulses follow routes or pathways through the nervous system. Some nerve pathways are relatively simple – the knee-jerk reflex involves just two neurones: a sensory neurone communicating directly with a motor neurone to connect receptor cells with effector cells. These simple nerve pathways are known as **reflex arcs** and are responsible for our **reflexes** – rapid, involuntary responses to stimuli. Figure 8.6 illustrates one reflex arc involving just three neurones.

However, most nerve pathways involve numerous neurones within the central nervous system. A sensory neurone connects to a range of neurones within the CNS and passes impulses to the brain to produce a coordinated response. Figure 8.6 shows how even in reflex arcs there are additional connections within the CNS to ensure a coordinated response. Some synapses with motor neurones will be inhibited to ensure that the desired response occurs.

Q8.1 What is the advantage of reflex pathways?

Q8.2 Describe the nerve pathways involved if instead of a candle, the hand was picking up a hot dinner plate, but the person was trying not to drop the plate.

The pupil reflex

When Kenge and his companion emerged from the trees they moved from deep shade to bright sunlight. Immediately a reflex arc caused a change in the diameter of their pupils. If you cover your eyes for a few minutes, and then uncover your eyes while looking in a mirror, you can see that the size of your pupils decreases and the size of the irises increases (see Figure 8.7).

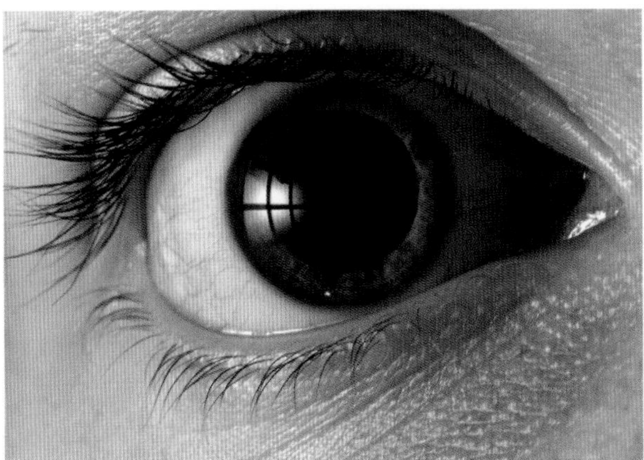

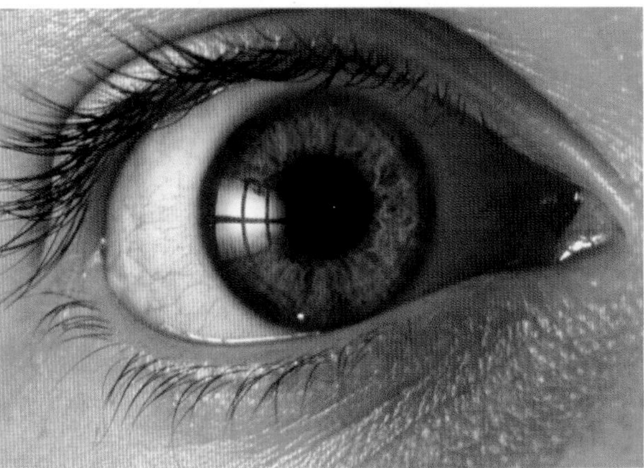

Figure 8.7 The pupil dilates or constricts in response to changing light intensities. Remind yourself about the location of the iris inside the eye by looking at Figure 8.24.

Q8.3 **a** Which of the eyes in Figure 8.7 is in low light?

b Compare the two photos and calculate the percentage increase in diameter.

How do the muscles of the iris respond to light?

The iris controls the size of the pupil. It contains a pair of antagonistic muscles: radial and circular muscles (Figure 8.8). These are both controlled by the autonomic nervous system (Figure 8.4). The radial muscles are like spokes of a wheel, and are controlled by a sympathetic reflex. The circular muscles are controlled by a parasympathetic reflex. One reflex dilates and the other constricts the pupil.

Q8.4 Which of the two sets of muscles will cause the pupil to dilate?

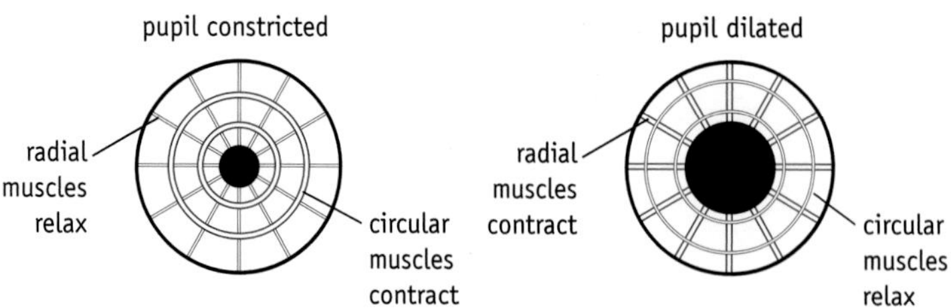

Figure 8.8 How the muscles act to constrict and dilate the pupil.

Controlling pupil size

High light levels striking the **photoreceptors** in the retina cause nerve impulses to pass along the optic nerve to a number of different sites within the CNS, including a group of coordinating cells in the midbrain. Impulses from these cells are sent along parasympathetic motor neurones to the circular muscles of the iris, causing them to contract. At the same time the radial muscles relax. This constricts the pupil, reducing the amount of light entering the eye. Figure 8.9 shows the reflex pathway involved.

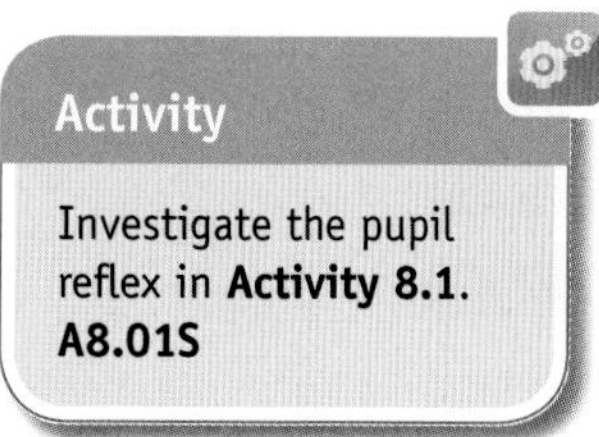

Activity
Investigate the pupil reflex in **Activity 8.1.** **A8.01S**

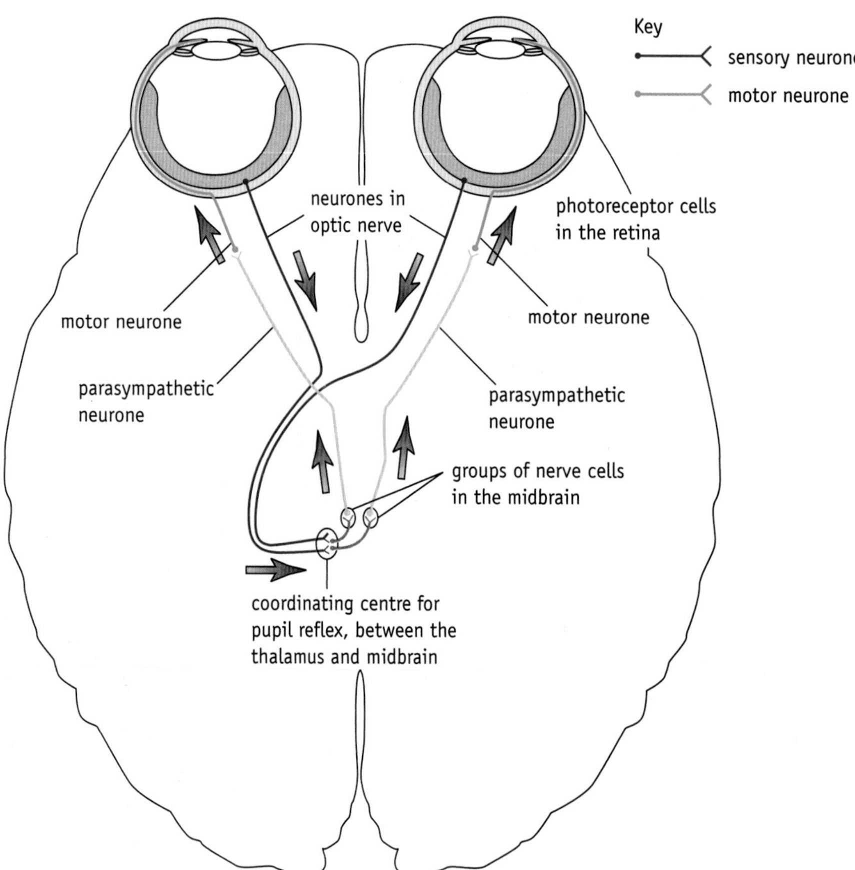

Figure 8.9 The reflex pathway involved in pupil constriction. Pupil dilation involves sympathetic neurones not shown here. Diagram not to scale.

Q8.5 **a** Name the components a–d involved in the pupil reflex in Table 8.1.

b Draw a flow chart to show how these components are linked in the pupil reflex pathway.

Part of the nervous system	Structure in the pupil reflex
receptors	**a**
sensory nerve fibres	**b**
coordinator	**c**
motor nerve fibres	oculomotor nerve
effector	**d**

Table 8.1 Components involved in the pupil reflex.

Q8.6 What is the purpose of the pupil reflex?

Q8.7 The pupil reflex response to increased light is very rapid. Why does this need to be the case?

Q8.8 How many synapses are there in the pupil reflex pathway shown in Figure 8.9?

Did you know?

Atropine

The plant deadly nightshade (*Atropa belladonna*) is the source of the drug atropine. Atropine was used in the Middle Ages by certain women to make their pupils dilate. This was thought to make them more attractive to men, hence 'belladonna', Latin for beautiful lady, in the species name.

Atropine inhibits parasympathetic stimulation of the iris, so the circular muscles of the iris relax. Today it is used to dilate the pupils for an eye examination. It is known as an **acetylcholine** antagonist. When you have completed all the work on the nervous system you should understand what this means!

How nerve cells transmit impulses

To understand how Kenge saw the buffalo, we must first understand how nerve cells transmit impulses, and how the receptor cells in the retina detect light, causing impulses to be sent to the brain where the signals are interpreted.

Much of the work done to establish what happens in a nerve fibre was carried out on the giant axons of the squid (Figure 8.10). Their large size makes them easier to work with. Hodgkin, Huxley and Eccles carried out this work in the 1940s and 1950s, and they eventually won a Nobel Prize for their efforts.

Weblink

Find out more about these Nobel Prize winners and their work on giant axons by visiting the Nobel Prize website.

Figure 8.10 The giant axons of the squid can be seen by the naked eye, making it possible to manipulate them in experiments.

Inside a resting axon

All cells have a potential difference (electrical voltage) across their surface membrane. Figure 8.11 shows an experimental setup designed to measure the potential difference across the membrane of an axon.

In Figure 8.11A, with both electrodes in the bathing solution, there is no potential difference. But if one of the electrodes is pushed inside the axon, as in Figure 8.11B, then the oscilloscope shows that there is a potential difference of around –70 millivolts. The inside of the axon is more negative than outside; the membrane is said to be polarised. The value of –70 mV is known as the **resting potential**.

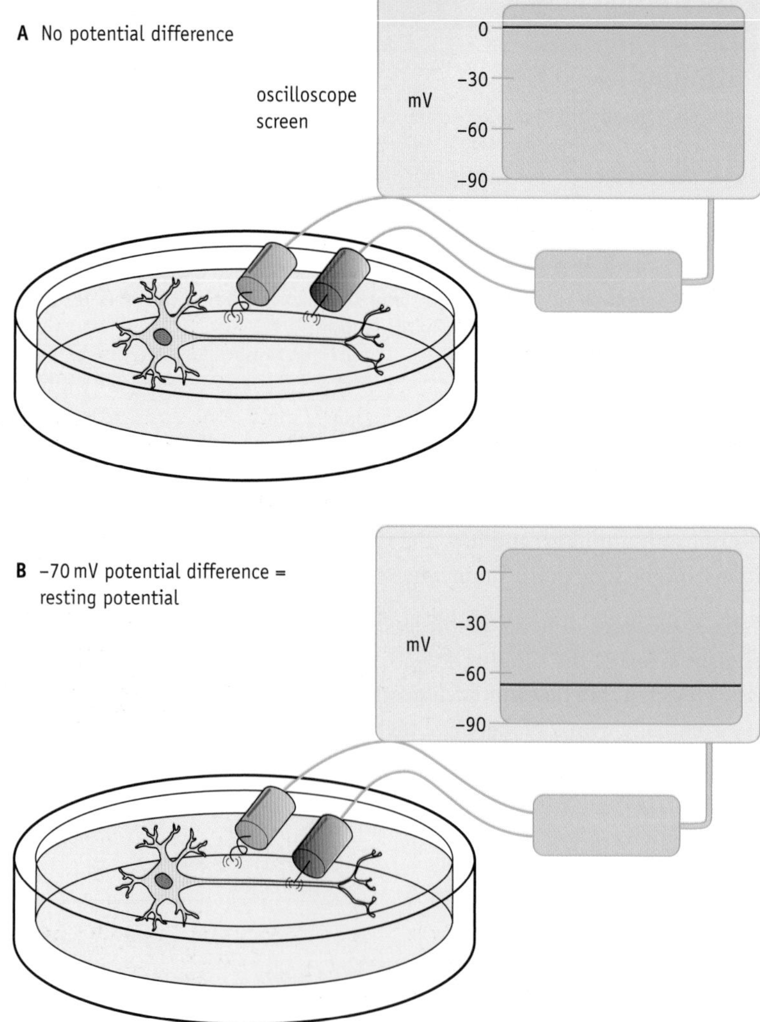

Figure 8.11 Measuring the potential difference across the axon membrane. The oscilloscope displays the potential difference between the two electrodes **A** when both electrodes are in the bathing solution and **B** across the axon membrane.

Key biological principle: Why is there a potential difference?

Table 8.2 shows the concentrations of some of the ions found in the solutions inside and outside a squid giant axon. The most obvious feature of this is that the distribution of the ions is far from equal.

Ion	Extracellular concentration /mmol kg^{-1}	Intracellular concentration /mmol kg^{-1}
K^+	30	400
Na^+	460	50
Cl^-	560	100
organic anions	0	370

Table 8.2 The approximate concentrations of ions inside and outside a nerve fibre.

This uneven distribution of ions across the cell surface membrane is achieved by the action of sodium-potassium pumps in the cell surface membrane of the axon. These carry Na^+ out of the cell and K^+ into the cell (Figure 8.12). These pumps act against the concentration gradients of these two ions and are driven by energy supplied by hydrolysis of ATP. The organic anions (e.g. negatively charged amino acids) are large and stay within the cell, so chloride ions move out of the cell to help balance the charge across the cell surface membrane.

The resting potential

Once the concentration gradients are established by the sodium-potassium pumps and there is no difference in charge between the inside and outside of the membrane, potassium ions diffuse out of the neurone down the potassium concentration gradient. The K^+ ions pass through potassium channels, making the outside of the cell surface membrane positive and the inside negative. The membrane is permeable to potassium ions but is virtually impermeable to sodium ions. There is some leakage of Na^+ into the neurone down the Na^+ concentration gradient. This movement of Na^+ does not balance the difference in charge across the membrane caused by the movement of K^+. The difference in charge caused by diffusion of K^+ causes a potential difference across the membrane, known as the resting potential (Figure 8.12).

Why is the axon resting potential –70 mV?

To understand why the axon resting potential is –70 mV we need to appreciate that there are two forces involved in the movement of the potassium ions. These result from:

- the concentration gradient generated by the Na^+/K^+ pump
- the electrical gradient due to the difference in charge on the two sides of the membrane resulting from K^+ diffusion.

Potassium ions diffuse out of the cell due to the concentration gradient. The more potassium ions that diffuse out of the cell, the larger the potential difference across the membrane. The increased negative charge created inside the cell as a consequence attracts potassium ions back across the membrane into the cell. When the potential difference across the membrane is around –70 mV, the electrical gradient exactly balances the chemical gradient. There is no net movement of K^+, and hence a steady state exists, maintaining the potential difference at –70 mV. An electrochemical equilibrium for potassium is in place and the membrane is polarised (Figure 8.12).

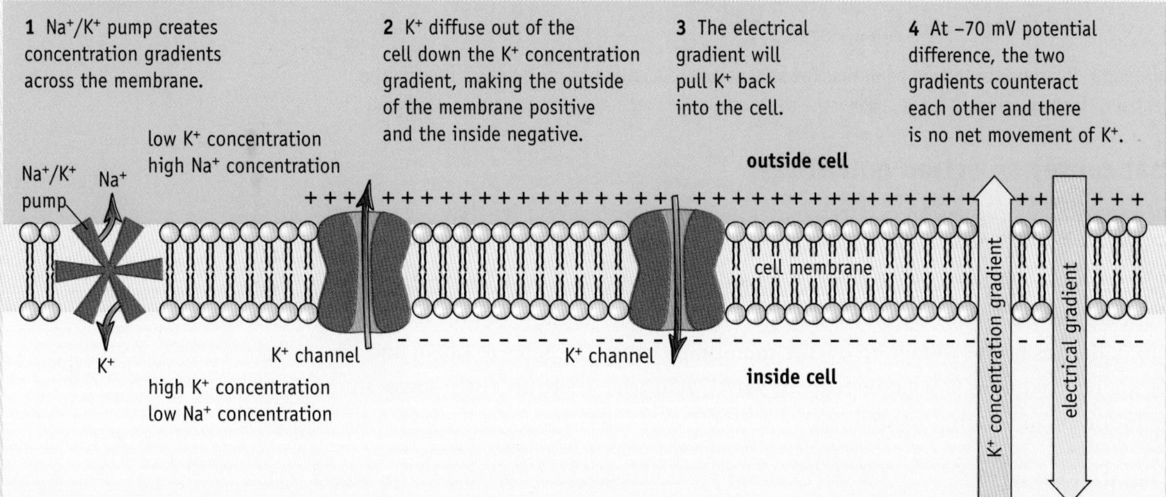

1 Na⁺/K⁺ pump creates concentration gradients across the membrane.
2 K⁺ diffuse out of the cell down the K⁺ concentration gradient, making the outside of the membrane positive and the inside negative.
3 The electrical gradient will pull K⁺ back into the cell.
4 At –70 mV potential difference, the two gradients counteract each other and there is no net movement of K⁺.
low K⁺ concentration
high Na⁺ concentration
Na⁺/K⁺ pump
Na⁺
K⁺
outside cell
cell membrane
K⁺ channel
K⁺ channel
inside cell
high K⁺ concentration
low Na⁺ concentration
K⁺ concentration gradient
electrical gradient

Figure 8.12 Movement of sodium and potassium ions across the cell surface membrane of the neurone leads to the formation of the resting potential. At –70 mV, further movement of K^+ out of the cell due to the concentration gradient is opposed by the electrical gradient across the cell surface membrane pulling K^+ back into the cell.

What happens when a nerve is stimulated?

Neurones are electrically excitable cells, which means that the potential difference across their cell surface membrane changes when they are conducting an impulse.

Figure 8.13 shows the effect of stimulating the axon by passing a small electric current through it. If an electrical current above a threshold level is applied to the membrane, it causes a massive change in the potential difference. The potential difference across the membrane is locally reversed, making the inside of the axon positive and the outside negative. This is known as **depolarisation**.

The potential difference becomes +40 mV or so for a very brief instant, lasting about 3 milliseconds (ms), before returning to the resting state, as shown by the oscilloscope trace (Figure 8.13). It is important that the membrane is returned to the resting potential as soon as possible in order that more impulses can be conducted. This return to a resting potential of –70 mV is known as **repolarisation**. The large change in the voltage across the membrane is known as an **action potential** (Figure 8.13).

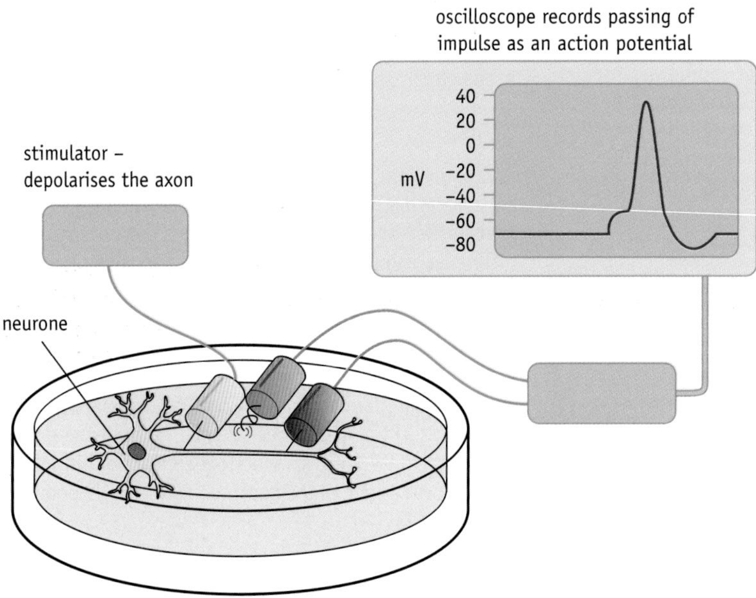

Figure 8.13 Measuring an action potential. The stimulator produces an electric current causing the potential difference across the axon membrane to reverse before returning to the resting potential.

What causes an action potential?

Once threshold stimulation occurs, an action potential is caused by changes in the permeability of the cell surface membrane to Na^+ and K^+, due to the opening and closing of voltage-dependent Na^+ and K^+ channels (Figure 8.14). At the resting potential, these channels are blocked by gates preventing the flow of ions through them. Changes in the voltage across the membrane cause the gates to open, and so they are referred to as voltage-dependent gated channels. There are three stages in the generation of an action potential.

1 Depolarisation

When a neurone is stimulated some depolarisation occurs. The change in the potential difference across the membrane causes a change in the shape of the Na^+ gate, opening some of the voltage-dependent sodium ion channels. As the sodium ions flow in, depolarisation increases, triggering more gates to open once a certain potential difference threshold is reached. The opening of more gates increases depolarisation further. This is an example of **positive feedback** – a change encourages further change of the same sort – and it leads to a rapid opening of all of the Na^+ gates. This means there is no way of controlling the degree of depolarisation of the membrane; action potentials are either there or they are not. This property is often referred to as **all-or-nothing**.

There is a higher concentration of sodium ions outside of the axon, so sodium ions flow rapidly inwards through the open voltage-dependent Na^+ channels, causing a build-up of positive charges inside. This reverses the polarity of the membrane. The potential difference across the membrane reaches +40 mV.

2 Repolarisation

After about 0.5 ms, the voltage-dependent Na^+ channels spontaneously close and Na^+ permeability of the membrane returns to its usual very low level. Voltage-dependent K^+ channels open due to the depolarisation of the membrane. As a result, potassium ions move out of the axon, down the electrochemical gradient (they diffuse down the concentration gradient and are also attracted by the negative charge outside the cell surface membrane). As potassium ions flow out of the cell, the inside of the cell once again becomes more negative than the outside. This is the falling phase of the oscilloscope trace in Figure 8.14.

3 Restoring the resting potential

The membrane is now highly permeable to potassium ions, and more ions move out than occurs at resting potential, making the potential difference more negative than the normal resting potential (Figure 8.14). This is known as **hyperpolarisation** of the membrane. The resting potential is re-established by closing of the voltage-dependent K^+ channels and potassium ion diffusion into the axon.

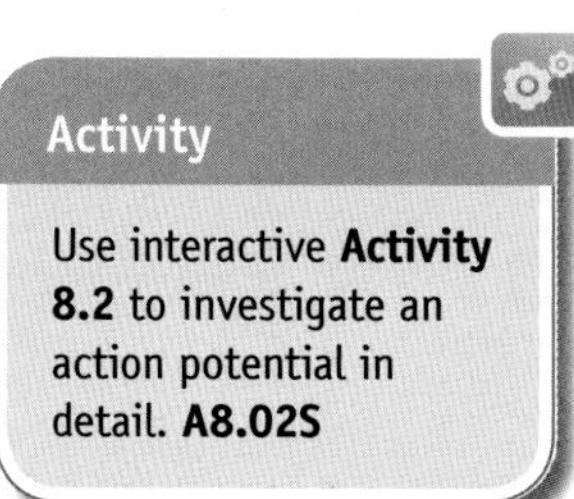

Activity
Use interactive **Activity 8.2** to investigate an action potential in detail. **A8.02S**

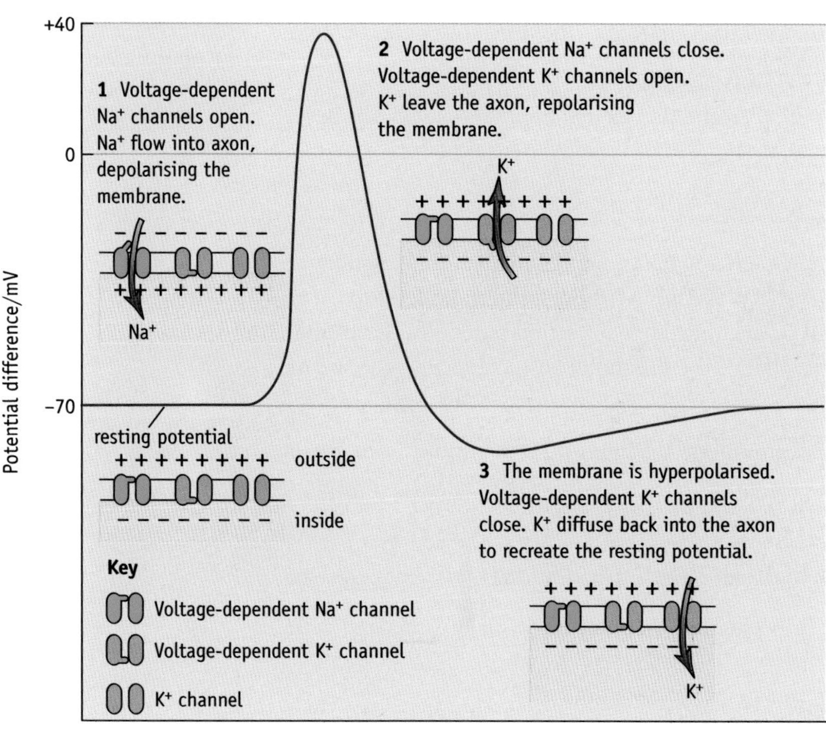

Figure 8.14 Voltage-dependent gates open and close to produce changes in potential difference during an action potential.

If lots (hundreds) of action potentials occur in the neurone, the sodium ion concentration inside the cell rises significantly. The sodium-potassium pumps start to function, restoring the original ion concentrations across the cell membrane (Table 8.2). If a cell is not transmitting many action potentials, these pumps will not have to be used very frequently. At rest there is some slow leakage of sodium ions into the axon. These sodium ions are pumped back out of the cell.

Checkpoint

8.2 Produce a bullet point summary of the membrane changes and ion movements that cause an action potential. You should aim to have at least ten bullet points.

Q8.9 Why is the Na^+ channel referred to as voltage-dependent?

Q8.10 Will it be possible for an action potential to be triggered in a dead axon? Give a reason for your answer.

How is the impulse passed along an axon?

When a neurone is stimulated, the action potential generated does not actually travel along the axon, but triggers a sequence of action potentials along the length of the axon. It is rather like pushing one domino to topple a whole line of standing dominoes. Figure 8.15 illustrates this propagation of the impulse along an axon.

As part of the membrane becomes depolarised at the site of an action potential, a local electrical current is created as the charged sodium ions flow between the depolarised part of the membrane and the adjacent resting region. The depolarisation spreads to the adjacent region and the nearby Na^+ gates will respond to this by opening as described earlier, triggering another action potential. These events are then repeated along the membrane. As a result, a wave of depolarisation will pass along the membrane. This is the nerve impulse.

Activity

Use **Activity 8.3** to help you understand how the nerve impulse is transmitted along the axon. **A8.03S**

A new action potential cannot be generated in the same section of membrane for about five milliseconds. This is known as the **refractory period.** It lasts until all the voltage-dependent sodium and potassium channels have returned to their normal resting state (closed) and the resting potential is restored. The refractory period ensures that impulses only travel in one direction.

Extension

Read about ion channels and episodic diseases in **Extension 8.1. X8.01S**

Q8.11 How does the refractory period ensure that an action potential will not be propagated back the way it came?

Are impulses different sizes?

A very strong light will produce the same size action potential in a neurone coming from your eye as does a dim light. A stimulus must be above a threshold level to generate an action potential. The all-or-nothing effect for action potentials means that the size of the stimulus, assuming it is above the threshold, has no effect on the size of the action potential.

Different mechanisms are used to communicate the intensity of a stimulus. The size of stimulus affects:

- the frequency of impulses
- the number of neurones in a nerve that are conducting impulses.

A high frequency of firing and the firing of many neurones are usually associated with a strong stimulus.

1

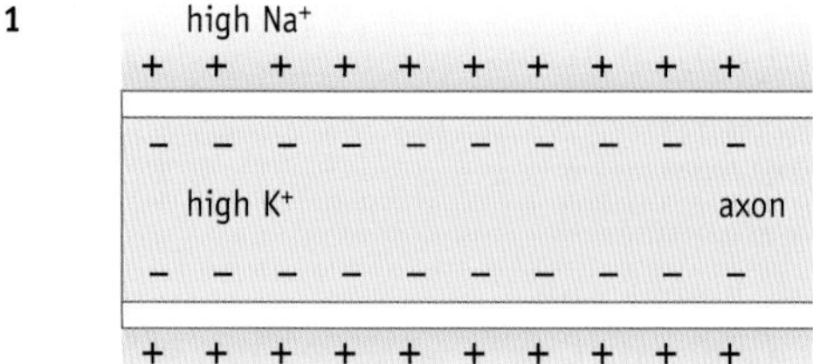

At resting potential there is positive charge on the outside of the membrane and negative charge on the inside, with high sodium ion concentration outside and high potassium ion concentration inside.

2

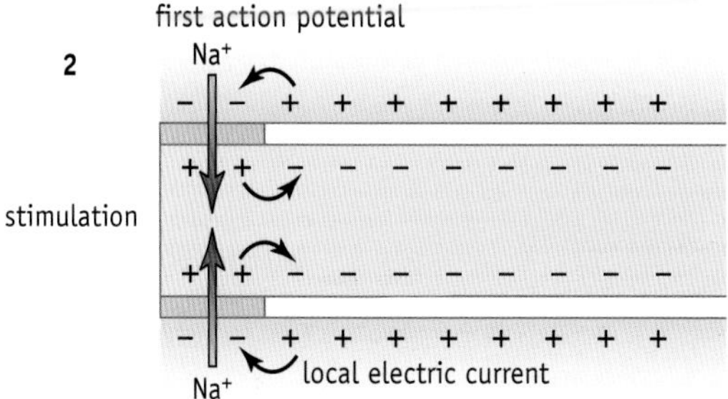

When stimulated, voltage-dependent sodium ion channels open, and sodium ions flow into the axon, depolarising the membrane. Localised electric currents are generated in the membrane. Sodium ions move to the adjacent polarised (resting) region causing a change in the electrical charge (potential difference) across this part of the membrane.

3

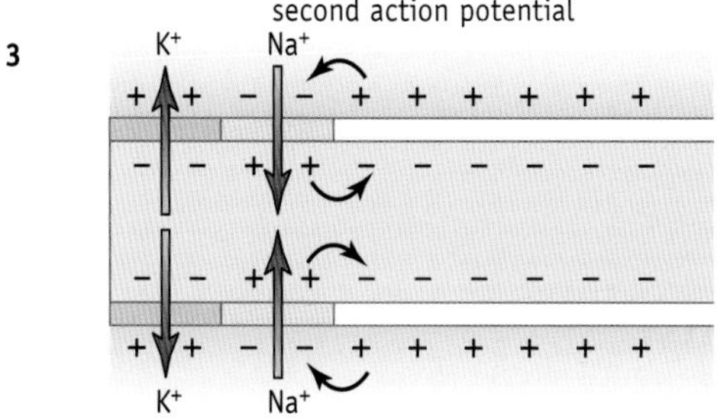

The change in potential difference in the membrane adjacent to the first action potential initiates a second action potential. At the site of the first action potential the voltage-dependent sodium ion channels close and voltage-dependent potassium ion channels open. Potassium ions leave the axon, repolarising the membrane. The membrane becomes hyperpolarised.

4

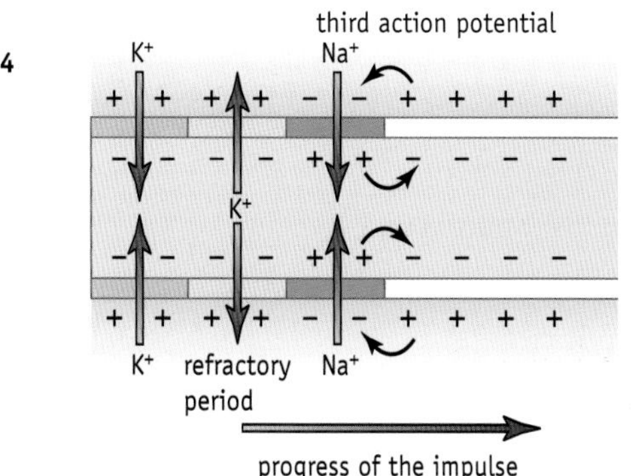

A third action potential is initiated by the second. In this way local electric currents cause the nerve impulse to move along the axon. At the site of the first action potential, potassium ions diffuse back into the axon, restoring the resting potential.

Figure 8.15 Propagation of an impulse along an axon.

Speed of conduction

The speed of nervous conduction is in part determined by the diameter of the axon. In general, the wider the diameter, the faster the impulse travels. The normal axons of a squid (diameter 1–20 μm) conduct impulses at around 0.5 ms^{-1}, whereas the giant axons (diameter up to 1000 μm, i.e. 1 mm) conduct at nearer 100 ms^{-1}. The nerve axons of mammals (diameter 1–20 μm) are much narrower than the squid giant axons, but impulses travel along them at up to 120 ms^{-1}. This apparent anomaly can be explained by the presence of the myelin sheath around mammalian nerve axons.

The myelin sheath acts as an electrical insulator along most of the axon, preventing any flow of ions across the membrane. Gaps known as **nodes of Ranvier** occur in the myelin sheath at regular intervals, and these are the only places where depolarisation can occur. As ions flow across the membrane at one node during depolarisation, a circuit is set up which reduces the potential difference of the membrane at the next node, triggering an action potential. In this way, the impulse effectively jumps from one node to the next. This is much faster than a wave of depolarisation along the whole membrane. This 'jumping' conduction, illustrated in Figure 8.16, is called **saltatory conduction**.

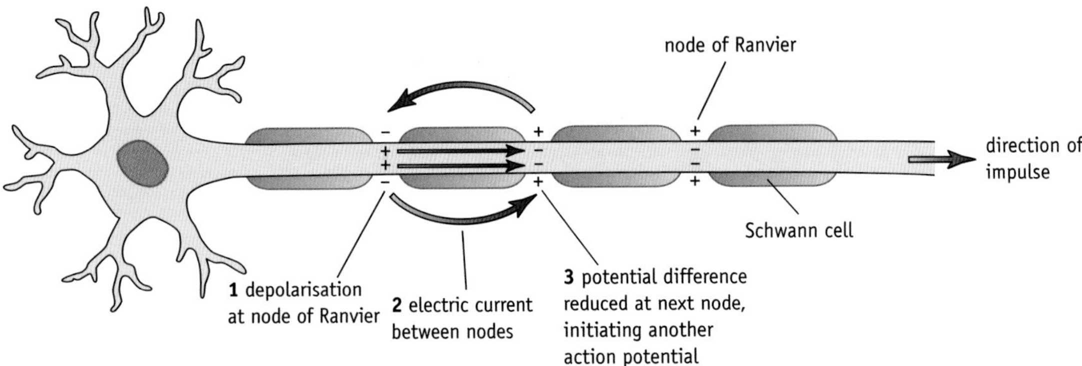

Figure 8.16 An impulse can move very quickly along the axon by jumping between the nodes of Ranvier.

How does a nervous impulse pass between cells?

Where two neurones meet is known as a **synapse**. The cells do not actually touch – there is a small gap, the **synaptic cleft**. So how does the nerve impulse, on which the function of the nervous system depends, get across this gap?

Synapse structure

A nerve cell may have very large numbers of synapses with other cells (Figure 8.3), as many as 10 000 for some brain cells. This is important in enabling the distribution and processing of information.

Figure 8.17 shows the structure of a typical synapse. Notice the synaptic cleft that separates the **presynaptic** membrane of the stimulating neurone from the **postsynaptic** membrane of the other cell. The gap is about 20–50 nm and a nerve impulse cannot jump across it. In the cytoplasm at the end of the presynaptic neurone there are numerous **synaptic vesicles** containing a chemical called a **neurotransmitter**.

Q8.12 Look at Figure 8.17B. Identify the presynaptic neurone – is it the left or right cell? Give a reason for your answer.

How does the synapse transmit an impulse?

The arrival of an action potential at the presynaptic membrane causes the release of the neurotransmitter into the synaptic cleft. The neurotransmitter diffuses across the gap, resulting in events that cause the depolarisation of the postsynaptic membrane, and hence the propagation of the impulse along the next cell. The presynaptic cell expends a considerable amount of energy to produce neurotransmitter and package it into vesicles ready for transport out of the cell.

A

1 An action potential arrives.

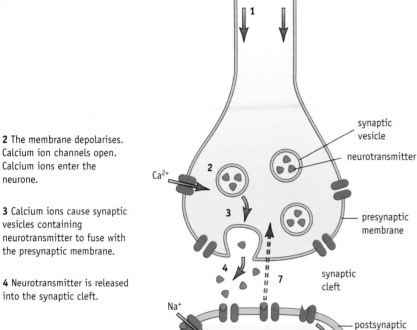

2 The membrane depolarises. Calcium ion channels open. Calcium ions enter the neurone.

3 Calcium ions cause synaptic vesicles containing neurotransmitter to fuse with the presynaptic membrane.

4 Neurotransmitter is released into the synaptic cleft.

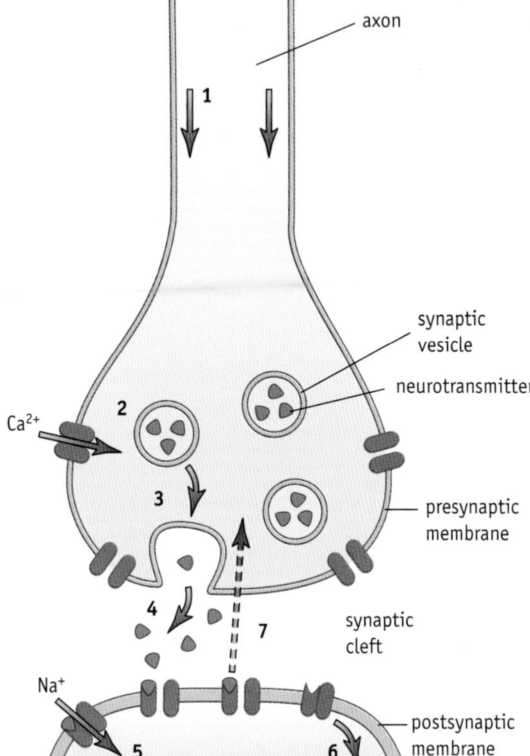

5 Neurotransmitter binds with receptors on the postsynaptic membrane. Cation channels open. Sodium ions flow through the channels.

6 The membrane depolarises and initiates an action potential.

7 When released the neurotransmitter will be taken up across the presynaptic membrane (whole or after being broken down), or it can diffuse away and be broken down.

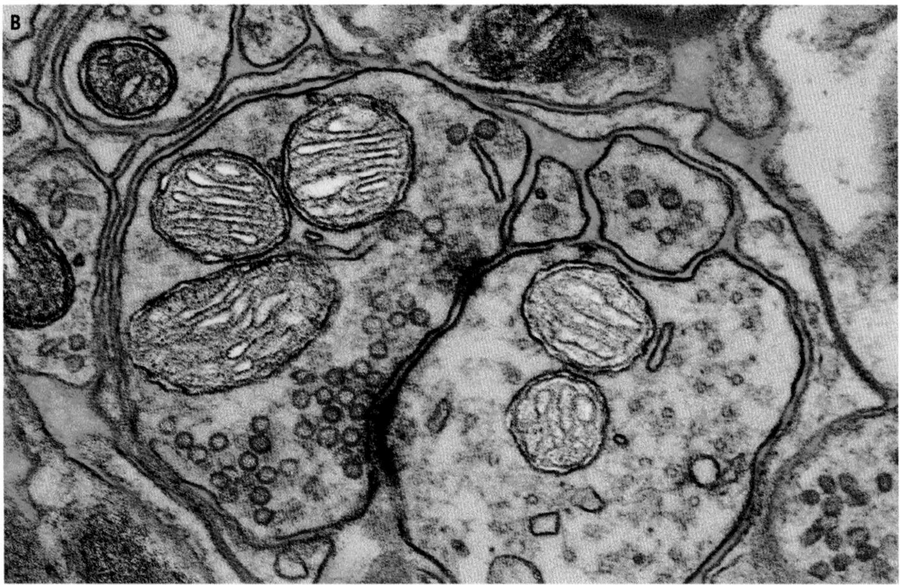

Figure 8.17 **A** The functioning of a synapse. **B** Coloured transmission electron micrograph showing a synapse between two neurones in the brain. The presynaptic membrane, postsynaptic membrane, synaptic vesicles and mitochondria are visible. Magnification × 61 000.

Many neurotransmitters have been discovered, with 50 identified in the human central nervous system. **Acetylcholine**, the first to be discovered, will be used here to describe the working of a synapse. Others will be considered later in the topic.

There are essentially three stages leading to the nerve impulse passing along the postsynaptic neurone:

- neurotransmitter release
- stimulation of the postsynaptic membrane
- inactivation of the neurotransmitter.

These stages are illustrated in Figure 8.17.

Neurotransmitter release

When the presynaptic membrane is depolarised by an action potential, channels in the membrane open and increase the permeability of the membrane to calcium ions (Ca^{2+}). These calcium ions are in greater concentration outside the cell, so they diffuse across the membrane and into the cytoplasm.

The increased Ca^{2+} concentration causes synaptic vesicles containing acetylcholine to fuse with the presynaptic membrane and release their contents into the synaptic cleft by exocytosis.

Stimulation of the postsynaptic membrane

The neurotransmitter takes about 0.5 ms to diffuse across the synaptic cleft and reach the postsynaptic membrane. Embedded in the postsynaptic membrane are specific receptor proteins that have a binding site with a complementary shape to part of the acetylcholine molecule. The acetylcholine molecule binds to the receptor, changing the shape of the protein, opening cation channels and making the membrane permeable to sodium ions. The flow of sodium ions across the postsynaptic membrane causes depolarisation, and if there is sufficient depolarisation, an action potential will be produced and propagated along the postsynaptic neurone.

The extent of the depolarisation will depend on the amount of acetylcholine reaching the postsynaptic membrane. This will depend in part on the frequency of impulses reaching the presynaptic membrane. A single impulse will not usually be enough; several impulses are usually required to generate enough neurotransmitter to depolarise the postsynaptic membrane. The number of functioning receptors in the postsynaptic membrane will also influence the degree of depolarisation.

Inactivation of the neurotransmitter

Some neurotransmitters are actively taken up by the presynaptic membrane and the molecules are used again. With others, the neurotransmitter rapidly diffuses away from the synaptic cleft or is taken up by other cells of the nervous system. In the case of acetylcholine, a specific enzyme at the postsynaptic membrane, **acetylcholinesterase**, breaks down the acetylcholine so that it can no longer bind to receptors. Some of the breakdown products are then reabsorbed by the presynaptic membrane and reused.

Activity

You can investigate the synapse in more detail using the animation in **Activity 8.4. A8.04S**

Checkpoint

8.3 Construct a flow chart to show the sequence of events that occur when a nerve impulse crosses a synapse.

What is the role of synapses in nerve pathways?

Control and coordination

Synapses have two roles:

- control of nerve pathways, allowing flexibility of response
- integration of information from different neurones, allowing a coordinated response.

The postsynaptic cell is likely to be receiving input from many synapses at the same time (Figure 8.18). It is the overall effect of all of these synapses that will determine whether the postsynaptic cell generates an action potential. Two main factors affect the likelihood that the postsynaptic membrane will depolarise:

- the type of synapse
- the number of impulses received.

Some synapses help stimulate an action potential. Others are inhibitory and make it less likely that the postsynaptic membrane will depolarise. A postsynaptic cell can have many inhibitory and excitatory synapses, and so whether or not an action potential results depends upon the balance of excitatory and inhibitory synapses acting at any given time.

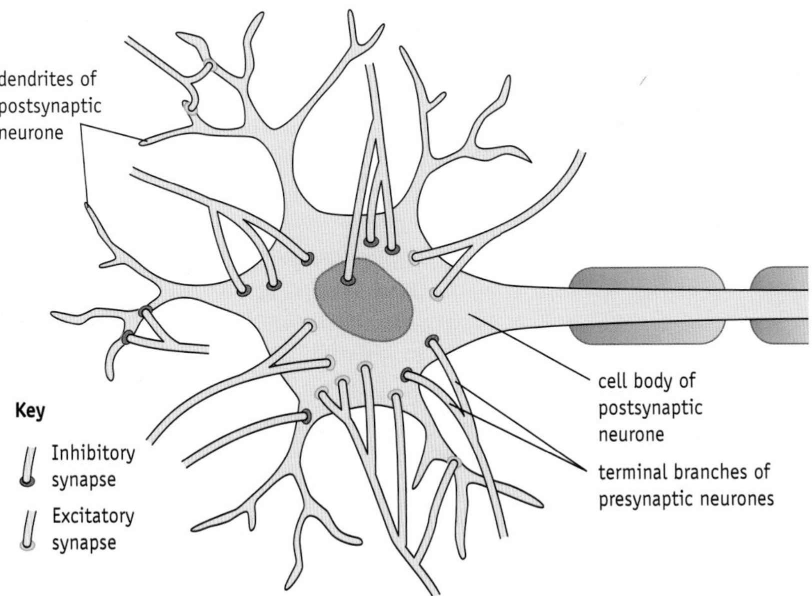

Figure 8.18 A postsynaptic neurone receives input from many excitatory and inhibitory synapses at the same time.

Q8.13 Before you read on, suggest how an inhibitory synapse might work to make it less likely that the postsynaptic membrane will depolarise.

Types of synapse

Excitatory synapses

Excitatory synapses make the postsynaptic membrane more permeable to sodium ions. A single excitatory synapse typically does not depolarise the membrane enough to produce an action potential, but several impulses arriving within a short time produce sufficient depolarisation via the release of neurotransmitter to produce an action potential in the postsynaptic cell. The fact that each impulse adds to the effect of the others is known as **summation**.

There are two types of summation:

- **Spatial summation** – here the impulses are from different synapses, usually from different neurones. The number of different sensory cells stimulated can be reflected in the control of the response (Figure 8.19A).
- **Temporal summation** – in this case several impulses arrive at a synapse having travelled along a single neurone one after the other. Their combined release of neurotransmitter generates an action potential in the postsynaptic membrane (Figure 8.19B).

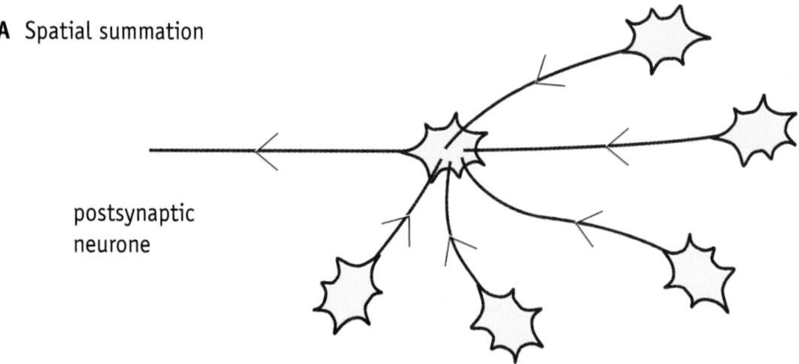
A Spatial summation
postsynaptic
neurone
Impulses from several different neurones produce an action potential in the postsynaptic neurone.

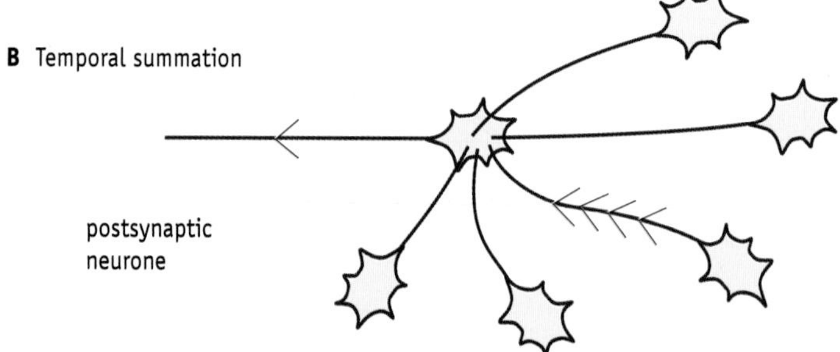
B Temporal summation
postsynaptic
neurone
Several impulses along one neurone produce an action potential in the postsynaptic neurone.

Figure 8.19 Spatial and temporal summation.

Q8.14 **a** You might not notice a tiny insect (e.g. a thrip) landing on your arm

Explain why you would be more likely to notice a larger insect.

b Explain why you would be very likely to respond to a butterfly crawling along your arm.

Inhibitory synapses

Inhibitory synapses make it less likely that an action potential will result in the postsynaptic cell. The neurotransmitter from these synapses opens channels for chloride ions and potassium ions in the postsynaptic membrane, and these ions will then move through the channels down their diffusion gradients (Figure 8.20). Chloride ions will move into the cell carrying negative charge, and potassium ions will move out carrying positive charge. The result will be a *greater* potential difference across the membrane as the inside becomes more negative than usual (about –90 mV), so-called hyperpolarisation. This makes subsequent depolarisation less likely. More excitatory synapses will be required to depolarise the membrane.

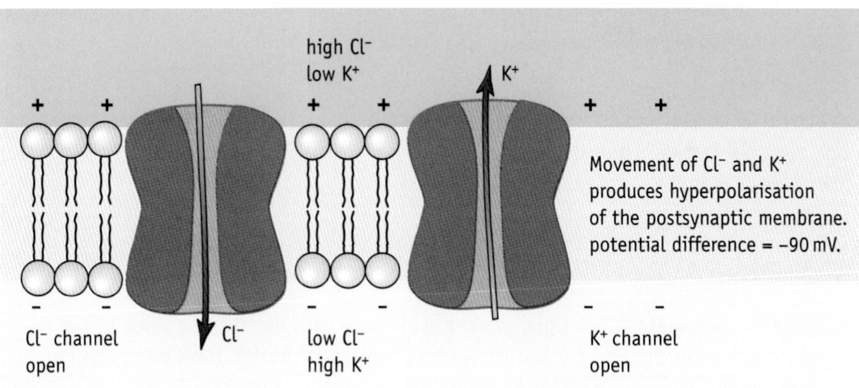

Figure 8.20 Neurotransmitters from inhibitory synapses open Cl^- and K^+ channels in the postsynaptic membrane.

Q8.15 Suggest the role of inhibitory synapses in the pupil reflex in Figure 8.9.

Key biological principle: Comparing nervous and hormonal coordination

The nervous system is not the only means by which the activities of the body can be coordinated. As you saw in Topic 7, hormones, which are secreted into the bloodstream by endocrine glands, act as a means of chemical communication with target cells (Figure 7.72 and Table 7.4).

Q8.16 Describe the ways that a hormone brings about a change in the activity of its target cell.

Many hormones are produced steadily over long periods of time to control long-term changes in the body, such as growth and sexual development. See, for instance, testosterone in Topic 7. Adrenaline, also encountered in Topic 7, is more short term in its action, but takes longer than the nervous system to produce a response.

Q8.17 Look back at Topic 7 and describe in detail the action of the hormone testosterone. Ensure your description includes:

- site of production
- method of transport
- location of target cells
- effect on target cells.

Checkpoint

8.4 Think up a list of key words to distinguish the features of **a** nervous and **b** hormonal control. If you can compose a mnemonic to help you remember, even better.

CONTINUED ▶

Table 8.3 contrasts nervous and hormonal control in animals.

Nervous control	Hormonal control
electrical transmission by nerve impulses and chemical transmission at synapses	chemical transmission through the blood
fast acting	slower acting
usually associated with short-term changes, e.g. muscle contraction	can control long-term changes, e.g. growth
action potentials carried by neurones with connections to specific cells	blood carries the hormone to all cells, but only target cells are able to respond
response is often very local, such as a specific muscle cell or gland	response may be widespread, such as in growth and development

Table 8.3 Nervous and hormonal control in animals.

Extension

In **Extension 8.2** read about how scientists link sex hormones with brain activities. **X8.02S**

Key biological principle: Coordination in plants

Animals use nervous and hormonal control to coordinate activity in the body. Plants lack a nervous system so must use chemicals to coordinate growth, development and responses to the environment. In some textbooks you will see these chemicals called hormones, but more often they are called plant growth regulators or **plant growth substances**. They are chemicals produced in the plant in very low concentrations and transported to where they cause a response.

There are several classes of plant growth substances but we will only consider **auxins**, the first to be discovered.

The discovery of auxins

Charles Darwin and one of his sons, Francis, completed experiments on phototropism (bending of plants toward a light source), which are considered to be some of the earliest work on the effects of auxin. Their experiments showed that an oat coleoptile (Figure 8.21) with its tip cut off stops bending towards the light (Figure 8.22B). Replacing the tip starts growth towards the light again. They concluded that 'some influence' was transmitted from the shoot tip to the lower part of the seedlings, causing them to bend. Later experiments by researchers such as Boysen-Jensen and Went identified the nature of this 'influence'.

It was shown that a chemical made in the tip passed down the coleoptile. This was demonstrated by removing the tip, placing it on a small block of agar jelly and putting the agar on top of the cut end of the coleoptile. The coleoptile started to grow again; a chemical produced by the tip had

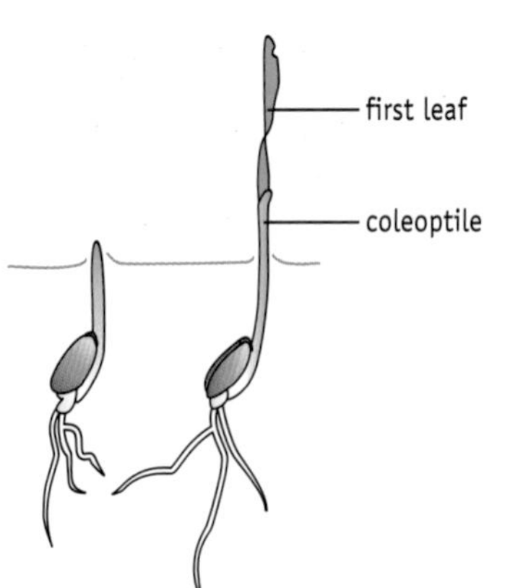

Figure 8.21 A coleoptile is the protective sheath that covers the first leaf in grasses and cereals.

CONTINUED ▶

diffused down through the agar jelly (Figure 8.22C). Went provided further evidence by placing the agar blocks on one side of the cut coleoptile tip in the dark; this caused the coleoptile to curve away from the side receiving the chemical messenger from the agar (Figure 8.22C). The chemical was eventually identified as the auxin, **indoleacetic acid** (**IAA**) and one of its major functions is to stimulate growth. The growth response is the result of cell elongation.

Why does the coleoptile curve towards the light when the tip is in place? Went measured the amount of chemical being produced on the shaded and unshaded side of the shoot and found that the total amount did not change compared to a shoot illuminated from all sides; instead more auxin had passed down the shaded side (Figure 8.22D). The increased concentration of auxin on the shaded side increased cell elongation; the reduced concentration on the illuminated side inhibited cell elongation. As a result the shoot grows towards the light. Similar interactions of auxins and gravity have been observed.

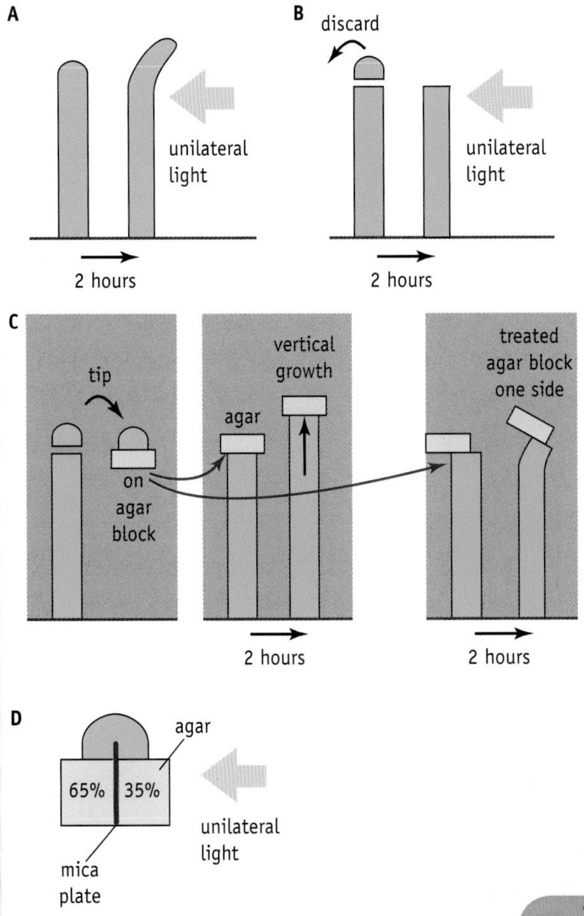

Q8.18 State what conclusions can be drawn from experiments B to D in Figure 8.22.

Q8.19 Sketch the result you would expect for experiment C if completed in uniform light.

Q8.20 Suggest a control for Went's experiments shown in Figure 8.22C.

Explaining growth curvature as resulting from the unequal distribution of auxin due to lateral transport of auxin is known as the Cholodny-Went model. The model has been widely criticised due to the small sample sizes and the difficulty of measuring the very small concentrations involved. However, many plant physiologists maintain that the basic features of the model still hold. Researchers today still investigate growth curvature but they now benefit from new technologies. New techniques being used to study tropisms include the use of genetically modified plants that produce fluorescent proteins in the presence of auxin, making it possible to visualise the location of the auxin.

Auxins are synthesised in actively growing plant tissues (known as meristems) such as shoot tips, developing leaves, seeds and fruits. The auxins are actively transported away from these sites to where they bring about a range of responses through their effect on cell elongation. By binding with receptors on the plasma membranes in the zone of shoot elongation, auxins produce second messenger signal molecules that bring about changes in gene expression. Transcription of genes coding for enzymes then results in metabolic changes. It is thought that the auxin causes acidification of the cell wall by indirectly stimulating the activity of proton pumps that move H^+ out of the cytoplasm. It is thought that the low pH affects an enzyme in the cell walls that causes bonds between cellulose microfibrils to break, allowing the cell wall to expand. The increased potential difference across the membrane enhances uptake of ions into the cell. This in turn causes uptake of water by osmosis, resulting in cell elongation.

Figure 8.22 A The curving growth of coleoptiles towards a unidirectional light source is an easy-to-observe response by plants to an environmental cue.
B–D Experiments that led to the understanding of the role of auxins in growth response to light. The mica plate is impermeable.

Activity

In **Activity 8.5** you can look at the effect of IAA on roots and shoots. **A8.05S**

Checkpoint

8.5 Draw up a table that summarises the similarities and the differences between the action of hormones in animals and growth substances in plants.

Did you know?

More than just shoot elongation

Auxins have many other effects in plants. They inhibit the growth of side branches down the plant; this is known as apical dominance. This effect can be observed if the growing tip at the top of a plant, its apical meristem, is removed. The side branches down the plant will start to grow. Auxins also initiate growth of lateral roots, stimulate fruit development, and are involved in leaf fall.

Many synthetic auxins have been produced and are widely used in agriculture. For example, 2,4-dichlorophenoxy acetic acid (2,4-D) is an effective herbicide. Monocotyledons, such as wheat and barley, inactivate synthetic auxins, such as 2,4-D, whereas in broad-leaved plants (dicotyledons) the auxins accumulate in cells, causing rapid growth that kills the plant. Spraying a lawn with synthetic IAA kills broad-leaved weeds such as dandelions and daisies but grass is unaffected.

Commercial fruit growers spray plants with synthetic auxins to induce fruiting. It means that if, for instance, a tomato plant is sprayed, fruits will be produced without the need for pollination, and the tomatoes will be seedless. Auxin is also used to help initiate rooting of cuttings for plant propagation (Figure 8.23). 'Agent orange', used by the United States as a defoliant for the rainforest in South Vietnam during the Vietnam War (1959–1975), was a mixture of synthetic auxins.

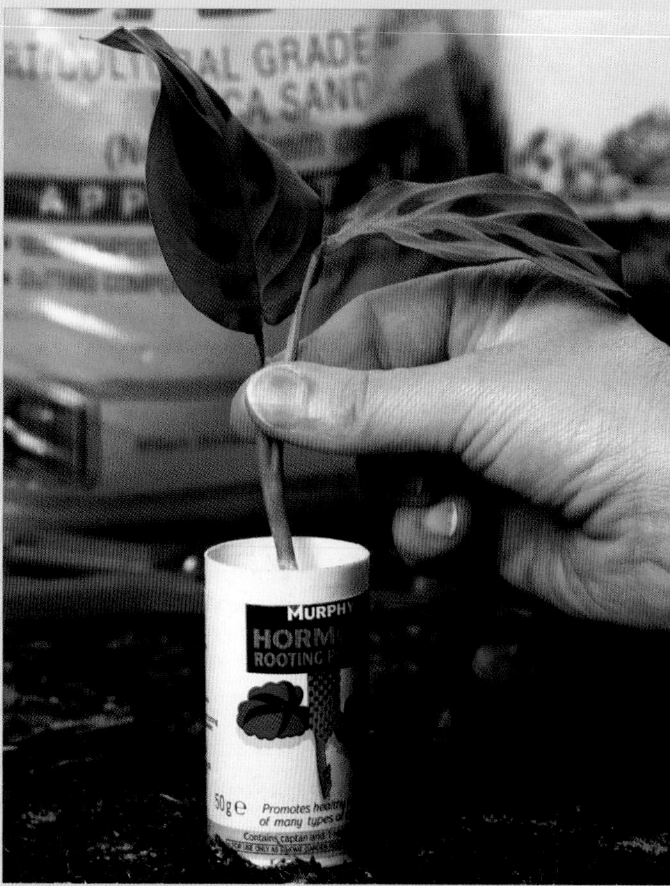

Figure 8.23 A plant cutting is dipped in rooting powder. This contains synthetic auxin, which will stimulate root development.

8.2 Reception of stimuli

How does light trigger nerve impulses?

When reflected light entered Kenge's eye, how was the light converted into electrical impulses that could be passed along the optic nerve to the brain?

Receptors

Stimuli (any changes that occur in an animal's environment) are detected by receptor cells that send electrical impulses to the central nervous system. Many receptors are spread throughout the body, but some types of receptor cells are grouped together into **sense organs**. Sense organs such as eyes help to protect the receptor cells and improve their efficiency; structures within the sense organ ensure that the receptor cells are able to receive the appropriate stimulus. The receptor cells that detect light are found in the eye. The lens and cornea refract (bend) the light so that it is focused on the retina where the photoreceptor cells are located.

Did you know?

Different types of receptors

Receptors allow us to perceive and respond to a wide variety of stimuli. The receptors can either be cells that synapse with a sensory neurone, or can themselves be part of a specialised sensory neurone, like the temperature receptors in the skin (see Figure 7.58). Four of the main types of receptor are shown in Table 8.4.

Type of receptor	Stimulated by	Examples of role in body
chemoreceptors	chemicals	taste, smell and regulation of chemical concentrations in the blood
mechanoreceptors	forces that stretch, compress or move the sensor	balance, touch and hearing
photoreceptors	light	sight
thermoreceptors	heat or cold	temperature control and awareness of changes in the surrounding temperature

Table 8.4 Four types of receptor and their roles.

All of the receptors, except the photoreceptors, work in a similar manner. At rest, the cell surface membrane has a negative resting potential. Stimulation of the receptor causes depolarisation of the cell. The stronger the stimulus, the greater the depolarisation. When depolarisation exceeds the threshold level, it triggers an action potential. This is either relayed across the synapse using neurotransmitters or passed directly down the axon of the sensory nerve.

Activity

Remind yourself about the structure and function of the different parts of the eye in **Activity 8.6. A8.06S**

Before addressing in detail the question 'How does light trigger nerve impulses?' remember what you've learned before about the way that the parts of the eye work together, using Figure 8.24 and the revision quiz in Activity 8.6.

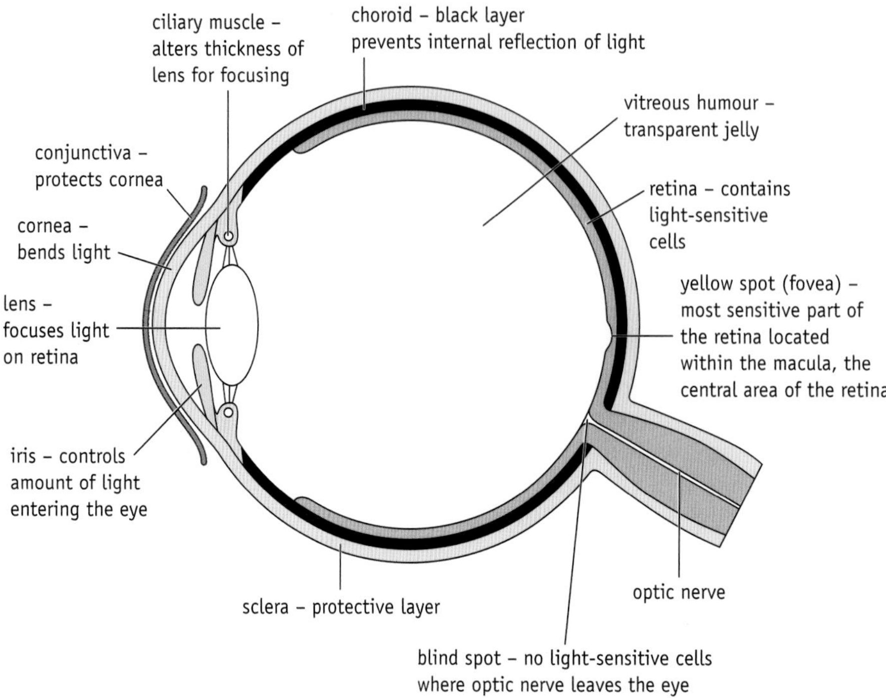

Figure 8.24 The structure of the eye.

Did you know?

The most common cause of blindness in the UK

The central part of the retina, known as the macula, receives light entering the eye from 'straight ahead'. The delicate cells of the macula sometimes become damaged causing progressive deterioration of sight. Age-related macular degeneration is the most common cause of blindness in the UK. The good news is that in 2008, a drug, Lucentis, was approved for treatment of the wet type of the condition. The drug binds to a protein growth factor, stopping the growth of new abnormal blood vessels under the retina that leak fluid and blood. Unfortunately only 10% of cases are the wet type. The more common dry type is caused by accumulation of fatty deposits beneath the retina which cause it to dry out.

Photoreceptors

The human retina contains two types of photoreceptor cells sensitive to light: **rods** and **cones** (see Figure 8.25). Cones allow colour vision in bright light; rods only give black and white vision but, unlike cones, work in dim light as well as in bright light. In the centre of the retina, in an area about the size of this 'o', there are only cones. This area allows us to pinpoint accurately the source and detail of what we are looking at. Over the remainder of the retina, rods outnumber cones by a factor of about 20 to 1.

In Figure 8.25 notice the arrangement of the three layers of cells that make up the retina. The rods and cones synapse with **bipolar neurone** cells, which in turn synapse with **ganglion neurones**, whose axons together make up the optic nerve. Light hitting the retina has to pass through the layers of neurones *before* reaching the rods and cones.

Extension

Read **Extension 8.3** to find out how the cones function to allow colour vision. **X8.03S**

Q8.21 Can you explain why some people describe the retina as *functionally* inside out?

How does light stimulate photoreceptor cells?

In both rods and cones, a photochemical pigment absorbs the light resulting in a chemical change. In rods the molecule is a purplish pigment called **rhodopsin**. In Figures 8.25 and 8.26 you can see that the rod cell has an outer and inner segment; these contain many layers of flattened vesicles. The rhodopsin molecules are located in the membranes of these vesicles.

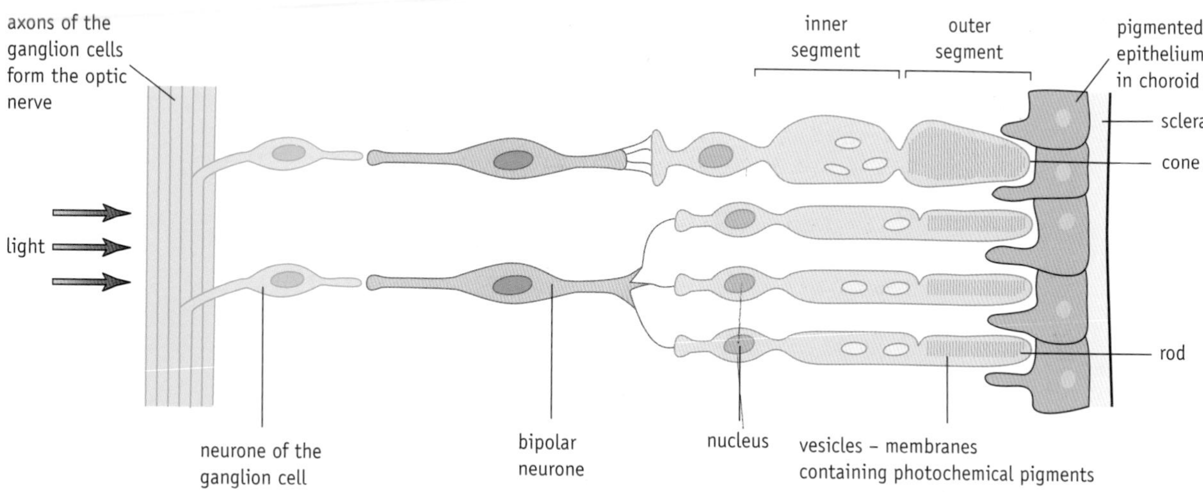

Figure 8.25 The structure of rods and cones within the retina.

Figure 8.26 An electron micrograph of rods shows the outer segments that contain the photochemical pigment. At the top right the choroid layer, normally attached to the retina, has peeled away. The rods synapse with the neurones on the left. Magnification × 1200.

Figure 8.27 summarises the processes that occur in the light and dark within the rod cells.

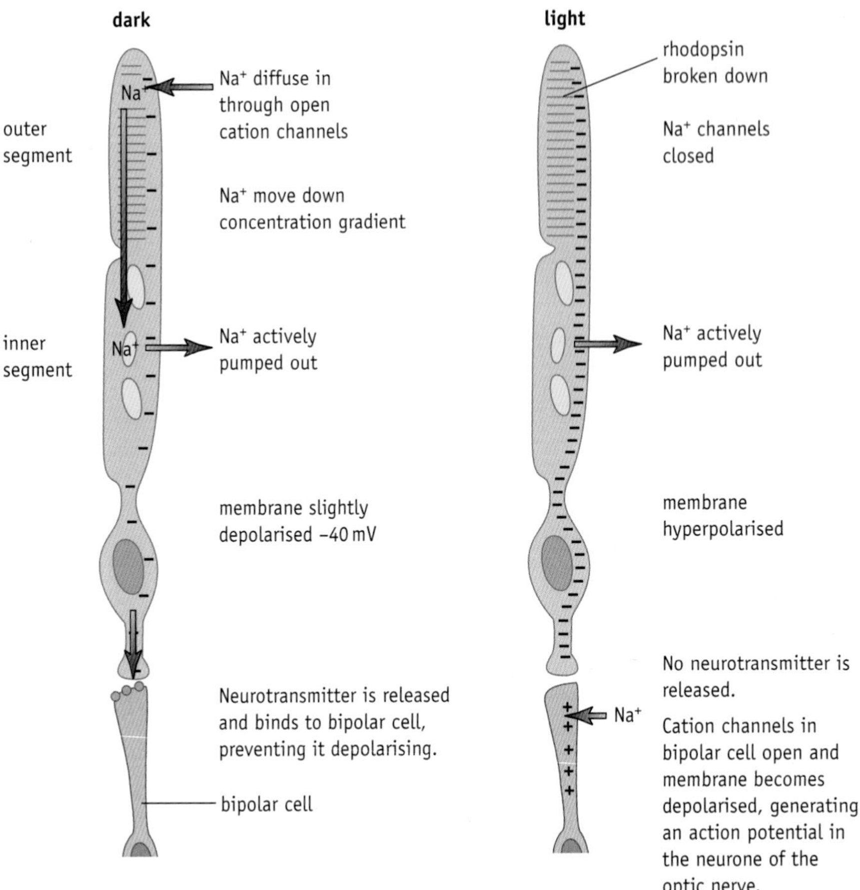

Figure 8.27 Rod cell in light and dark.

In the dark

In the dark, sodium ions flow into the outer segment through **non-specific cation channels**. The sodium ions move down the concentration gradient into the inner segment where pumps continuously transport them back out of the cell. The influx of Na^+ produces a slight depolarisation of the cell. The potential difference across the membrane is about –40 mV, compared to the –70 mV resting potential of a cell. This slight depolarisation triggers the release of a neurotransmitter, thought to be **glutamate**, from the rod cells. In the dark, rods release this neurotransmitter continuously. The neurotransmitter binds to the bipolar cell, stopping it depolarising.

In the light

When light falls on the rhodopsin molecule, it breaks down into retinal and opsin, non-protein and protein components. The opsin activates a series of membrane-bound reactions, ending in hydrolysis of a molecule attached to the cation channel in the outer segment. The breakdown of this molecule results in the closing of the cation channels. The influx of Na^+ into the rod decreases, while the inner segment continues to pump Na^+ out. This makes the inside of the cell more negative. It becomes hyperpolarised, and the release of the glutamate neurotransmitter stops.

The lack of glutamate results in depolarisation of the bipolar cell with which the rod synapses. The neurones that make up the optic nerve are also depolarised and respond by producing an action potential.

Checkpoint

8.6 Produce a series of statements that describe what happens in rod cells to enable light to generate an action potential in a bipolar cell. Get a fellow biology student to order the statements as a revision exercise.

Q8.22 **a** By what form of transport will sodium ions:

i be pumped out of the rod cell **ii** flow back into the cell?

b What do you think a 'non-specific cation channel' is?

c Why does the rod cell membrane become hyperpolarised in the light?

Once the rhodopsin has been broken down, it is essential that it be rapidly converted back to its original form so that subsequent stimuli can be perceived. Each individual rhodopsin molecule takes a few minutes to do this. The higher the light intensity, the more rhodopsin molecules are broken down and the longer it can take for all the rhodopsin to reform, up to a maximum of 50 minutes. This reforming of rhodopsin is called **dark adaptation**.

Activity

Try **Activity 8.7** to experience the effects of dark adaptation, and think through how the rod cells work. **A8.07S**

Did you know?

Why you should eat your carrots

Have you ever been told to finish your carrots so that you will be able to see in the dark? And did you believe it? Like many 'old wives' tales' there is a grain of truth in this one because carrots are a good source of vitamin A.

Poor night vision, sometimes called night blindness, has been known for many years to be one of the symptoms of the disease caused by a shortage of vitamin A in the diet.

Retinal, a derivative of vitamin A, is part of the rhodopsin found in the rods. A shortage of vitamin A leads to a lack of retinal and thus rhodopsin, which means poor vision in low light conditions.

Plants can also detect and respond to environmental cues

Plants, like all living things, detect and respond to stimuli (changes in their environment), adapting their growth and development to ensure their survival and reproductive success. Light is the most important environmental cue for plants, influencing many events in their growth and development.

Like animals, plants detect the quantity, direction and wavelength of light using photoreceptors, and respond to the changes in light conditions. Unlike animals, all messages in plants are chemical, and so all their responses are slower.

Plants contain several families of photoreceptors. We will focus on the most extensively studied photoreceptors, the **phytochromes**, which absorb red and far-red light. Five different phytochromes have been identified.

Phytochromes – plant photoreceptors

A phytochrome molecule consists of a protein component bonded to a non-protein light-absorbing pigment molecule. The five phytochromes differ in their protein component. The non-protein component exists in two forms, which are different isomers:

- $\mathbf{P_r}$ – phytochrome red; absorbs red light (660 nm)
- $\mathbf{P_{fr}}$ – phytochrome far-red; absorbs far-red light (730 nm).

These two isomers are photoreversible. Plants synthesise phytochromes in the P_r form; absorption of *red* light converts P_r into P_{fr}. Absorption of far-red light converts P_{fr} back into P_r. In sunlight P_r is converted into P_{fr}, and P_{fr} into P_r. The former reaction dominates in sunlight because more red than far-red light is absorbed. So P_{fr} accumulates in the light. In the dark, any P_{fr} present is slowly converted to P_r (see Figure 8.28).

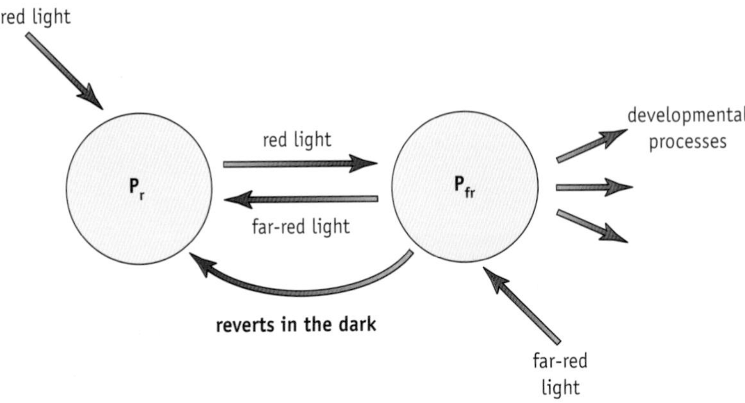

Figure 8.28 Phytochromes exist in two interconvertible forms, P_r and P_{fr}.

Q8.23 Suggest structural and functional similarities between phytochromes and the molecule rhodopsin found in the retina.

A wide range of responses are regulated by phytochromes, including seed germination, stem elongation, leaf expansion, chlorophyll formation and flowering.

Phytochromes trigger germination

Phytochromes were discovered through germination experiments. Gardeners have long known that a number of tiny seeds such as lettuce, coleus and mint require light to germinate. They have thin seed coats and few food reserves, and only germinate in optimum conditions, which includes being close enough to the soil surface. In the dark, they fail to germinate.

Experiments with lettuce indicate that a flash of red light will trigger germination, but if followed by a flash of far-red light, germination is inhibited. A further flash of red light again promotes germination, while a second flash of far-red once again inhibits germination. This suggests that the effects of red light and far-red light are reversible. The final flash of light determines whether germination occurs (Figure 8.29). Red light is particularly effective at triggering germination. Far-red light seems to inhibit germination.

Figure 8.29 Red light triggers germination, far-red light inhibits germination.

Before you read on, answer the question below to see if you can explain the effect of light on lettuce seed germination.

Q8.24 **a** Bearing in mind that red light stimulates germination of lettuce seeds and far-red light inhibits it, suggest which isomer of phytochrome, P_r or P_{fr}, needs to be present to stimulate germination.

b Describe the mechanism that prevents lettuce seeds from germinating in the dark.

When lettuce seeds are exposed to red light, P_r is converted to P_{fr}, stimulating responses that lead to germination. In lettuce seeds kept in the dark, *no P_r is converted to P_{fr}*. The seeds do not germinate because it is the appearance of P_{fr} that triggers germination. When exposed to far-red light, P_{fr} is converted back to P_r, inhibiting germination.

Activity

In **Activity 8.8** you can carry out your own investigation with coleus or lettuce seeds. **A8.08S**

Photoperiods, flowering and phytochromes

Plant species flower at a particular time of year (Figure 8.30), but what tells a plant when to flower? Why do chrysanthemums flower in autumn while strawberries flower in summer? The **photoperiod**, the relative length of day and night, is the environmental cue that determines time of flowering. The ratio of P_r to P_{fr} in a plant enables it to determine the length of day and night.

Long winter nights give ample time for P_{fr} to convert back to P_r, so that by sunrise all phytochrome will be P_r. Summer nights, on the other hand, may not be long enough to do so, so some P_{fr} may still be present in the morning.

Long-day plants only flower when day length exceeds a *critical value*. These plants, such as strawberries, oats, poppies and lettuce, flower when the period of uninterrupted darkness is *less than* (typically) 12 hours; they need P_{fr} to stimulate flowering.

Figure 8.30 Chrysanthemum is a typical short-day plant, flowering in the autumn. Strawberry is a long-day plant, flowering in the long days of summer.

Q8.25 Explain why a long-day plant can successfully flower in the summer.

Short-day plants like chrysanthemums and poinsettias tend to flower in spring or autumn when the period of uninterrupted darkness is *greater than* 12 hours. They need long hours of darkness in order to convert all P_{fr} present at sundown back to P_r. P_{fr} inhibits flowering in short-day plants. In most short-day plants, a flash of red light in the middle of the dark period negates the effect of the dark period.

Q8.26 Look at Figure 8.31.

a Work out the day length requirements for short-day and long-day plants.

b Explain why it might be better to call them long-night and short-night plants.

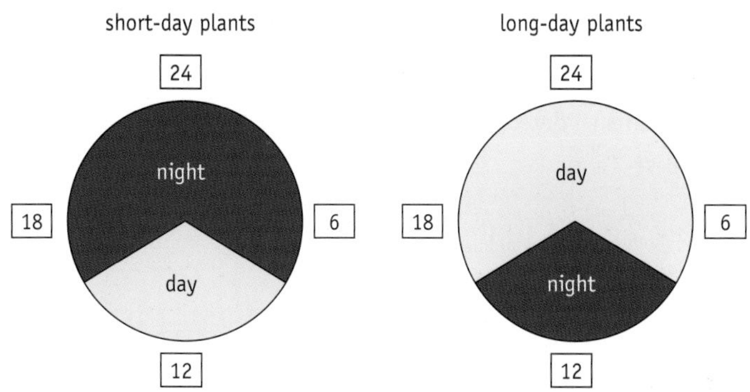

Figure 8.31 The pattern of light and dark throughout 24 hours required for flowering of a short-day plant (left) and a long-day plant (right).

Q8.27 A gardener adjusts the timer for lighting and heating in his greenhouse to give his prizewinning chrysanthemum plants longer and warmer days, thinking that this will make them produce bigger and better flowers. What would you say to the gardener to explain why his plants have failed to produce flowers?

Q8.28 Geraniums are said to be day-neutral. Suggest what this means with respect to flowering.

Activity

In **Activity 8.9** you can consider how light affects flowering. **A8.9S**

Phytochrome and greening

Once a shoot has broken through soil into sunlight, the plant undergoes profound changes in both its form and biochemistry (Figure 8.32). We call these changes **greening**. Once in the light, phytochromes promote the development of primary leaves, leaf unrolling and the production of pigments. They can also inhibit certain processes such as elongation of internodes (Figure 8.32).

Q8.29 **a** In Figure 8.32, the plant on the left has been grown in the dark. The plant on the right has been grown in the light. Describe the differences between the two plants.

b Suggest how the features you have described aid the survival of seedlings as they emerge.

Q8.30 Leaves absorb more red light than far-red light. Many species exhibit a shade avoidance response if their leaves are shaded by plants and thus exposed to high levels of far-red light.

a Explain how this response might be initiated by phytochromes.

b Suggest what type of response the plant might make to avoid the shade.

Figure 8.32 Two seedlings. The one on the right was grown in full light and shows the changes associated with 'greening'. The one on the left shows the typical features of etiolation, the result of being deprived of light.

How do phytochromes switch processes on or off?

Exposure to light causes phytochrome molecules to change from one form to another, bringing about a change in shape. It is thought that each activated phytochrome then interacts with other proteins (Figure 8.33); the phytochromes may bind to the protein or disrupt the binding of a protein complex. These signal proteins may act as transcription factors or activate transcription factors that bind to DNA to allow transcription of light-regulated genes. The transcription and translation of proteins result in the plant's response to light. For example, in seedlings, synthesis of the enzymes that control chlorophyll production will result in greening of the shoot.

Other photoreceptors

Scientists working with mutant *Arabidopsis* plants (a member of the cabbage family) have discovered at least three pigments used by plants to detect blue light, including phototropins that determine phototropic responses.

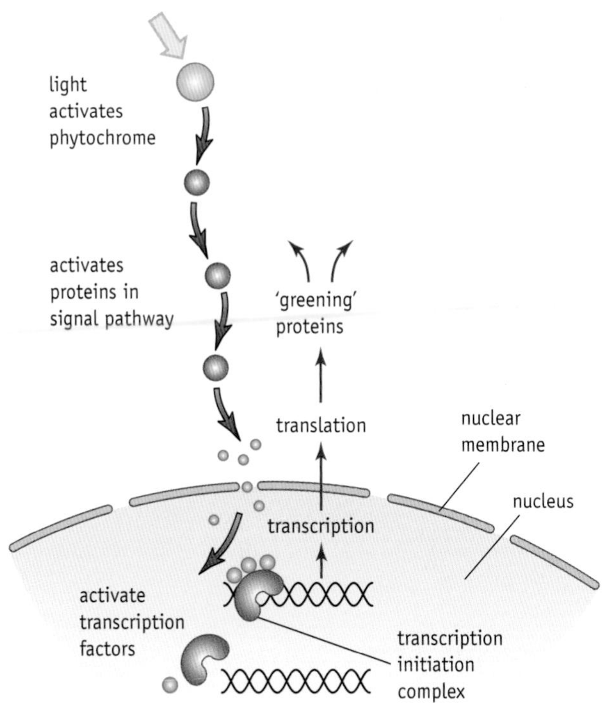

Figure 8.33 The action of phytochromes in greening.

Plants detect other environmental cues

Gravity

More than a short distance under the soil surface, light cannot be the cue for the shoot to grow upwards and the root to grow downwards. The stimulus for this is gravity and the response ensures that developing shoots reach the light while roots grow in the soil.

Touch and mechanical stress

Some plants are sensitive to touch and mechanical stress (Figure 8.34). Rubbing plant stems can result in shorter stems than in controls. It is thought that the mechanical stimulus activates signal molecules whose end result is the activation of genes that control growth.

Some plants have leaves that move rapidly in response to mechanical stimulation. When a leaf of *Mimosa pudica* is touched, it folds rapidly then collapses. The mechanism is that when touched, specialised cells lose potassium ions. Water follows by osmosis and the cells become flaccid, so no longer support the leaf and keep it upright.

Figure 8.34 Climbing plants such as vines and peas have tendrils that twist around supports: contact triggers the directional growth response.

8.3 The brain

The brain acts as the main coordinating centre for nervous activity, receiving information from sense organs, interpreting it, and then transmitting information to effectors. Different regions within the brain are involved in helping us respond to our external environment and in regulating our internal environment (see Figures 8.36 and 8.37). Which parts of the brain did Kenge use to interpret the view of the buffalo? At the end of this section you should know.

The cerebral hemispheres

Looking at the brain from the top down (Figure 8.35), you see the **cortex**. It is grey and highly folded, composed mainly of nerve cell bodies, synapses and dendrites. This outer layer of the brain is known as the **grey matter**.

The cortex, accounting for about two-thirds of the brain's mass, is the largest region of the brain. It is positioned over and around most other brain regions, and is divided into left and right **cerebral hemispheres**. Each hemisphere is composed of four regions called lobes (Figure 8.36):

- **frontal lobe**
- **parietal lobe**
- **occipital lobe**
- **temporal lobe**.

Activity

Try the interactive tutorial in **Activity 8.10** to work out the function of different areas of the brain. **A8.10S**

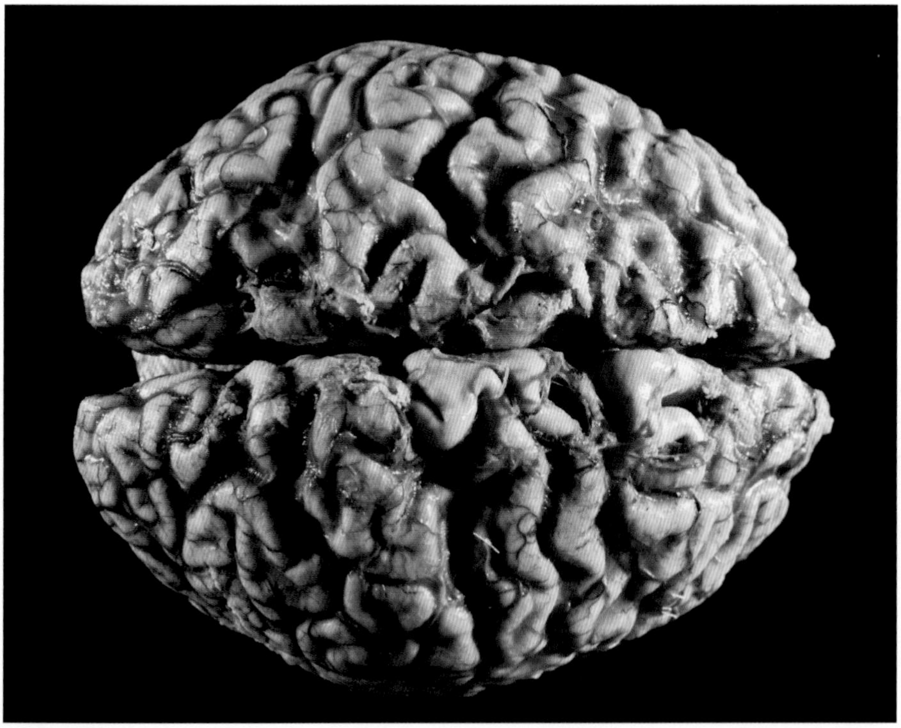

Figure 8.35 The two cerebral hemispheres can easily be seen from above.

Each lobe interprets and manages its own sensory inputs. The two cerebral hemispheres are connected by a broad band of **white matter** (nerve axons) called the corpus callosum.

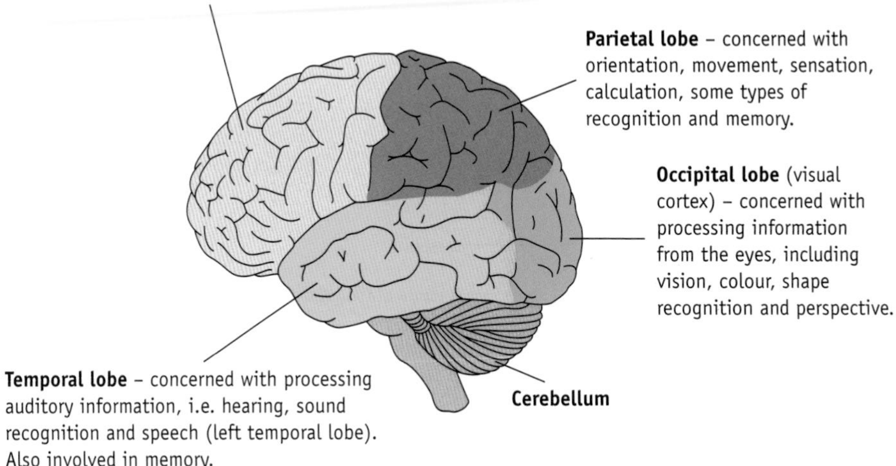

Frontal lobe (also referred to as the higher centres of the brain) – concerned with the higher brain functions such as decision making, reasoning, planning and consciousness of emotions. It is also concerned with forming associations (by combining information from the rest of the cortex) and with ideas. It includes the primary motor cortex which has neurones that connect directly to the spinal cord and brain stem and from there to the muscles. It sends information to the body via the motor neurones to carry out movements. The motor cortex also stores information about how to carry out different movements.
Parietal lobe – concerned with orientation, movement, sensation, calculation, some types of recognition and memory.
Occipital lobe (visual cortex) – concerned with processing information from the eyes, including vision, colour, shape recognition and perspective.
Temporal lobe – concerned with processing auditory information, i.e. hearing, sound recognition and speech (left temporal lobe). Also involved in memory.
Cerebellum

Figure 8.36 The regions of the cerebral hemispheres and their functions. The location of the cerebellum is also shown.

Q8.31 A blow to the back of your head may result in you seeing stars. Suggest why.

The structures lying directly below the corpus callosum include, among others, the thalamus, the hypothalamus and the hippocampus.

- The **thalamus** is responsible for routing all the incoming sensory information to the correct part of the brain, via the axons of the white matter.
- The **hypothalamus** lies below the thalamus and contains the thermoregulatory centre. This monitors core body temperature and skin temperature, and initiates corrective action to restore the body to its optimum temperature (see Topic 7, Section 7.4). Also located in the hypothalamus are other centres that control sleep, thirst and hunger. The hypothalamus also acts as an endocrine gland, secreting hormones such as antidiuretic hormone (this controls water reabsorption in the kidneys and hence controls blood concentration). The hypothalamus connects directly to the pituitary gland, which in turn secretes other hormones. See the Key biological principle box on hormonal coordination (page 214).
- The **hippocampus** is involved in laying down long-term memory.

The **basal ganglia** are a collection of neurones that lie deep within each hemisphere and are responsible for selecting and initiating stored programmes for movement.

The cerebellum and brain stem

The brain stem is, in evolutionary terms, the oldest part of the brain and is sometimes referred to as the reptilian brain. It lies at the top of the spinal column. The brain stem extends from the midbrain to the **medulla oblongata** (Figure 8.37).

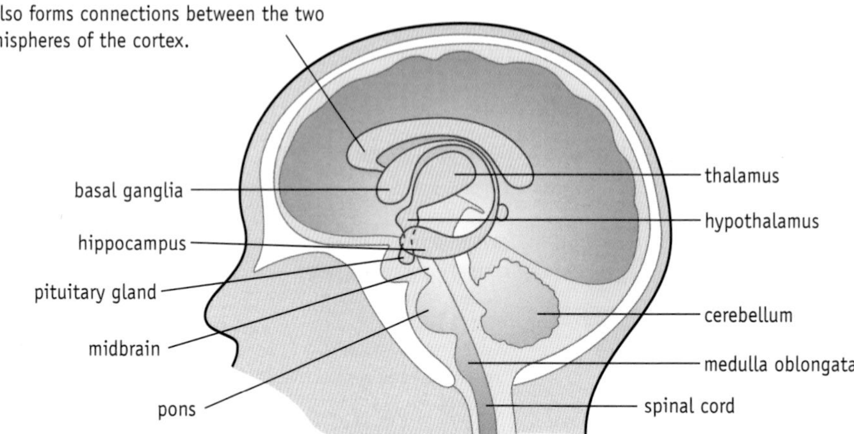
Corpus callosum – white matter composed mainly of axons, whose white myelin sheaths give it its characteristic appearance. It provides connections between the cortex and the brain structures below. It also forms connections between the two hemispheres of the cortex.
basal ganglia
hippocampus
pituitary gland
midbrain
pons
thalamus
hypothalamus
cerebellum
medulla oblongata
spinal cord

Figure 8.37 Structures below the corpus callosum.

Study Table 8.5, which provides information about the role of the lower regions in the brain, and then try answering the questions that follow.

cerebellum	• responsible for balance • coordinates movement as it is being carried out, receiving information from the primary motor cortex, muscles and joints • constantly checks whether the motor programme being used is the correct one, for example by referring to incoming information about posture and external circumstances
midbrain	relays information to the cerebral hemispheres, including auditory information to the temporal lobe, and visual information to the occipital lobe
medulla oblongata	regulates those body processes that we do not consciously control, such as heart rate, breathing and blood pressure

Table 8.5 The cerebellum and brain stem functions.

Q8.32 Imagine that you are whizzing downhill on a bike (Figure 8.38) and come across an unexpected sharp bend in the road. You need to apply the brakes or turn the handlebars to stop yourself falling off. Which regions of the brain, including those in the brain stem, are involved in your subsequent action?

Q8.33 Various diseases or conditions can give us insights into the functioning of the different areas of the brain. Research has shown that in Parkinson's disease neurones in a particular area of the brain have died. Parkinson's disease results in an inability to select and make appropriate movements. Suggest which lobe of the brain is damaged.

Q8.34 Suggest which parts of the brain were involved in Kenge seeing and interpreting the view of the buffalo.

Figure 8.38 Aaargh!

Discovering the function of each brain region

How do neuroscientists identify the different regions of the brain and know what each does? Until relatively recently, neuroscientists were only able to study the brain by looking at pathological specimens, by examining the effect of damage to particular areas of the brain, through studies using animal models (see Section 8.4) and, to some degree, by studying human patients during surgery.

Individuals with brain damage still provide valuable information in the study of the brain. However, neuroscientists now have a wide range of non-invasive imaging techniques for studying the function of the living brain.

Studies of individuals with damaged brain regions

By studying the consequences of accidental brain damage it is possible to determine the functions of certain regions of the brain, as the examples that follow illustrate. Researchers have also studied the consequences of injuring or destroying neurones to produce lesions (areas of tissue destruction) in non-human animal 'models', and the consequences of the removal of brain tissue.

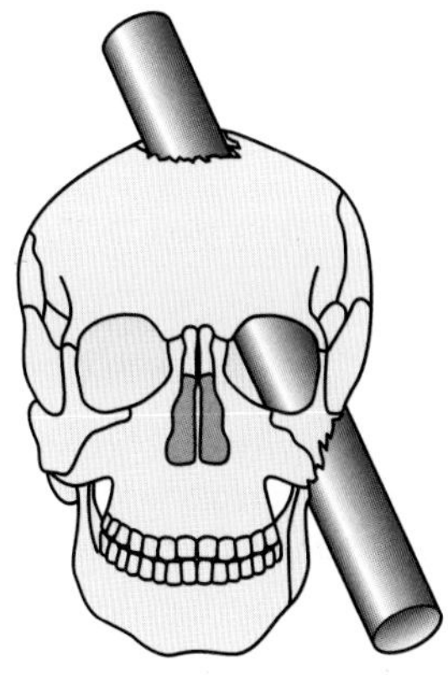

Figure 8.39 The iron bar travelled behind Gage's left eye and flew out through the top of his skull.

The story of Phineas Gage

Phineas Gage was the foreman of a railway construction company; a hard-working, fit, popular and responsible man. One day in 1848 he was working with dynamite when an explosion propelled a three-and-a-half-foot long iron bar through his head (Figure 8.39). Amazingly Gage didn't die and, although most of the front part of the left-hand side of his brain was destroyed, he could still walk and talk.

But after the accident Gage's personality changed: he became nasty, foul-mouthed and irresponsible. He was also impatient and obstinate and was unable to complete any plans for future action.

Phineas Gage died twelve years later. Researchers at the Harvard University Medical School have since combined photographs and X-rays of Gage's skull with computer graphics to determine the areas of his brain that were damaged by the bar. It is highly probable that the accident severed connections between his midbrain and frontal lobes. Gage's reduced ability to control his emotional behaviour after the accident was related to damage at this site.

Weblink

Listen to Lincoln Holmes on the BBC News website. Damage to an area of his temporal lobe left him 'face-blind'.

The strange case of Lincoln Holmes

Imagine what it would be like to never be able to put a name to a face. That's what it's like for Lincoln Holmes. He finds recognising a face impossible. Thirty years ago a car accident left him with damage to an isolated part of his temporal lobe and he is now 'face-blind'. Even when shown a photograph of himself, he has to be prompted before he realises he is staring at his own image. Lincoln can see facial features, but they all appear as a jumble and he is unable to put all the component parts together. Lincoln's case has revealed that recognition of faces is at least partly carried out by a specific face recognition unit that is in the temporal lobe.

Activity

In **Activity 8.11** you can identify regions of the brain by considering symptoms that occur after damage to that area. **A8.11S**

Q8.35 Physical damage is one obvious cause of brain damage. Can you think of any others?

The effects of strokes

Brain damage caused by a stroke (Figure 8.40) can cause problems with speaking, understanding speech, reading and writing. In the nineteenth century Paul Broca studied several post-mortems of patients who could not speak due to strokes. He concluded that lesions in a small cortical area in the left frontal lobe (subsequently known as Broca's area) were responsible for deficits in language production.

Some patients can recover some abilities after a stroke, showing the potential of neurones to change in structure and function. This is known as **neural plasticity**. The structure of the brain remains flexible even in later life and can respond to changes in the environment. Brain structure and functioning is affected by both nature and nurture.

Activity

In **Activity 8.12** you use brain images to identify the different regions of the brain and their functions. Visit some of the websites associated with the activity to see lots of CT and MRI scans. **A8.12S**

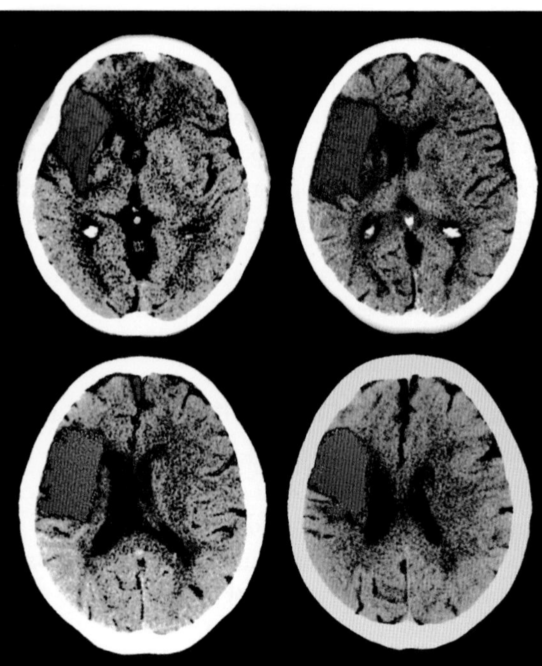

Figure 8.40 Four brain CT images from a patient who has had a stroke. A massive brain lesion in the left hemisphere affects the language areas. Surrounding areas are still intact, allowing the patient to sing but not to speak.

Brain imaging

CT scans

Computerised Axial Tomography (**CT** or **CAT**) imaging was developed in the 1970s to overcome the limitations of X-rays. Standard broad-beam X-rays cannot be used for imaging soft tissue as they are only absorbed by denser materials such as bone.

CT scans use thousands of narrow-beam X-rays rotated around the patient to pass through the tissue from different angles. Each narrow beam is attenuated (reduced in strength) according to the density of the tissue in its path. The X-rays are detected and are used to produce an image of a thin slice of the brain on a computer screen in which the different soft tissues within the brain can be distinguished (Figure 8.40).

CT scans give only 'frozen moment' pictures. They look at structures in the brain rather than at functions, and are used to detect brain disease and to monitor the tissues of the brain over the course of an illness. They have only limited resolution so small structures in the brain cannot be distinguished.

Techniques that do not rely on harmful X-rays and can therefore be used more frequently have been developed, including magnetic resonance imaging.

Magnetic resonance imaging

Magnetic Resonance Imaging (MRI) uses a magnetic field and radio waves to detect soft tissues. When placed in a magnetic field the nuclei of atoms line up with the direction of the magnetic field, in much the same way as a compass needle aligns itself to the Earth's magnetic field. Hydrogen atoms in water are monitored in MRI imaging because there is such a high water content in the tissues under investigation and hydrogen has a strong tendency to line up with the magnetic field.

In an MRI scanner the magnetic field runs down the centre of the tube in which the patient lies. Another magnetic field is superimposed on this, which comes from the magnetic component of high frequency radio waves. The combined fields cause the direction (axis) and frequency of spin of the hydrogen nuclei to change, taking energy from the radio waves to do so. When the radio waves are turned off, the hydrogen nuclei return to their original alignment and release the energy they absorbed. This energy is detected and a signal is sent to a computer, which analyses it to produce an image on the screen (Figure 8.41).

Different tissues respond differently to the magnetic field from the radio waves and so produce contrasting signals and distinct regions in the image. MRI examines tissues in small sections, normally thin 'slices', which when put together give three-dimensional images.

Nowadays MRI is widely used in the diagnosis of tumours, strokes, brain injuries and infections of the brain and spine. MRI can be used to produce finely detailed images of brain structures, as shown in Figure 8.41, with better resolution than CT scans for the brain stem and spinal cord. In Topic 1, MRI scans helped to diagnose Mark's stroke.

Extension

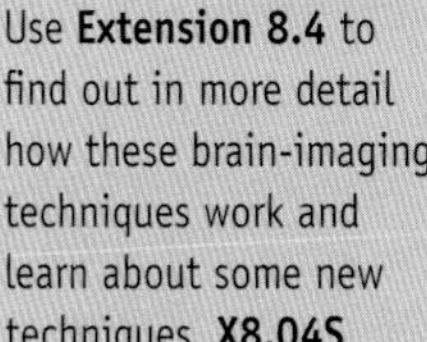

Use **Extension 8.4** to find out in more detail how these brain-imaging techniques work and learn about some new techniques. **X8.04S**

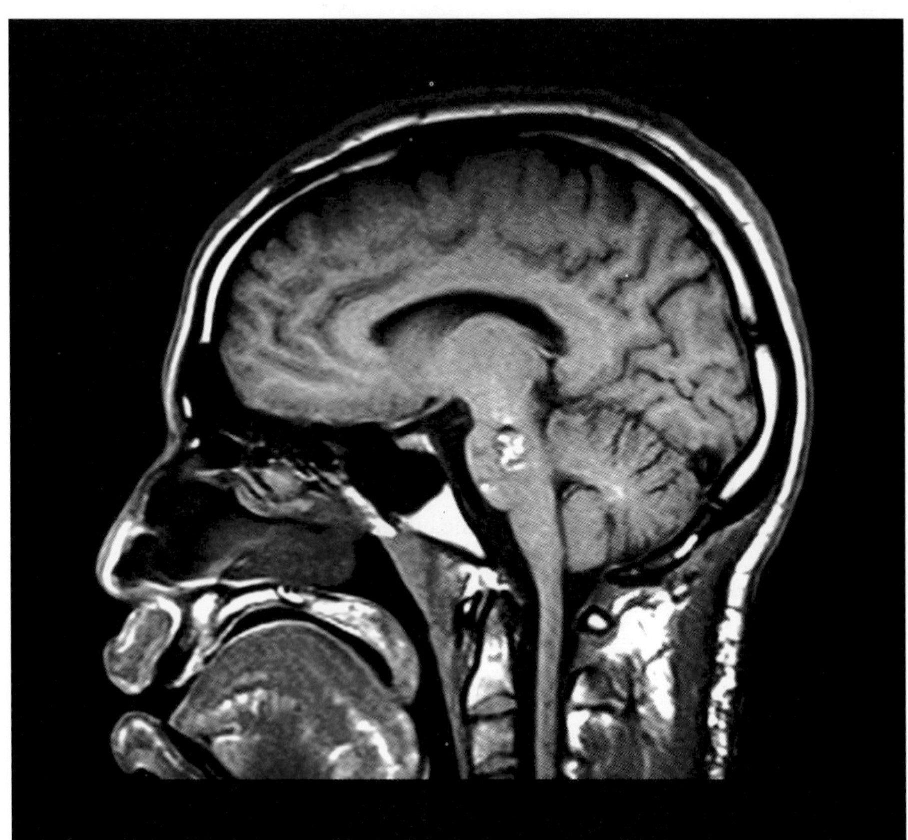

Figure 8.41 A typical MRI image showing good resolution of soft tissues.

Functional magnetic resonance imaging

Functional Magnetic Resonance Imaging (**fMRI**) is a particularly useful and exciting tool for the neuroscientist, as it can also provide information about the brain in action. Using this technique it is possible to study human activities, such as memory, emotion, language and consciousness.

fMRI is used to look at the functions of the different areas of the brain by following the uptake of oxygen in active brain areas. This is possible as deoxyhaemoglobin absorbs the radio wave signal, whereas oxyhaemoglobin does not. Increased neural activity in a brain area results in an increased demand for oxygen, and hence an increase in blood flow. Although there is a slight increase in oxygen absorption from the blood, overall there is a large increase in oxyhaemoglobin levels in the enhanced blood flow so less signal is absorbed. The less radio signal there is absorbed, the higher the level of activity in a particular area, so different areas of the brain will 'light up' according to when they are active (Figure 8.42). You can see some fMRI images on the websites linked to Activity 8.12.

fMRI can produce up to four images per second, so the technique can be used to follow the sequence of events over quite short time periods. In a typical fMRI experiment, images are collected continually while the subject alternates between resting and carrying out some task, such as object recognition, listening or memorising number sequences.

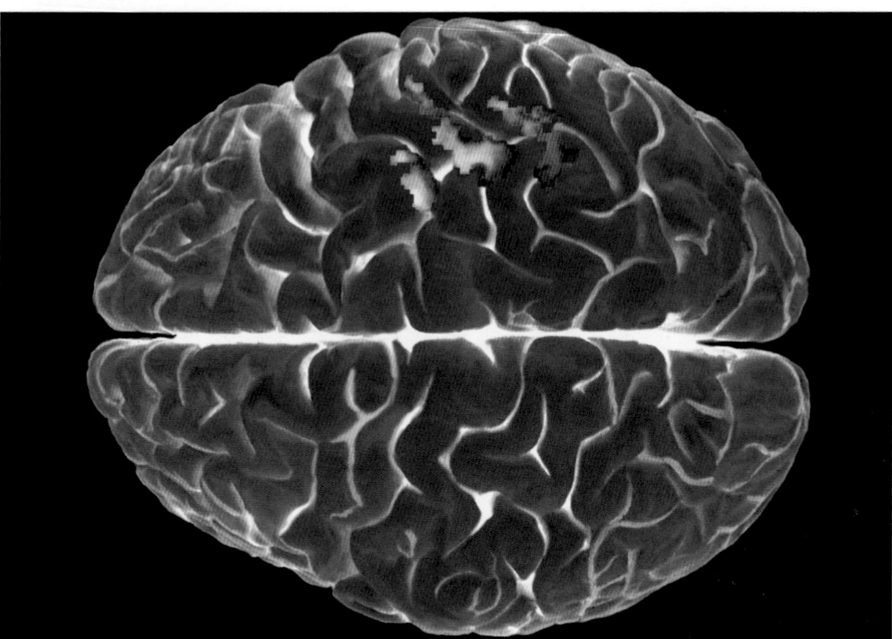

Figure 8.42 Active areas of the brain appear as coloured areas on an fMRI scan due to the high levels of oxyhaemoglobin present.

Q8.36 A neuroscientist conducts an fMRI experiment to investigate brain activity when subjects perform voluntary actions like pressing a lever. Look back at the section on regions of the brain and decide which part of the brain you would expect to be active when the lever is being pressed.

Q8.37 A study of a group of London taxi drivers showed that the right hippocampus is involved in recalling a well-developed mental map of London. A second study examined brain scans of 16 London taxi drivers and found that the only areas of their brains that were different from 50 control subjects were the left and right hippocampus. A particular region, the posterior hippocampus, was found to be significantly larger in the taxi drivers, whilst the front of the hippocampus was smaller than in the control subjects.

a What imaging method could have been used in: **i** the first and **ii** the second investigation?

b What do you think the scientists were able to infer from their results?

From the eye to the brain

Which parts of the brain were involved in Kenge seeing and interpreting what he saw? The axons of the ganglion cells that make up the optic nerve pass out of the eye and extend to several areas of the brain, including a part of the thalamus as shown in Figure 8.43. The impulses are then sent along further neurones to the primary visual cortex where the information is processed further.

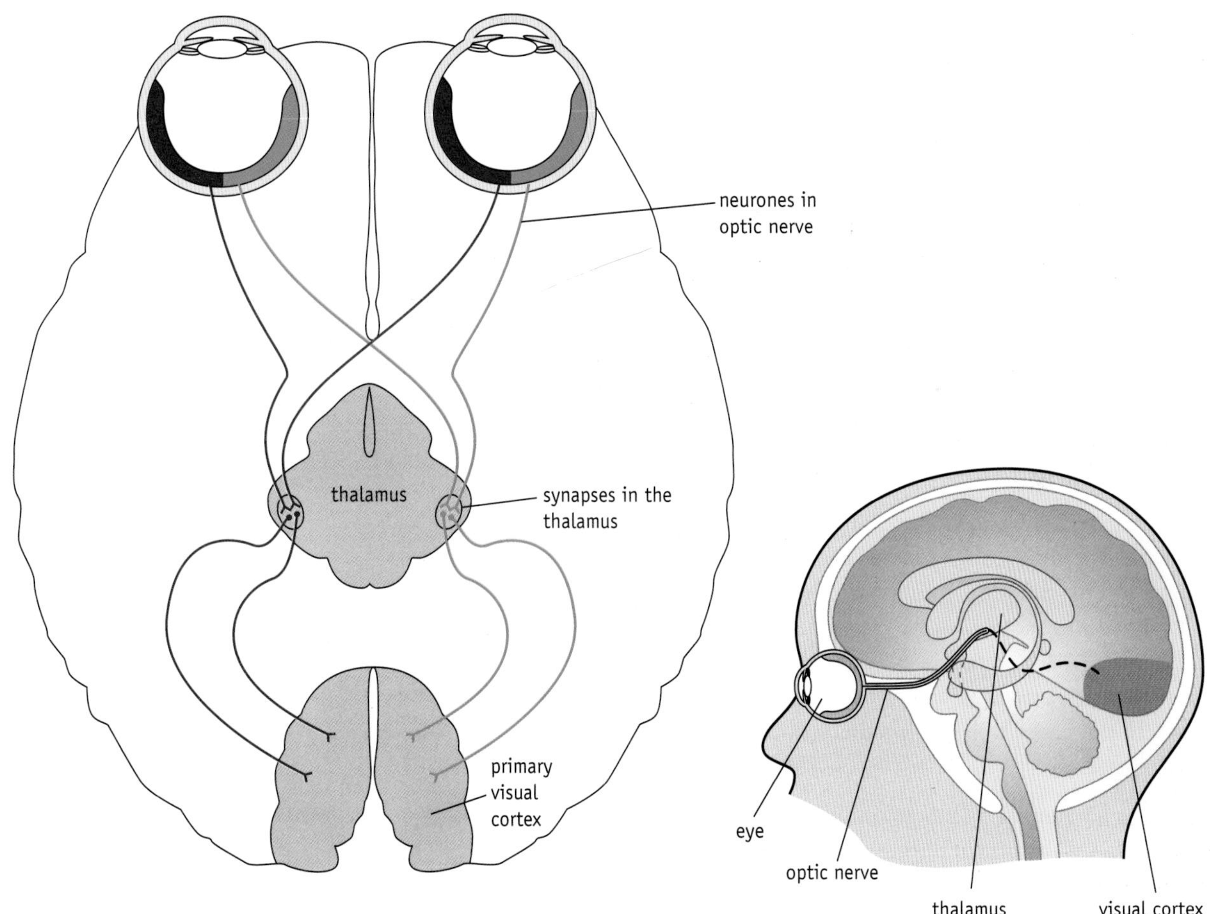

Figure 8.43 The visual pathway. The right side of the brain interprets input from the right side of both eyes, that is, the left side of the field of view. The left side of the brain interprets information from the left side of both retinas.

Before reaching the thalamus, some of the neurones in each optic nerve branch off to the midbrain, where they connect to motor neurones involved in controlling the pupil reflex (Figure 8.9) and movement of the eye. Audio signals also arrive at the midbrain so we can quickly turn our eyes in the direction of a visual or auditory stimulus.

8.4 Visual development

How does the visual cortex develop? What about the connections from the eye to the visual cortex? Was Kenge's visual development in some way faulty, causing him to see the buffalo as insects?

The human nervous system begins to develop soon after conception. By the 21st day, a neural tube has formed. The front part of the neural tube goes on to develop into the brain, while the rest of the neural tube develops into the spinal cord. The rate of brain growth during development is truly astonishing. At times, 250 000 neurones are added every minute!

A baby arrives in the world with about 100 000 million neurones. There is no huge increase in the number of brain cells after birth, though there is a large postnatal increase in brain size. This is caused by several factors, principally elongation of axons, myelination and the development of synapses. By six months after birth the brain will have grown to half its adult size. By the age of two years, the brain is about 80% of the adult size.

Once neurones have stopped dividing, the immature neurones migrate to their final position and start to 'wire themselves'. Axons lengthen and synapse with the cell bodies of other neurones. Neurones must make the correct connections in order for a function such as vision to work properly.

Axon growth

Axons of the neurones from the retina grow to the thalamus where they form synapses with neurones in the thalamus in a very ordered arrangement. Axons from these thalamus neurones grow towards the visual cortex in the occipital lobe.

Staining techniques and studies using electrical stimulation show that the visual cortex is made of columns of cells. Axons from the thalamus synapse within these columns of cells. Adjacent columns of cells receive stimulation from the same area of the retina in the right and left eye, with the pattern repeated across the visual cortex (see Figure 8.44). In this way a map of the retina is created within the visual cortex.

It used to be thought that these columns of cells in the visual cortex were formed during a **critical period** for visual development after birth, the result of nurture rather than nature. This is now known *not* to be the case. Working with ferrets, Crowley and Katz showed by injecting labelled tracers that the columns in the visual cortex are formed before the critical period for development of vision (Figure 8.45). Columns are also seen in newborn monkeys, suggesting that their formation is genetically determined and not the result of environmental stimulation.

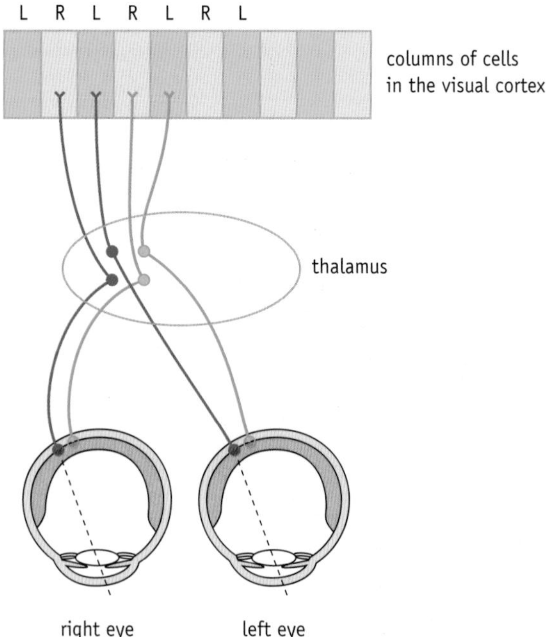

Figure 8.44 The relationship between the cells in each retina and the cells in the visual cortex.

However, periods of time during postnatal development have been identified when the nervous system must obtain specific experiences to develop properly. These are known as critical periods, **critical windows** or sensitive periods.

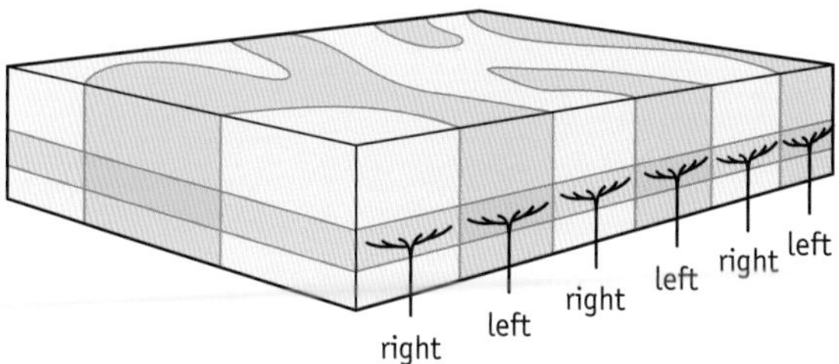

Figure 8.45 Radioactive label moves from one eye and is concentrated into distinct bands in the visual cortex, showing the columns of cells that receive input from that eye. These banding patterns have been observed in animals that have received no visual stimulation.

Evidence for a critical period in visual development

The way in which the environment affects the wiring of the nervous system has been extensively studied using the visual system. The evidence for a critical period in visual development comes from several sources, including medical observations of people with eye conditions affecting sight and results of experiments using animals.

Medical observations

One well-known case is that of a young Italian boy who, as a baby, had a minor eye infection. As part of his treatment, his eyes were bandaged for just two weeks. When the bandage was removed he was left with permanently impaired vision.

Studies of people born with cataracts have also contributed to our understanding of critical periods in development. You may know that a cataract is the clouding of the lens of the eye (Figure 8.46); this affects the amount of light getting to the retina. If cataracts are not removed before the child is ten years old, they can result in permanent impairment of the person's ability to perceive shape or form, including difficulties in face recognition.

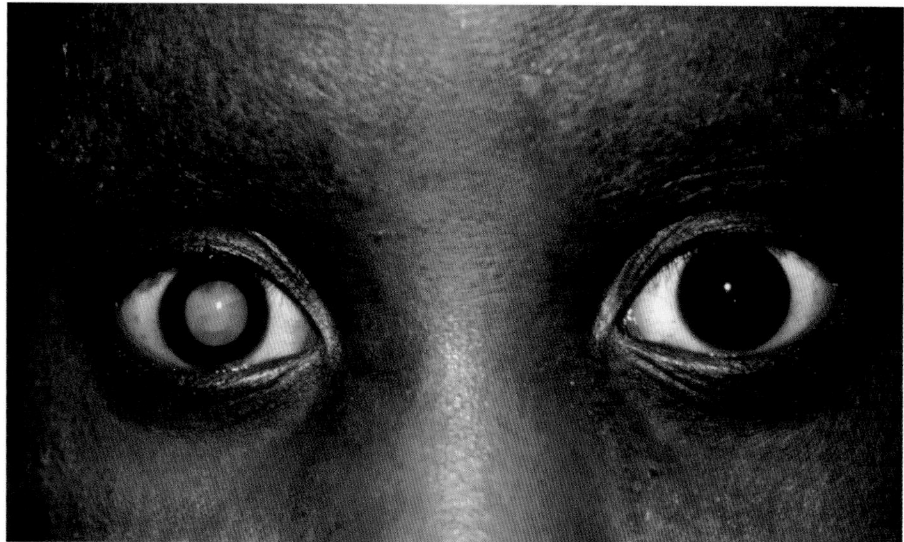

Figure 8.46 A cataract is the result of clouding of the lens.

By contrast, elderly people who develop cataracts later in life and have them for several years report normal vision if they are removed. This suggests that there is a specific time in development when it is crucial for a full range of light stimuli to enter the eye. Cataract removal in children is now carried out at a much earlier stage of development.

Research using animal models

Much of our knowledge and understanding of the visual system, and evidence for critical periods, comes from studies using animals. Animals are used extensively in the study of biological processes, including brain development and function. Because most research is conducted on just a few types of animals, a wealth of information is available about them. They are known as **animal models**.

Most animal models are easy to obtain, easy to breed, have short life cycles and a small adult size. In Topic 3 we saw fruit flies, *Drosophila*, being used for studying the links between genes and development. Nematode worms, chickens, mice, frogs and zebrafish are also used in this area of research. Mice are used extensively in the study of cancer and disease. In the study of visual development kittens and monkeys have been used, because of their similarity to humans.

Research using animals raises ethical issues – see Section 8.6.

Studies of newborn animals

In one study, one group of newborn monkeys was raised in the dark for the first three to six months of their lives, and another was exposed to light but not to patterns. When the monkeys were returned to the normal visual world, researchers found that *both* groups had difficulty with object discrimination and pattern recognition.

Q8.38 What does this suggest is required for visual development in these monkeys?

In a series of studies, Hubel and Wiesel raised monkeys from birth to six months, depriving them of any light stimulus in one eye. This is known as **monocular deprivation**. After six months the eye was exposed to light. On exposure to light it was clear that the monkey was blind in the light-deprived eye. Retinal cells in the deprived eye *did* respond to light stimuli, but the cells of the visual cortex did *not* respond to any visual input from the formerly deprived eye. Deprivation for only a single week during a certain period after birth produced the same result, with the deprived eye's visual cortex cells failing to respond to light. Deprivation in adults had no effect. Interestingly, visual deprivation of both eyes during this critical window has much less effect than when just one eye is deprived.

Q8.39 Hubel and Wiesel tested kittens for the effects of monocular deprivation at different stages of development and for different lengths of time. They found:

- deprivation at under three weeks had no effect
- deprivation after three months had no effect
- deprivation at four weeks had a major effect – even if the eye was closed for merely a few hours.

Bearing in mind that kittens are born blind, can you explain the results above?

Q8.40 Young doves and chaffinches raised without exposure to adult song will sing the adult song perfectly the first time they try it – the behaviour is inborn. Most other bird species need exposure to the adult song during a critical period in order to learn the proper song. Comment on the main influence on the development of the part of the brain associated with bird song: is it nature or nurture?

What is happening during the critical period for development of vision?

If the columns in the cortex are created before birth, what is happening during the critical period that can result in impaired vision if one eye is deprived of light? There must be further visual development.

At birth in monkeys there is a great deal of overlap between the territories of different axons (Figure 8.47A). In adults there is less overlap (Figure 8.47B), even though the mass of the brain is greater with more dendrites and synapses. Are these changes the result of visual experience or are they genetically determined?

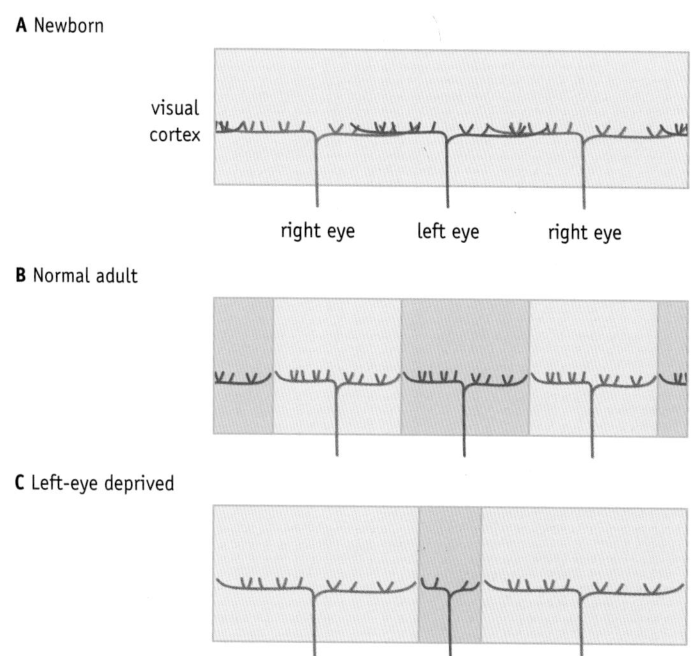

Figure 8.47 **A** Overlapping columns in the visual cortex are present at birth. **B** Refinement during the critical period produces the distinctive pattern of columns driven by the left and right eye. **C** Columns that receive input from a light-deprived eye become much narrower.

After light deprivation in one eye, columns with axons from the light-deprived eye are narrower than those for the eye receiving light stimulation (Figure 8.47C). Dendrites and synapses from the light-stimulated eye take up more territory in the visual cortex. This suggests that visual stimulation is required for the refinement of the columns, and so for full development of the visual cortex.

Axons compete for target cells in the visual cortex. Every time a neurone fires onto a target cell, the synapses of another neurone sharing the target cell are weakened, and they release less neurotransmitter. If this happens repeatedly, the synapses that are not firing will be cut back (Figure 8.48). When one eye is deprived of light, the axons from that eye will not be stimulated. In the area of the visual cortex where neurones from both eyes overlap, only the synapses on axons receiving light stimulation will fire. The synapses from the light-deprived eye will be weakened and eventually lost (Figure 8.47).

Throughout the nervous system, many more neurones are produced than are required, and numerous synapses and axons are pruned back. In the retina up to 80% of the original neurones die.

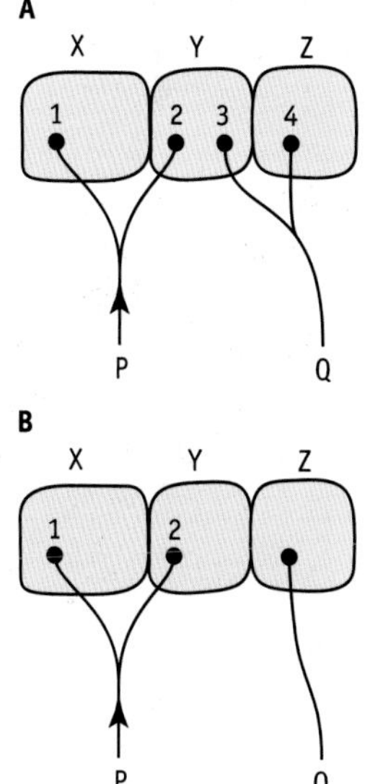

Figure 8.48 **A** Impulses passing along neurone P cause synapses 1 and 2 to fire. Synapse 3 releases less neurotransmitter and is eventually lost if there are many more impulses passing along P than Q. **B** Neurone Q no longer synapses with cell Y.

Q8.41 Rewrite these sentences in the correct order, to fit together the pieces of this jigsaw of ideas.

1 There is a lack of visual stimulation in one eye.
2 Inactive synapses are eliminated.
3 Axons from the non-deprived eye pass impulses to cells in the visual cortex.
4 Synapses made by active axons are strengthened.
5 Axons from the visually deprived eye do not pass impulses to cells in the visual cortex.

Activity

In **Activity 8.13** you can look at the evidence for a critical period for development of vision. **A8.13S**

Did you know?

Using stem cells to restore vision

Mike May (Figure 8.49), a successful businessman from San Francisco, lost his sight in a chemical explosion when he was three years old. Forty-three years later, in 2000, he underwent pioneering surgery on his cornea in a bid to restore some vision. Donor stem cells were transplanted into the eye. To be successful the cells needed to multiply and bind to the original cornea cells.

When the bandages came off, to his surprise a limited amount of Mike's vision had been restored. However, Mike's biggest problem is understanding what he's seeing. His main method of perception is still through touch. He has to learn how to interpret what he sees. He will never learn to see as clearly as most people. However, he still holds the speed record for downhill skiing by a totally blind person (65 mph). He is married to Jennifer and they have two sons.

Figure 8.49 Mike May, who is learning to see again.

8.5 Making sense of what we see

If synapses are made and maintained during the critical period, neurones in the visual cortex are able to respond to the information from the retina. Individual neurones in the columns of cells within the visual cortex respond in different ways to the information from the retina, and to different characteristics of the object being viewed. Some neurones, called simple cells, respond to bars of light. Others, called complex cells, respond to edges, slits or bars of light that move; others to the angle of the edge; still others to contours, movement or orientation.

However, visual perception is not simply the creation of an image of what is being viewed, but involves knowledge and experience as the brain interprets the sensory information received from the retina to create our visual experience of the world.

Kenge laughed when he was told that the animals on the horizon were buffalo. He could see the animals on the plain perfectly well. His visual development was not a problem; impulses were successfully sent from the retina to the cortex where cells were stimulated. Although he had seen buffalo up close before, he did not recognise them from a distance. Not knowing that a buffalo in the distance appears smaller, he failed to make sense of what he saw.

Depth perception

When we look at any object we can make a judgement about how far away the object is. The brain does this in different ways for close and distant objects.

Close objects

For objects less than 30 m away from us, we depend on the presence of cells in the visual cortex that obtain information from both eyes at once. The visual field is seen from two different angles, and cells in the visual cortex let us compare the view from one eye with that from the other. This is called stereoscopic vision and allows the relative position of objects to be perceived.

Q8.42 Why might a child who had had an eye patch during visual development never develop stereoscopic vision?

Activity

Check your own depth perception using **Activity 8.14. A8.14S**

Distant objects

For objects that are more than 30 m distant, the images on our two retinas are very similar, so visual cues and past experiences are used when interpreting the images. Look at the beach in Figure 8.50. What visual cues might help in depth perception? Lines converge in the distance giving perspective, the impression of distance. The stones further away are smaller and, because from experience we know that the stones along a beach are usually roughly the same size, the small ones in the picture are perceived as being further away. We know the size of buildings, so the ones in the background are not smaller, just further away.

Figure 8.50 Visual clues in the photograph allow depth perception.

Overlaps of objects and changes of colour also help in judging depth. Even if an object becomes smaller on the retina, for example when a car drives away into the distance, we perceive it as moving further away not getting smaller.

Usually we are unaware of the clues being used. It is only when they are not present, or cause the brain to misinterpret the image, that we become aware of them, such as when viewing optical illusions – see Figure 8.51.

Figure 8.51 Are these two people different heights? The person standing further away may look taller, but the lines that imply depth fool the brain. If you measure the people with a ruler you will see that they are actually the same height.

Is depth perception innate or learned?

It is possible that Kenge had not developed depth perception over long distances because he had had no experience of seeing long distances on an open plain, having always lived in the enclosed forest. Studies into how people with different cultural experiences and newborn children respond to optical illusions help answer this question.

Cross-cultural studies

What is culture? It has been defined in a number of different ways. In this course we will view culture as a system of beliefs that are shared among a group of people. It thus shapes experience and behaviour. People from different cultures may not share the same beliefs, and they may show different behaviours.

According to the **carpentered world hypothesis**, those of us who live in a world dominated by straight lines and right angles tend to perceive depth cues very differently from those who live in a 'circular culture'. In the carpentered world we are surrounded by buildings with right angle corners so unconsciously from an early age we tend to interpret images with acute and obtuse angles (so long as they don't differ too greatly from 90°) as right angles (Figure 8.52A). People who live in a 'circular culture' with few straight lines or right angle corners, such as the Zulu people of Africa who have circular houses and no roads, have little experience of interpreting acute and obtuse angles on the retina as representations of right angles. Studies have shown they are rarely fooled by visual illusions such as the Müller-Lyer illusion (Figure 8.52B).

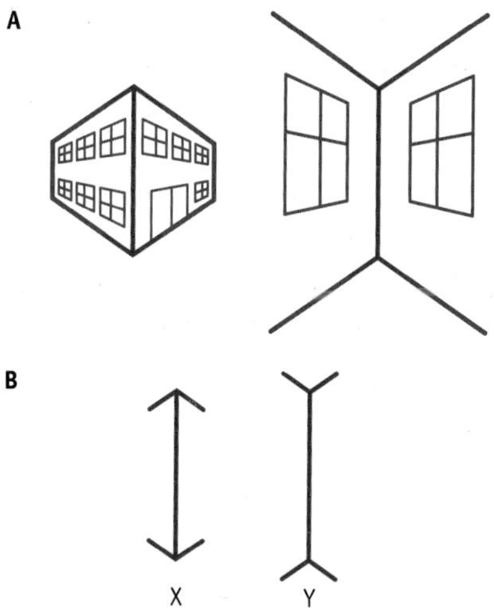

A

B

X Y

Figure 8.52 **A** Most of us will automatically interpret images with acute and obtuse angles as right angles. **B** Decide which of the two vertical lines X or Y is longer. Most people think they differ in length, but they are the same length.

In the Müller-Lyer illusion the angles between the line and the arrow heads are interpreted as right angles, providing depth clues to the image. We unconsciously perceive line X to be shorter because the depth clues make us think it is further away and Y is closer to us.

Some researchers think that lack of susceptibility to the Müller-Lyer illusion is not the result of different experience, but is due to genetic differences in pigmentation between individuals. They suggest that individuals who find it harder to detect contours are less susceptible to the Müller-Lyer illusion. They link poor contour detection to higher retinal pigmentation. In light-coloured people, with low retinal pigmentation, contour detection is good so such people, according to the argument, are more easily caught out by the illusion.

Q8.43 Why might researchers be cautious about the pigment and contour detection evidence supporting the idea that susceptibility to the Müller-Lyer illusion is genetic rather than environmental?

Depth cues in other pictures

In another cross-cultural study, individuals from a number of different cultures were shown pictures with depth cues such as object size, overlap of objects and linear perspective, similar to those in Figure 8.53.

It was found that all young children had difficulty perceiving the pictures as three-dimensional. They would have said that the man in Figure 8.53 was pointing his finger at the elephant not the antelope – they failed to interpret the depth cues. By the age of 11 years, almost all the European children interpreted the pictures in three dimensions, but some Bantu and Ghanaian children still did

Figure 8.53 Pictorial depth cues tell us that the man must be pointing at the antelope not the elephant.

not, and nor did non-literate adults, both Bantu and European. They had had less experience in the interpretation of depth cues in pictures.

What does seem clear is that the depth cues in pictures, which most of us take for granted, are not **innate** (inborn); they have to be learned. This suggests that visual perception is, in part at least, learned.

Q8.44 What depth cues are used to decide that the man is pointing at the antelope in Figure 8.53 and not the elephant?

Q8.45 Suggest reasons why caution should be used when interpreting some of these cross-cultural studies.

Activity

In **Activity 8.15** you can investigate the Müller-Lyer illusion using the interactive tutorial, before examining some data from cross-cultural studies. **A8.15S**

Studies with newborn babies

The fact that babies are born with a range of characteristic behaviours suggests that these are determined by genes; for example crying, walking movements and grasping are all present from birth. Within 24 hours of birth newborn babies can distinguish human faces and voices from other sights and sounds, and prefer them. These types of inborn capacities exhibited by a newborn baby are used as evidence for the role of genes in determining the hard wiring of the brain before birth.

Although born short-sighted, babies can see people and items clearly at a distance of about 30 cm. It is suggested that their preference for stripes and other patterns shows they are imposing order on their perceptions in early infancy. Long before they can crawl, they can tell the difference between a happy face and a sad one. They imitate people's expressions, and by the time they're old enough to pick up a phone they can mimic what they have seen others doing with it.

The visual cliff

In a classic experiment babies are encouraged to crawl across a table made of glass or Perspex, below which is a visual cliff. Patterns placed below the glass create the appearance of a steep drop, as shown in Figure 8.54. If the perception of depth is innate, then babies should be aware of the drop even if they have not previously experienced this stimulus themselves.

Q8.46 If the baby does have a perception of depth, how will they react when invited to crawl over the 'edge' of the cliff?

Young babies were very reluctant to crawl over the 'cliff', even when their mothers encouraged them to do so.

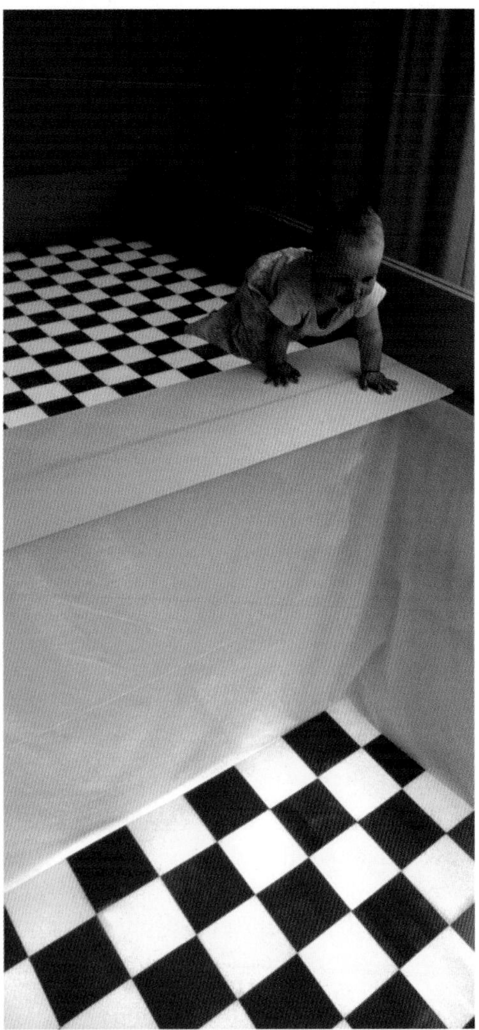

Figure 8.54 The visual cliff.

Q8.47 **a** This reaction is presumed to indicate that depth perception is innate. Explain why.

b Argue the case for this not illustrating that depth perception is innate.

This experiment is only possible with babies who have learned to crawl. It is likely that a six-month-old human may already have *learned* depth perception. Therefore the experiment was repeated with animals that can walk as soon as they are born such as chicks, kids (young goats) and lambs. They too refused to cross the cliff.

Q8.48 Explain if this supports or does not support the idea that this type of perception is innate.

Q8.49 Having read Section 8.5, hypothesise why Kenge, a forest dweller, thought that the buffalo on the horizon were insects.

Did you know?

Pattern recognition

Look at the list of letters on the right. You should easily recognise the letters unless you have dyslexia. Some theories of perception suggest that there is a template stored in our long-term memory. The template may not be an exact representation of the letters but a model that has all the key features, so when patterns are viewed their key features are matched with those in the long-term memory, allowing us to recognise and interpret what we see. There is still a lot about dyslexia that we don't know but it clearly has something to do with problems in recognising certain patterns. For example 'b' and 'd' are often confused.

CQOOCG
QQDCOQ
GGCOQD
CGQSSU
GCOQUC
ODQUCO
GHOQGD
DGOQDS

If you look quickly down the list of letters on the right you will probably be able to quickly pick out the single letter H with its straight vertical and horizontal lines – the other letters all have curved features. You do not have to look at every single letter separately. Cells that respond to the horizontal and straight lines are used; once stimulated they can check that the letter has the features of an H, using a stored template.

When reading we do not have to recognise every single letter. If we did it would take much longer to read this sentence, in the way a small child must sound out each letter in turn. Instead, for all but very long or rarely encountered words, we recognise complete words or even groups of words.

Recognition of objects probably works in a similar way.

Q8.50 Look at the pictures in Figure 8.55, and decide what each is.

Extension

In **Extension 8.5** and **Extension 8.6** you can investigate pattern recognition. **X8.05S X8.06S**

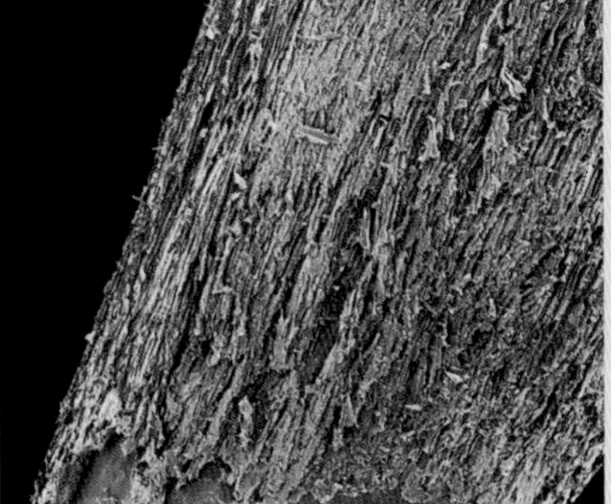

Figure 8.55 Can you recognise these objects? The answer to Question 8.50 will let you check if you were correct.

8.6 Learning and memory

Kenge knew what buffalo looked like even though he did not perceive them correctly from a distance. He had learned what they looked like and could recall this information from his memory. What is learning and how do we store information in our memory?

Learning occurs throughout our lives, and is any relatively permanent change in behaviour or knowledge that comes from experience. For learning to be effective, you must be able to remember what you have learned, and studies of learning and memory have always gone hand in hand. Throughout our lives, the memory stores vast amounts of information, from sights and sounds to emotions and skills such as riding a bike or texting. You saw earlier the plasticity of the nervous system, with changes occurring in our network of neurones, often by the modification of synapses. It is also changes to the synapses that underpin learning and memory.

Extension

In **Extension 8.7** you can investigate types of learning, including conducting Pavlov's dog conditioning experiment yourself, using the animation on the Nobel Prize website, which is in the weblinks that accompany this extension. **X8.07S**

Where memories are stored

Memory is not localised in one part of the brain. It is distributed throughout the cortex with different sites for short- and long-term memory. Different types of memory are controlled by different parts of the brain. This is clearly demonstrated by looking at cases where people have lost the use of particular parts of the brain, as the following case study illustrates.

One of the most famous patients in the study of the brain, HM, suffered from severe epilepsy, so doctors in 1953 removed those areas of the brain that appeared to be causing the problem, and immediately caused amnesia. HM's long-term memories from before the operation were unaffected, but he could no longer form new long-term memories, and found it difficult to remember what he had done just a few minutes before. By contrast, his memory for how to do everyday things was still intact.

Q8.51 Look back at the section of this topic on regions of the brain and find out the parts of the brain that are involved in memory.

Q8.52 HM's surgery affected his medial temporal lobes and hippocampus. What does this suggest about the function of these parts of the brain in forming and recalling memories?

How memories are stored

In the brain, every neurone connects with many other neurones to make up a complex network. HM's story shows that making memories is an active process. Memories can be created in two ways, by altering:

- the pattern of connections
- the strength of synapses.

Sea slugs and habituation – changing synapse strength

Eric Kandel, amongst others, studied the molecular biology of learning in the giant sea slug (*Aplysia*, Figure 8.56) to help understand learning in humans. He shared a Nobel Prize in 2000 for his work on *Aplysia*.

There are no fundamental differences between the nerve cells and synapses of humans and those of lower animals such as the sea slug. However, with only 20 000 neurones, the neurobiology of a sea slug is much simpler than that of a human. Sea slugs have large accessible neurones (Figure 8.57) so those involved in particular behaviours can be identified. Sea slug behaviour can be modified by learning, and the effects on neurones and synapses studied.

Figure 8.56 The giant sea slug (*Aplysia californica*). This species may grow to be 30 cm in length and weigh 1 kg.

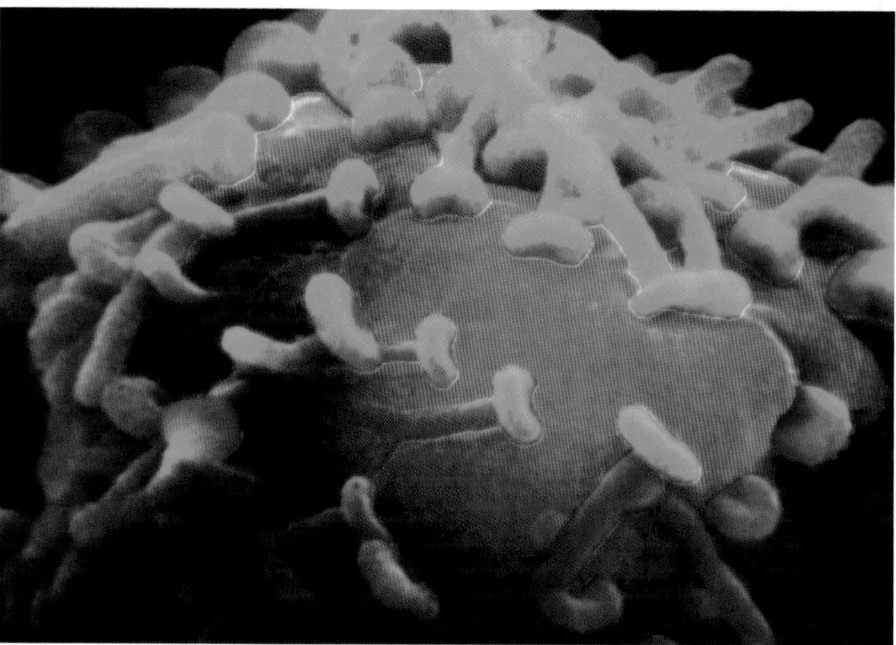

Figure 8.57 Scanning electron micrograph of synapses in the giant sea slug. Magnification × 2350.

Q8.53 **a** Sea slug neurones cannot pass impulses by saltatory conduction. Suggest a reason for this.

b Suggest how the sea slug neurones can nevertheless transmit impulses rapidly.

The giant sea slug breathes through a gill located in a cavity on the upper side of its body. Water is expelled through a siphon tube at one end of the cavity. If the siphon is touched, the gill is withdrawn into the cavity (Figure 8.58). This is a protective reflex action similar to removal of a hand from a hot plate.

Because they live in the sea, *Aplysia* are frequently buffeted by the waves and learn not to withdraw their gill every time a wave hits them. They become habituated to waves. **Habituation** is a type of learning. When you put your socks on you feel them at first, but after a few minutes you no longer notice them. You have become habituated to the feeling of the socks on your feet, even though they are still providing a touch stimulus.

Habituation allows animals to ignore unimportant stimuli so that limited sensory, attention and memory resources can be concentrated on more threatening or rewarding stimuli.

Activity

In **Activity 8.16** you can test whether snails can also become habituated to a stimulus. **A8.16S**

A Gill withdraws when siphon stimulated.

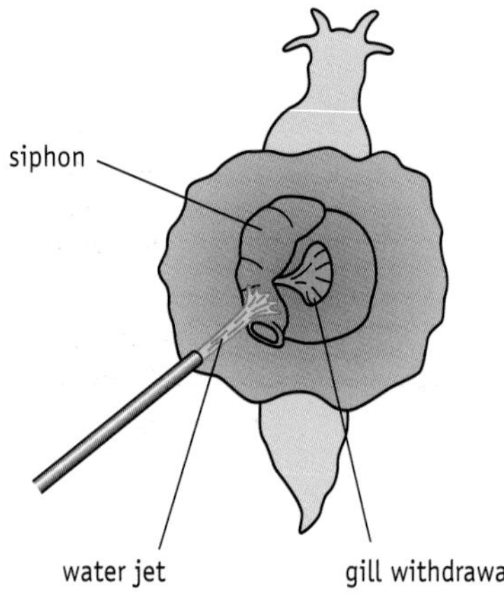
siphon
water jet
gill withdrawal

B After several minutes of repeated stimulation of the siphon the gill no longer withdraws.

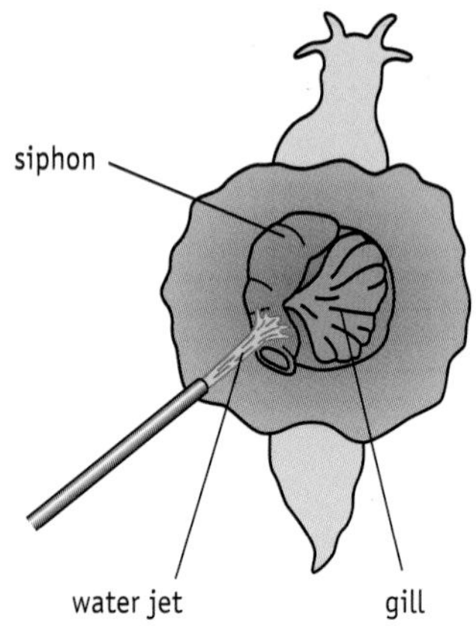
siphon
water jet
gill

C How habituation is achieved.

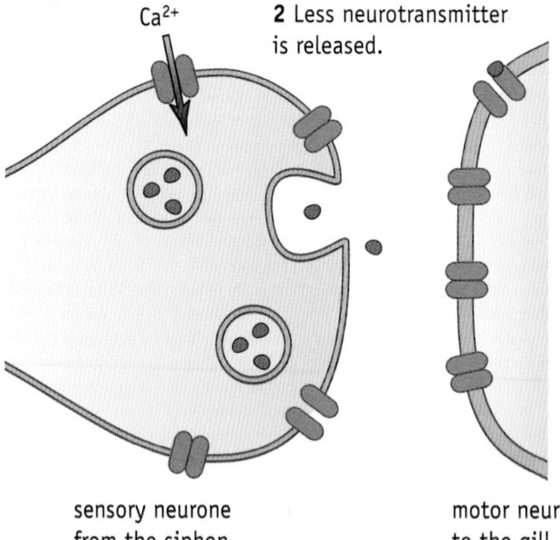
1 With repeated stimulation, Ca^{2+} channels become less responsive so less Ca^{2+} crosses the presynaptic membrane.
Ca^{2+}
2 Less neurotransmitter is released.
3 There is less depolarisation of the postsynaptic membrane so no action potential is triggered in the motor neurone.
sensory neurone from the siphon
motor neurone to the gill

Figure 8.58 Stimulating the siphon causes the gill of the sea slug to be withdrawn.

Q8.54 Sketch a diagram to show the reflex arc that is involved in withdrawal of the sea slug gill.

What is happening during habituation?

Kandel stimulated the sea slug's siphon repeatedly with a jet of water. The response gradually faded away, until the gill was not withdrawn any more. The neurones involved in the reflex were identified, and Kandel found that the amount of neurotransmitter crossing the synapse between the sensory and motor neurones decreased with habituation. With repeated stimulations, fewer calcium ions move into the presynaptic neurone when the presynaptic membrane is depolarised by an action potential; fewer neurotransmitter molecules are then released (Figure 8.58).

Activity

In **Activity 8.17** you can investigate what is happening at synapses during habituation. **A8.17S**

More connection – longer memory

Long-term memory storage involves an increase in the number of synaptic connections. Repeated use of a synapse leads to creation of additional synapses between the neurones.

Did you know?

Sensitisation

Sensitisation is the opposite of habituation. It happens when an animal develops an enhanced response to a stimulus. Humans can learn by sensitisation. Imagine you are at home alone late at night, and you hear a loud crash from outside. For a short period after the crash, any small noises (which you previously had not noticed because you were habituated to them) seem very loud and provoke a similar enhanced response. You have become sensitised to these noises.

If a predator attacks *Aplysia*, the sea slug becomes sensitised to other changes in its environment and responds strongly to them. In experiments studying sensitisation, Eric Kandel gave an electric shock to the sea slug tail before stimulating the siphon again with a jet of water. This provoked an enhanced gill withdrawal response that lasted from several minutes to over an hour. Use Figure 8.59 to help you to understand how sensitisation occurs.

A A shock to the tail enhances the gill withdrawal due to the water jet.

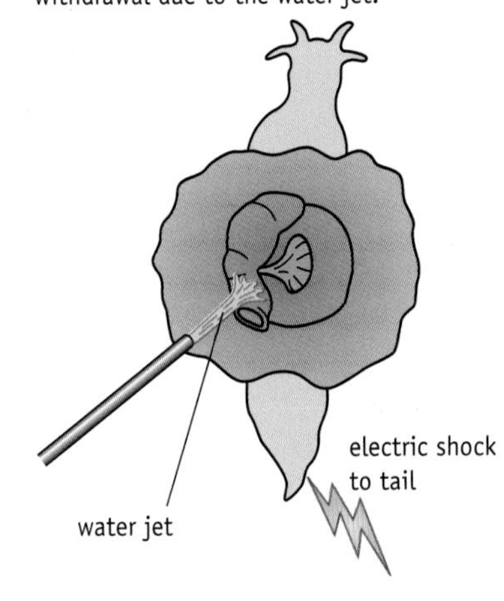

B How sensitisation is achieved.

1 impulse due to electric shock to tail
2 serotonin released
3 greater calcium ion uptake
4 impulse passes along sensory neurone
5 more neurotransmitter released
6 greater depolarisation
7 higher frequency of action potentials
8 enhanced gill withdrawal response

Figure 8.59 Sensitisation in sea slugs.

The ethics of using animals in medical research

The work of Kandel and others on sea slugs has helped our understanding of learning. However, people vary greatly in their views about whether or not it is acceptable to use animals in medical research (Figure 8.60) – or for other purposes such as farming or as pets.

First of all there are those who believe in **animal rights**. People don't any longer consider it morally acceptable to have human slaves, so, the argument goes, why is it acceptable to keep animals captive in laboratory cages or on farms? If you believe that humans have certain rights, it is quite difficult to explain why animals have none.

Of course, this doesn't mean that animals such as chimpanzees or dogs would have a right to vote or join a trade union – such rights are as meaningless for them as they are for a three-year-old child. Some people hold that animal rights mean that we should never keep animals because this would be to enslave them; others hold that the important thing is to respect their rights to such things as food, water, veterinary treatment and the ability to express normal behaviours.

Figure 8.60 The use of animals in medical research is strictly regulated in the UK and many other countries, but remains controversial.

The importance of consent

From the point of view of medical research, accepting that animals have rights would mean that we could only use animals that consented to participate in medical experiments, just as we only use humans in medical experiments if they give their informed consent. In practice, except perhaps for such things as feeding trials, this would bring an end to the use of animals in medical research.

Animal welfare rather than animal rights

A much more widespread position than believing that animals have rights is the belief that humans should treat animals well so far as is possible. Here the emphasis is on **animal welfare**. This is pretty much the position in European law. No country in the European Union is allowed to use vertebrates in medical experiments if there are non-animal alternatives. If there are not, animals can be used provided the research warrants it and strict guidelines are followed.

Animal suffering and experience of pleasure

Both the animal rights approach and the animal welfare approach assume that animals can suffer and experience pleasures. This seems pretty obvious to anyone who has had a pet cat or dog. But what of fish? Can they suffer or experience pleasures? There is genuine scientific and philosophical debate over this – though the balance is turning in favour of the fish. And what of spiders and insects? Most experts reckon they can't suffer. If you spend your weekends pulling the wings off flies it probably means that you are an inconsiderate and unpleasant person (or have suffered an abusive childhood), but it probably doesn't increase the amount of animal suffering in the world.

A utilitarian approach to the use of animals

One ethical framework we have introduced in this course is **utilitarianism** – the belief that the right course of action is the one that maximises the amount of overall happiness or pleasure in the world (see AS Topic 2). A utilitarian framework allows certain animals to be used in medical experiments *provided* the overall expected benefits are greater than the overall expected harms. Suppose, for example, to oversimplify greatly, that it takes the lives of 250 000 mice used in medical experiments to find a cure for breast cancer and that 50 000 of these mice are in pain for half their lives. There could still be a utilitarian argument for using the mice.

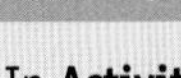

Activity

In **Activity 8.18** you investigate what people think of using animals for medical research. **A8.18S**

Q8.55 What would be the utilitarian argument for using 250 000 mice to find a cure for breast cancer?

Activity

Activity 8.19 lets you summarise the methods used to compare the contributions of nature and nurture to brain development discussed throughout the topic. **A8.19S**

8.7 Problems with the synapses

The functioning of the nervous system is completely dependent on synapses passing impulses. In everything we do, with every sensation, action and thought, neural pathways are stimulated. The control and coordination of all this activity in the brain relies on synapses. The synapses in turn depend on neurotransmitters. Unfortunately things can go wrong, with adverse consequences for health. Imbalances in the naturally-occurring brain chemicals can cause problems, as can drugs when they cross the blood-brain barrier. The endothelial cells forming the capillaries in the brain are more tightly joined together than those in capillaries elsewhere in the body. They form a barrier to control the movement of substances into the brain and protect the brain from the effects of changes in blood ionic composition and from toxic molecules.

There are about 50 different neurotransmitters in the human central nervous system, and they need to be present in controlled amounts. The function of two examples, dopamine and serotonin, are discussed here, together with the consequences of abnormal levels. The effects of ecstasy are used to illustrate how drugs can affect synapses.

Activity

In **Activity 8.20** you can refresh your memory of how synapses work, and review how they are affected by chemical imbalances. **A8.20S**

Parkinson's disease

Dopamine and Parkinson's disease

Dopamine is a neurotransmitter secreted by neurones, including many located in part of the midbrain. The axons of the neurones in this area extend throughout the frontal cortex, the brain stem and the spinal cord. In people with Parkinson's disease, dopamine-secreting neurones in the basal ganglia die (Figure 8.37). These neurones normally release dopamine in the motor cortex. Parkinson's patients' motor cortexes receive little dopamine and there is a loss of control of muscular movements. The main symptoms of the disease are:

- stiffness of muscles
- tremor of the muscles
- slowness of movement
- poor balance
- walking problems.

Other problems that may arise include depression and difficulties with speech and breathing. Between 1% and 2% of people in the UK over fifty are affected by Parkinson's disease, although the onset can be well before that age.

Treatment for Parkinson's disease

Recent developments in drug treatments for Parkinson's have made it much easier for some people to live with the effects of the disease. Some of the treatments for Parkinson's disease are outlined below.

- Slowing the loss of dopamine from the brain, with the use of drugs such as selegiline. This drug inhibits the enzyme monoamine oxidase, which is responsible for breaking down dopamine in the brain.
- Treating the symptoms with drugs. Dopamine itself cannot be given to treat Parkinson's because it cannot cross into the brain from the bloodstream, but L-dopa, a precursor in the manufacture of dopamine, can be given. Once in the brain L-dopa is converted into dopamine, increasing the concentration of dopamine and controlling the symptoms of the disease.
- Use of dopamine **agonists**. Dopamine agonists are drugs that activate the dopamine receptor directly. These drugs mimic the role of dopamine in the brain, binding to dopamine receptors at synapses and triggering action potentials. They can be particularly useful in the treatment of Parkinson's disease, since they avoid higher than normal levels of dopamine in the brain. Abnormally high dopamine levels can have unpleasant side effects (see the 'Did you know' box below).
- Results of gene therapy trials in animals and phase I trials in humans show promise. Genes for proteins that increase dopamine production and that promote the growth and survival of nerve cells are inserted into the brain. Cell therapy in which the proteins themselves are injected is also being trialled.
- New surgical approaches are being trialled, some of which are generating encouraging results.

Q8.56 Suggest why non-peptide compounds that can induce proteins that aid survival of neurones are being developed.

Weblink

Visit the Birmingham Department of Clinical Neuroscience website to see some Parkinson's video clips showing the effect of treatment with L-dopa.

Did you know?

Schizophrenia, hyperactivity and the effect of taking cocaine

Excess dopamine in the brain is believed to be a major cause of schizophrenia. This excess dopamine in the brain can be treated with drugs that block the binding of dopamine to its postsynaptic receptor sites. These drugs are usually similar to dopamine in structure, but are unable to stimulate the receptors. This reduces the effect of the dopamine in triggering postsynaptic action potentials. A side effect in some patients taking these drugs is symptoms of Parkinson's disease.

Ritalin is a drug used to treat childhood hyperactivity associated with low levels of dopamine. The effect of the drug is to prevent dopamine from being taken back up by the presynaptic membrane; more dopamine is left in the synaptic cleft, to stimulate postsynaptic receptor molecules.

Cocaine also prevents dopamine reuptake, by binding to proteins that normally transport dopamine. Not only does cocaine bind to the transport proteins in preference to dopamine, it remains bound for much longer than dopamine. As a result, more dopamine remains to stimulate neurones; this causes prolonged feelings of pleasure and excitement. Amphetamines also increase dopamine levels. Again, the result is over-stimulation of these pleasure-pathway nerves in the brain. Long-term amphetamine use can induce schizophrenia-like illness.

Depression

Serotonin and depression

The neurotransmitter **serotonin** plays an important part in determining a person's mood. Neurones that secrete serotonin are situated in the brain stem. Their axons extend into the cortex, the cerebellum and the spinal cord, targeting a huge area of the brain (Figure 8.61).

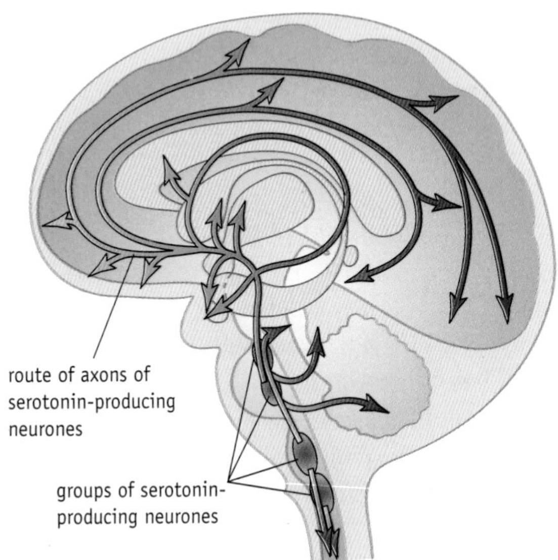

Figure 8.61 The neurones that secrete serotonin are located in the brain

A lack of serotonin has been linked to depression (Figure 8.62). Depression is associated with feelings of sadness, anxiety and hopelessness. Loss of interest in pleasurable activities and reduced energy levels are common, as are insomnia, restlessness and thoughts of death.

The causes of depression are not completely understood. There may be a genetic element, since it runs in families. It is probable that depression is a **multifactorial** condition; several genes may be involved but they probably simply confer a susceptibility to the condition, with environmental factors also contributing. For some people, traumatic or stressful events, such as bereavement, illness, or job or money worries, can be the trigger.

One gene, 5-HTT, known to influence our susceptibility to depression, codes for a transporter protein that controls serotonin reuptake into presynaptic neurones. People with the 'short' version of the 5-HTT gene are more likely to develop depression after a stressful life event. Two people who inherit the same susceptibility genes may not both develop depression; it will depend on environmental factors acting as triggers to bring about the symptoms of the disease.

Figure 8.62 Clinical depression is a worryingly common condition, involving far more than feeling a bit glum. It can last for months or years and can have a profound effect on work and relationships. It is under-diagnosed, particularly in children.

Q8.57 In the case of depression, alleles at many loci are thought to be involved in inheritance of the condition.

a State the name of this pattern of inheritance.

b Suggest two other conditions that show this pattern of inheritance.

A number of neurotransmitters may have a role to play in depression, including dopamine and noradrenaline, but reduced serotonin levels seem to be most commonly involved. However, it is not fully understood whether the lack of serotonin is a cause of depression or a result of it.

When someone is depressed, fewer nerve impulses than normal are transmitted around the brain, which may be related to low levels of neurotransmitters being produced. Pathways involving serotonin have a number of abnormalities in people with depression. The molecules needed for serotonin synthesis are often present in low concentrations, but serotonin-binding sites are more numerous than normal, possibly to compensate for the low level of the molecule.

Twin studies allow determination of the genetic component of a disease. Identical twins, produced by division of the same egg, are genetically identical. The degree of similarity between the twins is a measure of the influence of the genes on that characteristic. 99% of identical twins have the same eye colour, and 95% the same fingerprint ridge count. Where the environment has a greater effect the similarity falls. Twin studies have been useful in suggesting that there may be specific genes that cause clinical depression to develop in certain families, but they also indicate that there are a number of other genetic factors that seem to confer a vulnerability to depression.

Drug treatment for depression

One of the most effective types of drug used to treat the symptoms of depression inhibits the reuptake of serotonin from synaptic clefts. A drug of this type is a Selective Serotonin Reuptake Inhibitor (SSRI), meaning that it blocks only the uptake of serotonin. One of the more common drugs of this type is Prozac, which maintains a higher level of serotonin, and so increases the rate of nerve impulses in serotonin pathways. This has the effect of reducing some of the symptoms of depression.

How drugs affect synaptic transmission

Synapses have a number of features that can be disrupted by interference from certain chemicals (Figure 8.63). For example, a chemical with a similar molecular structure to a particular neurotransmitter is likely to bind to the same receptor sites, and perhaps stimulate the postsynaptic neurone. Other chemicals may prevent the release of neurotransmitter, block or open ion channels, or inhibit a breakdown enzyme such as acetylcholinesterase or monoamine oxidase (see AS Topic 3). In all of these cases the normal functioning of the synapse will be disrupted, with consequences that depend on the nerve pathway involved.

The effect of ecstasy

The drug known as **ecstasy** affects thinking, mood and memory. It can also cause anxiety and altered perceptions (similar to but not quite the same as hallucinations). The most desirable effect of ecstasy is its ability to provide feelings of emotional warmth and empathy.

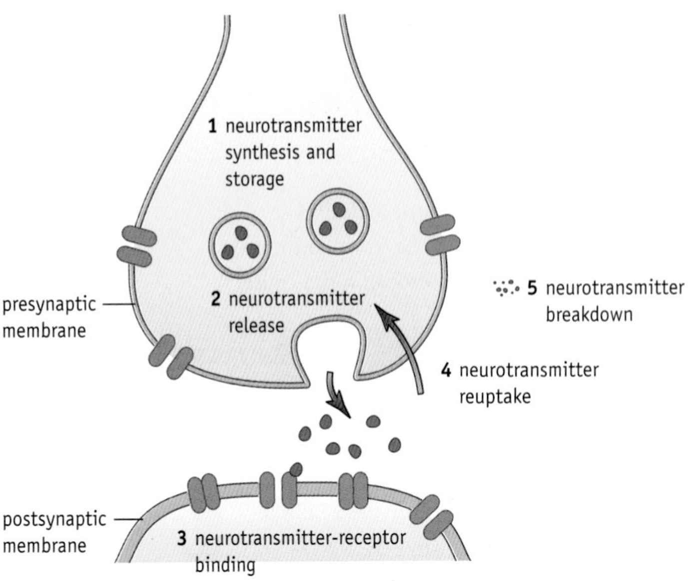

Figure 8.63 Five different stages in synaptic transmission that can be affected by drugs. Ecstasy binds to serotonin transport molecules, preventing uptake back across the presynaptic membrane.

Ecstasy is a derivative of amphetamine. Its chemical name is 3,4-methylenedioxy-N-methylamphetamine (MDMA). The short-term effects of ecstasy include changes in behaviour and brain chemistry. The long-term effects include changes in behaviour and brain structure.

How ecstasy affects synapses

Ecstasy increases the concentration of serotonin in the synaptic cleft. It does this by binding to molecules in the presynaptic membrane that are responsible for transporting the serotonin back into the cytoplasm. This prevents its removal from the synaptic cleft. The drug may also cause the transporting molecules to work in reverse, further increasing the amount of serotonin outside the cell. These higher levels of serotonin bring about the mood changes seen in users of the drug. It is possible that the ecstasy has a similar effect on molecules that transport dopamine as well.

Effects of using ecstasy

Users report feelings of euphoria, well-being and enhanced senses. There are, however, side effects and unpredictable consequences of using the drug. Some people experience clouded thinking, agitation and disturbed behaviour. Also common are sweating, dry mouth (thirst), increased heart rate, fatigue, muscle spasms (especially jaw-clenching) and hyperthermia, since ecstasy can disrupt the ability of the brain to regulate body temperature. Repeated doses or even a single high dose of ecstasy can cause hyperthermia, high blood pressure, irregular heartbeat, muscle breakdown and kidney failure. This can be fatal.

There is growing evidence of long-term effects including insomnia, depression and other psychological problems. It is possible that ecstasy can have an effect on normal brain activity even when the drug is no longer taken. Because the drug has stimulated so much serotonin release, the cells cannot synthesise enough to meet demand once it has gone, resulting in a feeling of depression.

Checkpoint

8.7 Construct a spider diagram showing the ways in which drugs can affect synaptic transmission. Include examples of the type of drugs that produce each effect.

Better treatments

The sheer misery and human trauma caused by conditions such as Parkinson's disease, depression, Huntington's disease, Alzheimer's disease, schizophrenia and bipolar disorder (manic depression) mean that the search for effective treatments for such conditions will be one of the major targets for 21st century biology.

The deciphering of the base sequence in the human genome as part of the **Human Genome Project** (HGP) means that we are now gaining a much better understanding of the way genes control our phenotype. This is leading to major advances in our understanding of diseases with a genetic basis and to improvements in treatment. It is hoped that it will also put us in a better position to avoid the risk factors specific to our own genetic make-up, and continue to improve the help given to people with these sorts of conditions.

How the outcomes of the Human Genome Project help

As you saw in AS Topic 3, a **genome** is all the DNA of an organism (or species), including the genes that carry all the information for making the array of proteins required by the organism (or species). It is these proteins that help determine all the characteristics of the organism, from individual biochemical pathways to its overall appearance.

Finding the sequence of bases in DNA

In 1977, Fred Sanger invented the first DNA sequencing process. In this process DNA is used as a template to replicate a set of DNA fragments, each differing in length by one base. The fragments are separated according to size using gel electrophoresis and the base at the end of each fragment is identified. This allows the sequence of bases in the whole DNA chain to be determined.

Following this it became possible to work out the entire sequence of bases in the human genome, and determine the location of all our genes. This was a massive undertaking, as Figure 8.64 illustrates.

Figure 8.64 The human genome contains a total of some three thousand million (three billion, i.e. 3×10^9) bases. It's hard to imagine such a huge number, so try this: 15 four-drawer filing cabinets, each with drawers full of A4 paper, printed on both sides with the letters A, C, G and T, typed (with no spaces or punctuation) in 12 point text.

In 1986 the Human Genome Project was officially started, with the USA and UK being full partners. It was thought that the task of deciphering the whole of the human genome would take over 20 years, but with the rest of Europe and Japan joining the project in 1992, and the development of sophisticated computerised systems (Figures 8.65 and 8.66), progress was more rapid.

Figure 8.65 DNA sequencing is highly automated, with computers collating the sequence.

Figure 8.66 DNA sequence on a computer screen. Each of the four bases is shown by a different colour, the order of colours giving the base sequence.

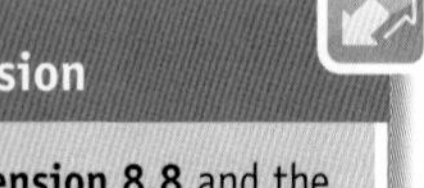

Extension

In **Extension 8.8** and the associated weblinks you can find out more about DNA sequencing and the Human Genome Project. **X8.08S**

Despite its name the project did not sequence only the human genome. The complete DNA sequence of yeast was published in 1996 and this was followed in 1998 by that of the nematode worm *Caenorhabditis elegans*. Other organisms used in biological and medical research, such as *Drosophila,* zebrafish, the mouse and the rat have also been sequenced.

In 1999 chromosome number 22 became the first human chromosome to be fully sequenced. A working draft of the whole human genome was published in 2001. The draft contained gaps and had an error rate of about 1 in 1000 bases. Work continues, with sections being sequenced up to ten times to produce a high quality finished sequence without gaps and with 99.99% accuracy. The policy of the publicly funded project was to immediately publish the sequence on the Internet, making it a freely available resource for biologists across the world.

Using the sequence biologists are gaining a better understanding of the genome itself, identifying new genes, working out how they are controlled, and discovering what products they code for. The sequence also has a major part to play in research into the role of genes in disease and the development of new diagnostic techniques and treatments.

Detailed information about the genome

The human genome is now thought to be 3 200 000 000 bases in length. It is estimated that our genome contains 20 000 to 25 000 genes, which are more complex than those of simpler organisms and give rise to a large number of more complex proteins.

The average human gene is now known to consist of about 3000 bases, but sizes vary greatly. The largest known human gene codes for dystrophin; it is 2.4 million bases long. The dystrophin protein is faulty in individuals with the muscle-wasting disease Duchenne muscular dystrophy. However, the functions of most of our genes are still unknown.

Scientists have identified about 1.4 million locations where single-nucleotide polymorphisms (SNPs – pronounced 'snips') occur in humans. A SNP is a DNA sequence variation that involves a change in a single nucleotide.

Non-coding sequences ('junk DNA') make up at least 50% of the human genome. Once considered to have no direct function, at least some of them are now known to be important in gene regulation.

Identification of new genes

Many disease genes have been identified, including the breast cancer gene BCRA2 and the total colour blindness genes CNGA3 and CNGB3 (these code for subunits of a protein channel in photoreceptor cells in the retina). Several genes associated with Parkinson's disease have been identified, including parkin, DJ1 and PINK1.

It is now possible to locate a candidate gene (a gene that may cause a particular disease) on our DNA and then screen this gene for mutations in affected individuals. Analysing DNA sequence patterns in humans side by side with those in well-studied model organisms (such as yeast or *Drosophila*) has become one of the most powerful strategies for identifying human genes and interpreting their functions.

Six genes have been identified as increasing susceptibility to Alzheimer's disease, a form of progressive dementia. Alzheimer's is a multifactorial disease; age, genes and environment are all factors that can increase the risk of developing the disease. Symptoms include confusion and progressive severe memory loss. In a few families, a genetic fault on chromosome 21 in the APP gene for production of the precursor of a particular protein (known as an amyloid protein) has been implicated in the development of Alzheimer's. Many cases of Alzheimer's have been linked to possession of alleles of a gene known as ApoE. There are three common ApoE alleles (ApoE2, ApoE3 and

ApoE4), and having two alleles of ApoE4 seems to increase the risk of getting the disease tenfold. The ApoE gene controls production of a lipoprotein used in the repair of cell membranes in damaged neurones. The ApoE4 allele produces a variant of this, an amyloid protein that is deposited in insoluble plaques in the brain.

Unfortunately how the genes and environmental factors interact to cause Alzheimer's disease is still unclear. It is currently being debated whether many of the dementia conditions are single diseases triggered by a number of genes, or a number of diseases whose symptoms are very similar. You may recall from AS Topic 1 that ApoE alleles are also involved in cardiovascular disease.

Identification of new drug targets

A drug target is a specific molecule that a drug interacts with to bring about its effect. Before the entire draft human DNA sequence was published, there were only 500 drug targets. The identification of disease genes and their products is allowing biologists to find new drug targets. By the time the draft sequence was published in 2000, scientists had already searched for DNA sequences similar to those for existing drug target proteins, and found 18 new sequences that may potentially provide new targets. The identification of drug targets is growing; estimates of the number of potential drug targets are in the thousands. For example, in 2008 a study using the nematode *C. elegans* identified 80 genes that influence the formation of the protein alpha-synuclein, which accumulates in the brains of Parkinson's disease patients. These genes and the proteins they code for may be important in providing new drug targets for the disease.

Preventative medicine and improved drug treatment

Some drugs work very well for one person with few side effects, whereas in other patients the same drug is ineffective and may have major side effects. It is thought that these different responses may often be due to slight variations in each individual's genome, depending on which of the 1.4 million SNPs a person possesses. Prescribing the best drug for a patient is currently largely trial and error – at best. It is hoped that information about a person's genome will enable doctors to prescribe the right drug at the correct dose. Some single-nucleotide polymorphisms have been associated with differences in drug targets, and the effects of drugs.

If a person knew that they carried mutations associated with a particular disease, they might be able to make changes to their lifestyle that would reduce the risk of the disease, or opt for preventative treatment. For example, if we knew the gene mutations associated with depression and we knew the environmental factors that trigger attacks, people with the mutations might be able to avoid the environmental factors that trigger the condition.

Ethical dilemmas

The human genome sequence offers great hope for the future, but there is also considerable concern about the potential use of this information. In 1991, shortly after the Human Genome Project got underway, the Director of the National Institute of Health (NIH) in America filed applications for patents on 6122 fragments of DNA. As a result, James Watson resigned from his position as leader of the American branch of the Human Genome Project. The President of the USA and the UK Prime Minister issued a statement in March 2000 declaring that all genetic data should be released into the public domain and not restricted by patent. However, patents on human DNA are still being taken out.

Did you know?

Other uses of the DNA sequence

Understanding basic biology

In addition to medical applications, the DNA sequence contributes to research in many areas of cell biology and physiology. For example, researchers investigating the molecular basis for why some people cannot detect bitter tastes used the DNA sequence, which was freely available. By identifying the sections of DNA involved and searching the DNA sequence, they discovered receptor proteins that are found in the taste buds of people who can detect bitter tastes. Experiments using cultured cells confirmed that the receptors do respond to the bitter taste.

Investigating evolution of the human genome

By comparing our genome with those of other animals it is possible to look at the evolutionary history of the genome, determining when particular genes first appeared and seeing how mutations accumulate over time. The same genes are almost always found in other vertebrate species, strongly suggesting a common ancestor in evolution. The occurrence of SNPs will also provide a useful tool for investigating human evolution.

One of the main project goals of the Human Genome Project is to address the ethical, legal and social issues that may arise from the project. A proportion of the budget (3–5%) has been set aside for this purpose. However, many issues are still to be addressed adequately:

- Testing for genetic predisposition has many implications. Would it be acceptable, for example, for insurers to have this information about people who are applying for health insurance?
- Who should decide about the use of genetic predisposition tests, and on whom they should be used?
- Making and keeping records of individual genotypes raises acute problems of confidentiality.
- Many medical treatments made possible through the development of genetic technologies will initially be very expensive. Their restricted availability will add considerably to the problems faced by the health services in deciding who is eligible for such treatments.

Genetic testing and discrimination

Individuals identified as possessing genes associated with inherited diseases may face discrimination in employment or when applying for insurance. Individuals may be faced with higher premiums or be unable to get certain types of cover even though the possession of the gene does not mean that the person will necessarily develop the condition. This is particularly true in the US, where the majority of medical treatment is paid for through health insurance. In 2003 the US Senate unanimously voted for the Genetic Information Nondiscrimination Act 2003. This Act prohibits the use of genetic information by insurers or employers, and became law in 2008. In October 2001 the UK Government announced a five-year moratorium on the use of genetic test results in assessing applications for some insurance policies. The Association of British Insurers have extended the moratorium until 2014. However, at the time of writing (2008) there is no Act of Parliament prohibiting this type of discrimination.

Q8.58 Suggest reasons why carriers of inherited disease genes should not face discrimination when applying for jobs or insurance.

Using genetically modified organisms to produce drugs

Genetic modification

With the Human Genome Project enabling the identification of genes, it has been possible to genetically modify non-human organisms to produce specific human proteins, such as human growth hormone, insulin and collagen – a protein found in the skin and connective tissues.

The artificial introduction of genetic material from another organism, through **genetic modification**, produces a **transgenic** or **genetically modified organism** (**GMO**). Figure 8.67 shows genetically modified mice. Genetic modification is also known as **genetic engineering** or genetic manipulation or recombinant DNA technology; it is easy to get confused!

Figure 8.67 These genetically modified mice glows green under blue light. A gene from jellyfish that codes for a green fluorescent protein has been incorporated into the mouse DNA.

Modifying microorganisms

The first success in genetic engineering was with bacteria. Bacteria contain simple DNA structures, **plasmids**, which can be transferred from one cell to another. Using restriction enzymes, the circular plasmid can be cut, and using another set of enzymes a piece of DNA from another species can be inserted into it. The plasmid is inserted back into the bacteria, which are then allowed to multiply in a fermenter. The protein produced is extracted from the culture.

Figure 8.68 shows how this process is used for producing the very valuable human protein, insulin. In 1982 this was the first commercial product of genetic modification. The first vaccine was produced four years later, to protect against hepatitis B.

Today many biopharmaceutical industries use genetically modified microorganisms to produce therapeutic proteins and vaccines. This is sometimes referred to as 'pharming' (drug production).

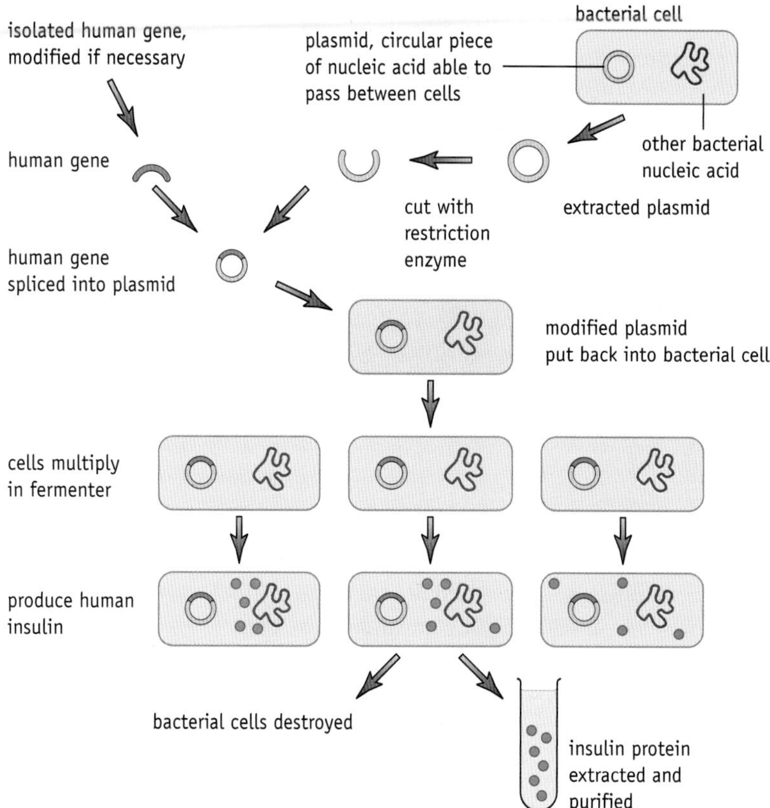

Figure 8.68 Using bacteria to produce human insulin.

? Did you know?

DNA technology is used to make hard cheese

During cheese-making a number of substances are added to sour milk. One of them is rennet. Rennet is a crude extract of enzymes including chymosin. These enzymes act on a milk protein called casein. Their effect is to cause the milk to form a soft curd, also known as junket. Without rennet, most cheeses cannot be made.

Traditionally rennet has been obtained from the stomachs of young calves (or piglets, kids, lambs or water buffalo calves). Rennets can be of vegetable origin, but until recently by far the most important source was young calves. Calves' stomachs were ground up in salt water, ten of them being required for one gallon of rennet. Calves' stomachs contain a lot of stuff in addition to rennet, so the purity wasn't very high and, of course, vegetarians objected.

Chymosin was the first enzyme produced from a GMO licensed for use in food. The DNA encoding the protein chymosin is isolated from calf cells. A copy of this gene is inserted into yeast cells. As the yeast cells reproduce, the modified plasmids are copied and chymosin is made by the yeast.

Genetically modified plants

Throughout history, humans have selected particular plants for cultivation as crops, because they were edible, produced good fibres, contained useful chemicals or just looked ornamental. For centuries farmers have picked out the hardiest and most prolific plants from their crops and have saved the seeds from these plants for sowing the following year. In this way crops have steadily improved. This process is called **artificial selection**. It selects alleles for characteristics that are agriculturally valuable.

Plant breeding is a slow process, and from identifying a potential new variety to commercial production takes many years. When genetic engineers introduce new genes with alleles for desired characteristics into a plant's DNA, the resulting **genetically modified plants** (Figure 8.69) can be produced in a timescale of months rather than the years required for traditional methods.

Figure 8.69 Genetically modified maize. The plants have been modified by inserting a bacterial gene that makes the plants herbicide resistant. When the crop is treated with herbicide, the maize is unaffected but the weeds between the rows of maize are killed.

Genetically modified crops such as maize or soya have the potential to mass produce medicines and other chemicals cheaply and efficiently. Plants in crop trials have been genetically engineered to manufacture proteins for healing wounds and for treating conditions such as cirrhosis of the liver, anaemia and cystic fibrosis (CF). Other trials are exploring the possibilities of using genetically engineered plants to produce antibodies to fight cancer, and vaccines against rabies, cholera and foot-and-mouth disease.

Foreign genes are inserted into plant cells. This can be achieved using a range of methods:

- A bacterium that infects many species of plant, such as the soil-inhabiting bacterium *Agrobacterium tumefaciens*, can be used. When the bacteria invade plant cells, genes from plasmid DNA become incorporated into the chromosomes of the plant cells. Scientists have developed a technique using this natural system. They insert the desired genes into a plasmid, which then 'carries' these genes into the plant DNA.
- Minute pellets that are covered with DNA carrying the desired genes are shot into plant cells using a particle gun.
- Viruses are sometimes used. They infect cells by inserting their DNA or RNA (depending on the type of virus). They can be used to transfer the new genes into the cell.

> **Activity**
>
> Use the interactive tutorial in **Activity 8.21** to produce your own summary explaining how plants can be genetically modified. **A8.21S**

Gene insertion is never 100% successful. Scientists therefore need a method of screening to find out which plant cells actually have the new gene. This has generally been done by incorporating a gene for antibiotic resistance, often called a **marker gene**, along with the new desired gene. The plant cells are then incubated with the antibiotic, which kills off any unsuccessful cells that have not taken up the new genes. The only cells to survive are those that have successfully incorporated the new genes and are resistant.

The genetically modified plant cells can then be 'cultured' in agar with nutrients and plant growth substances to produce new plants. This **micropropagation** can multiply a single cell to form a callus (a mass of plant cells), which then differentiates to form plantlets (tiny plants) (Figure 8.70 and 8.71) and finally novel plants.

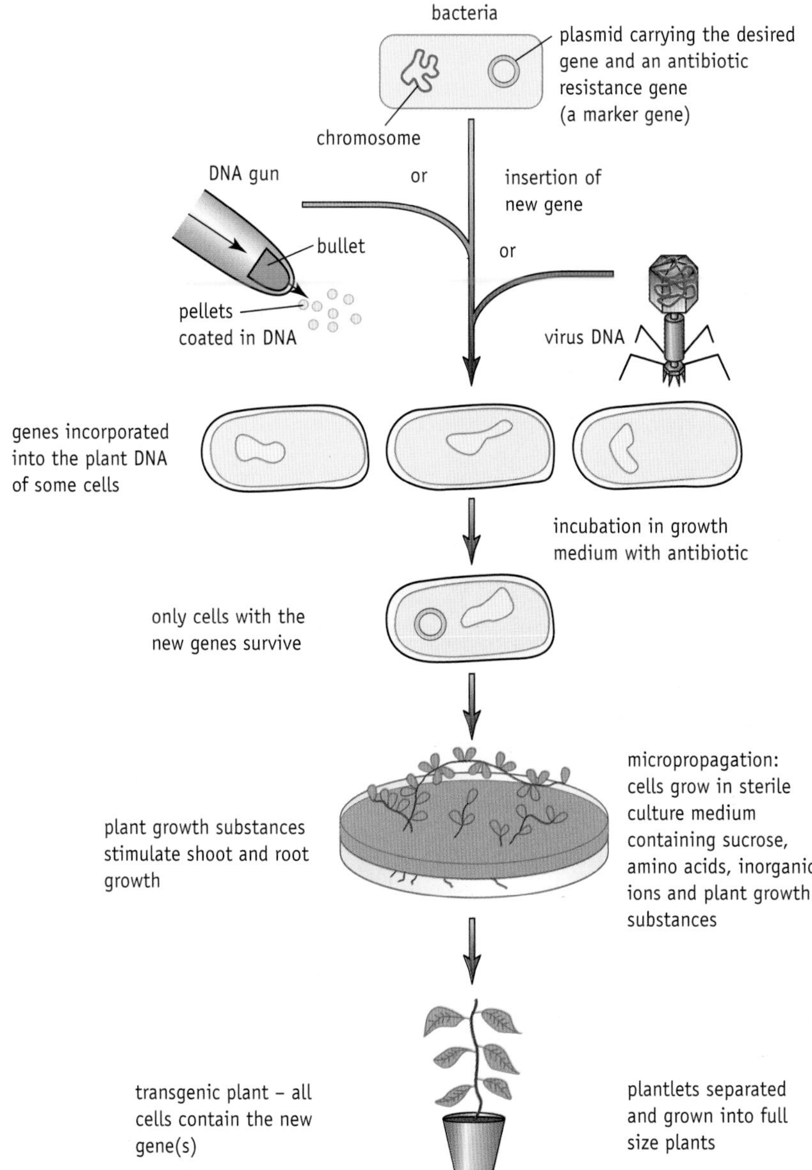

Figure 8.70 Production of genetically modified plants.

Figure 8.71 A callus of parenchyma cells is treated with growth substances that enable cells to differentiate into shoot and root tissue.

A wide range of genetically modified crops have already been produced. Some examples are described in Extension 8.9.

Extension

Read about some examples of genetically modified crops in **Extension 8.9. X8.09S**

Genetically modified animals

Several techniques can be used to artificially introduce genes into animal cells. You have already met the use of liposomes in gene therapy (AS Topic 2). A number of methods are used for the creation of genetically modified animals. These include injecting DNA directly into the nucleus of a fertilised egg. The egg is implanted into a surrogate female. The DNA is only successfully incorporated into the genome of about 1% of the treated embryos (higher success is achieved with mice). Retroviruses have also been used to introduce new genes into fertilised eggs. This type of virus incorporates its DNA into the host's DNA.

Q8.59 Suggest reasons why it is much more difficult to introduce genes in eukaryotes.

In 1991, Tracey was the most famous sheep in the world (Figure 8.72); she was the first transgenic sheep. Her DNA contained the human gene for the protein AAT (alpha-1-antitrypsin).

Figure 8.72 Tracey, the first genetically modified ewe.

AAT is normally made by our liver cells and inhibits the enzyme elastase. Elastase is released from neutrophils, the white blood cells that fight infection (Topic 6). It is a protease that digests damaged or ageing lung cells, foreign particles and bacteria. AAT prevents elastase attacking normal tissue.

In the inherited disease A1AD (alpha-1-antitrypsin deficiency), a mutation in the gene coding for AAT results in no protein or a non-functioning protein being produced. A lack of AAT can cause lung disease such as emphysema, as the elastase attacks normal lung tissue. The abnormal protein also collects in the liver where it contributes to liver cirrhosis and hepatitis.

AAT produced and secreted in the milk of transgenic ewes like Tracey can be extracted and used for treatment of A1AD. A flock of 3000 transgenic ewes were produced in Scotland but the financial backers withdrew funding, and so although trials are ongoing, there is no commercial production.

Q8.60 AAT is a possible treatment for cystic fibrosis. Suggest how AAT helps the condition.

Q8.61 The drug produced by genetically modified ewes is undergoing trials on animals. Recall the stages of drug testing and describe the phases it still has to go through before it can be prescribed for human use.

Scientists have successfully produced several proteins in the milk of genetically modified cows, sheep, goats and rabbits. There is now the prospect of using chickens as 'biofactories'. At the Roslin Institute, the research centre that created Tracey, and Dolly the cloned sheep, scientists have bred genetically modified chickens that lay eggs containing miR24, a monoclonal antibody with the potential to treat malignant melanoma (AS Topic 3). Other chickens have been modified to produce a form of beta interferon used to treat multiple sclerosis. The proteins secreted into the whites of the eggs are extracted and purified. However, large doses of such drugs are needed over long periods, so the challenge is to increase yield in the egg whites.

Q8.62 What advantages may chickens offer over other animals for such 'pharming'?

Genetically modified organisms are extensively used in research. For example, human disease genes can be incorporated into the genome of model animals. Transgenic mice have been produced that develop conditions such as Alzheimer's disease, CF and tumours.

Concerns about genetic modification

So far so good. There are potential benefits of GM technology, but as with the advent of any major new technology (such as nuclear power), we also need to examine carefully the risks associated with its use.

Health

The main health concerns that have been raised are:

- transfer of antibiotic-resistance genes to microbes
- formation of harmful products by new genes
- transfer of viruses from animals to humans.

Antibiotic resistance

GM plants often contain not only the gene (for pesticide resistance, a higher yield, or whatever) that codes for the desired product but also a marker gene to select for the new plants. These, as discussed on page 266, are sometimes antibiotic-resistance genes. For example, Bt corn contains a marker gene that gives resistance to ampicillin and other penicillin-type antibiotics. When eaten the gene could potentially be transferred to microbes in the gut, which could build up resistance to certain antibiotics used in medical treatments.

In 1996 the EU banned Ciba-Geigy's Bt corn on the grounds that it contained a gene for ampicillin resistance. (This outright ban has since been replaced by a requirement for specific labelling.) In 1999 the British Medical Association urged a ban on all use of antibiotic-resistance genes in GM foods, stating that 'the risk to human health from antibiotic resistance developing in microorganisms is one of the major public health threats that will be faced in the twenty-first century'.

Since antibiotic-resistance marker genes are there only to assist the scientists in developing the new crop, and do not benefit anyone once the crop is on the market, it seems unnecessary to leave them in the products. Routes for their removal are already available and increasingly being used.

Harmful products from new genes

It can be difficult to extract and purify proteins from GMOs to the standard required for medicinal use. Could the substances made by the new genes in GMOs be harmful in any way? It is difficult to guarantee that any activity in life is risk free – as we saw back at the very beginning of the AS course (AS Topic 1). So far there are no reported cases of ill health resulting from the consumption of GMOs. However, biochemical changes to oils, proteins and other substances might conceivably result in toxic compounds or new allergens.

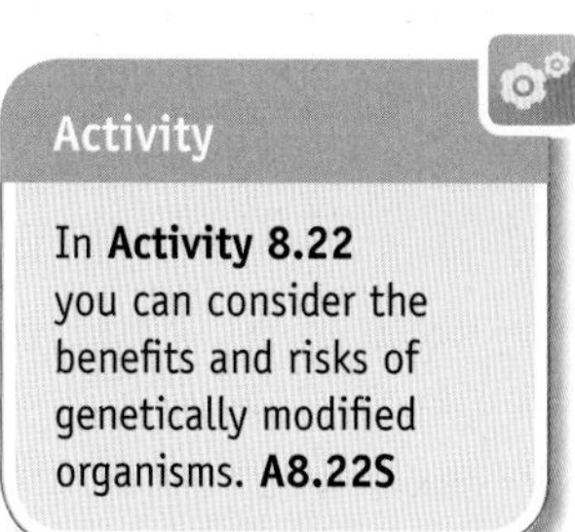

Activity
In **Activity 8.22** you can consider the benefits and risks of genetically modified organisms. **A8.22S**

Transfer of viruses from animals to humans

There are concerns that viruses that infect animals could be transferred to humans in products from genetically modified animals.

Choice

Most people have freedom to eat or use drugs provided they are safe and legal. If genes are changed in a GMO, and the product is deemed by the authorities to be safe, shouldn't people be allowed to choose whether or not they would like to eat food made from it or the products extracted from it?

Environmental issues

The main environmental concerns about GMOs are:

- transfer of genes to non-GM plants
- increased chemical use in crops.

Cross-pollination, i.e. transfer of pollen from one plant to another, can occur over quite long distances by wind or insects. Some of the crops we grow are related to wild plants living nearby and can cross with them (Figure 8.73). Furthermore, GM crops can cross with conventional crops of the same species growing in nearby fields. This means that genes introduced into a GM crop will almost inevitably spread to conventional crops – and this has been shown to occur. They may also be transferred to wild plants. Once they have 'escaped', such genes cannot be recalled.

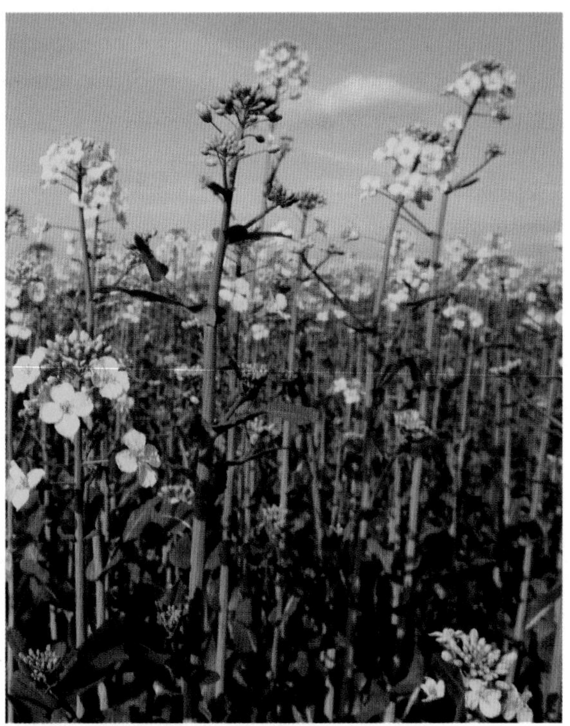

Figure 8.73 Oil seed rape is related to wild mustard.

Most of our crops cannot survive in the wild – they are not fit enough. Wild species would soon overrun our fields if crops were not given a helping hand by pesticides, fertilisers, herbicides and mechanical weeding. Therefore any escaped transgenic crop plants might be presumed to disappear before long. But suppose the introduced gene makes the crop or any plants it is transferred to much fitter, for example capable of surviving drought, frost or insect attack. Then it could have an advantage, even over its wild relatives, and we might have 'superweeds'.

We already have examples of 'superweeds' – without genetic engineering – due to the introduction of plants, such as rhododendrons and Japanese knotweed, from other countries. Some such plants have proved very well suited to survive in our environments and have caused considerable ecological havoc.

One solution to the problem of GM superweeds would be to ensure that the outcrosses (crosses between GM plants and other plants of a different variety) are not fertile and cannot proliferate. An alternative is the development of technology whereby the pollen does not contain the modified gene so it cannot spread. Both approaches are being investigated and look to be feasible.

In the US, wild mustard that is resistant to Roundup herbicide has invaded fields of Roundup Ready oil seed rape, and resistant mare's-tail has been found amongst Roundup Ready soybean crops. Resistant rye grass has been reported in Australia and resistant goose grass in Malaysia. Organisations that campaign against GM technology, such as Greenpeace and Friends of the Earth, argue that far from reducing the extent of pesticide use, GM crops will end up causing us to use even more chemicals to control resistant weeds and insects than we would otherwise have done.

Who owns these new organisms?

Even before genetic engineering was possible, plant breeders could protect a new variety under the Plant Breeders' Rights Act. They 'owned' the seed, but many farmers kept back seed from each harvest for the next year to avoid buying more. The advantage to the farmers was that they saved money. One disadvantage was that saved seed is sometimes not as good as the original. Biotechnology companies now patent the new technology used in gene modification, the introduced genes and, indeed, new varieties. Biotechnologists point out that they have spent time and money 'inventing' the new crop and argue that they deserve revenue from patents just like any other inventor.

Developing countries are unlikely to be able to afford to pay for expensive licences to use the new technology to grow GM crops with characteristics suited to local conditions, such as drought resistance or salt tolerance. Ownership of this powerful new technology by a handful of huge multinational corporations may threaten developing societies.

Farmers with enough capital may be able to benefit from GM technology, producing high-yield crops that they can then sell at competitive prices. Most of these farmers live in relatively affluent countries that rarely have food shortages. In contrast, small farmers living in less developed areas where there are often food shortages often cannot afford to invest in the pricey seeds and necessary agrochemical backup. Moreover, when they come to sell their traditional crops they may find that they have to compete with cheaper or higher quality products from GM varieties. As a result, they may well get even poorer, fall into debt and be forced to sell off their small plots of land. It is rarely easy to predict the social consequences of scientific developments.

We began the course in AS Topic 1 by looking at the effects of lifestyle on health; we end it by considering the use of modern biotechnology, which might include producing products to overcome some of the problems associated with our lifestyles. To a certain extent each of the topics in Salters-Nuffield Advanced Biology has covered a great range of material. Now that you have survived to the end of the course we hope you feel that it has all come together. We are confident that this approach will help you to understand current issues in biology whether you go on to study biology further or not.

Checkpoint

8.8 List the key benefits and risks of using GMOs.

Activity

Use **Activity 8.23** to check your notes using the topic summary provided. **A8.23S**

Review

Now that you have finished Topic 8, complete the end-of-topic test. Congratulations – you have completed the A2!

Answers to in-text questions Topic 5

Q5.1 **a** Brine shrimps; algae; microorganisms; **b** Salt water; mineral substrate; air; light; temperature;

Q5.2 Just about all of them can be: for example, humans can affect the light that reaches a forest floor by cutting down trees; they can affect climate through the greenhouse effect; they can affect parasitism by using veterinary products, and so on;

Q5.3 Wave action; salinity; temperature; light; desiccation due to exposure at low tide; geology; topography; oxygen availability;

Q5.4 Small ears, small tail, long thick fur and thick layer of fat reduce heat loss; long nose with several passages warms air entering the lungs; mating in the summer, remaining in the den in winter and cubs emerging from the den in the spring helps the bear survive the harsh winter conditions and increases the chance of the cubs surviving;

Q5.5 **a** It grows very slowly; **b** The caribou migrate over vast areas so only have a small effect on any one area;

Q5.6 Being prickly; stinging; mimicry – the white deadnettle gets left alone because it looks like a nettle even though it doesn't sting; being unpalatable – the common field buttercup has an unpleasant taste and so cows leave it alone; the mat grass on moorland has silica in its leaves which make it rough so the sheep eat more tender grasses instead;

Q5.7 **a** Leaves grow close to the ground, which reduces their chance of being eaten or mown down; if the leaves are grazed or cut during mowing, the growing point at the base is unlikely to be affected and leaves will be replaced;

b The low growing daisy cannot compete with tall plants for light, so it becomes less abundant in a lawn or meadow if the grass is allowed to grow tall;

Q5.8 Large ears; large back legs; white tail;

Q5.9 Producer: *Laminaria*; *Chondrus crispus*; toothed wrack; *Corallina*; bladder wrack; channel wrack; spiral wrack; lichens;

Primary consumer: blue rayed limpet; flat periwinkle; limpet; rough periwinkle; black periwinkle;

Secondary consumer: dog-whelk;

any appropriate adaptation for each example selected;

Q5.10 **a** Seeds could be carried to the island by birds; **b** Breaking up of rock by weathering; addition of some organic material from animals visiting island or from plants that can cope with conditions;

Q5.11 **a** Nitrogen and sulphur; **b** Nitrogen and phosphorus;

Q5.12 **a** H_2O; **b** Light; **c** Reduced NADP; **d** ATP;

Q5.13 **a** $H_2O \rightarrow 2H^+ + 2e^- + \frac{1}{2}O_2$; **b** $2H^+ + 2e^- + NADP \rightarrow$ reduced NADP

Q5.14 **a** RuBP; **b** GALP; **c** 12 GP are needed to make 12 GALP; this provides the two GALP to make glucose and the ten to recreate RuBP;

Q5.15 **a** Algae; **b** Brine shrimps; **c** Flamingo; **d** Eagle;

Q5.16 Phytoplankton → zooplankton (animal plankton) → fish → seal → polar bear;

Q5.17 The blue and red parts of the spectrum;

Q5.18 **a** Water; **b** Carbon dioxide; **c** Carbon dioxide; **d** Water;

Q5.19 **a** Little effect unless rainfall changed;

b and **c** Increase growth rate as carbon dioxide is limiting;

d Little effect, as even if rainfall increased the rocky substrate holds little water;

Q5.20 **a i** 23 140 – 17 820 = 5320 kJ m^{-2} y^{-1};

ii 5100 – 1960 = 3140 kJ m^{-2} y^{-1};

iii GPP = NPP + R = 26 000 + 8000 = 34 000 kJ m^{-2} y^{-1};

GPP in the tropical rainforest is very high due to the high incident light, high temperatures and constant availability of water; the high temperature also means that the respiration rate is high and this means that NPP is not as high as the high rate of GPP might suggest;

GPP in the pine forest is much lower than the rainforest, partly because it is colder; however, although GPP is less than a quarter of the tropical rainforest GPP, the NPP is over half of the tropical rainforest NPP, because respiration is much less than in a rainforest;

Maize is very efficient at trapping light energy and so has a very high GPP; it has been bred to have a high NPP;

b % efficiency = $\frac{34\,000}{2 \times 10^6} \times 100 = 0.017 \times 100 = 1.7\%$

c Most of the light is the wrong wavelength; and is not absorbed (by the chlorophyll/chloroplast); some is reflected by the leaves; some passes through the leaves without being absorbed; some warms the leaves; some is lost during the light-dependent reactions of photosynthesis;

d A desert has very low NPP due to lack of water limiting photosynthesis; and sparse vegetation;

Q5.21 **a** 125 kJ; **b** 4%

Q5.22 **a** 12.5%; **b** 20%

Q5.23 A straight line shows a gradual increase of about 1.8 °C between the mid-seventeenth century and the present day; a curve following trends shows fluctuations between 1700 and 1900, then a gradual rise until the present day;

Q5.24 Absence of oxygen and presence of acidic conditions reduce the activity of microorganisms; fewer survive and enzyme activity in those that do may be affected by low pH;

Q5.25 The climate was wet at the time;

Q5.26 Large amount of birch; replaced by Scots pine, elm and oak; the Scots pine disappeared; lime and alder appeared; elm numbers decreased leaving high percentage of oak and alder and a small amount of lime and elm up until the present day;

Q5.27 **a** If temperature extremes were becoming more pronounced with colder winters and hotter summers; this would not be shown by an average for the whole year;

b Two cold periods 18 000–13 000 and 11 000–10 000 years ago; trend towards higher temperatures in February but with very wide fluctuations;

Q5.28 The youngest ring is at the outside and as you go deeper into the tree trunk the wood gets older;

Q5.29 **a** 900 AD and 1450 AD; **b** 1150 AD;

Q5.30 Nitrogen, oxygen and argon – negligible; carbon dioxide 3.8×10^{-2}; methane 4.5×10^{-3}; nitrous oxide 9.5×10^{-3}; chlorofluorocarbon CCl_3F 1.2×10^{-4};

Q5.31 **a** Carbon dioxide 35%; methane 153%; nitrous oxide 18%; CFCs infinity % (was originally zero); **b** CFCs;

Q5.32 **a** The rate of increase in global carbon dioxide concentration is rising;

b There can be significant variations from year to year owing to climatic events such as volcanic eruptions; ten year average evens out this background 'noise' in the data;

Q5.33 Carbon dioxide emissions decreased for about ten years but have recently increased again; methane and nitrous oxide emissions have been greatly reduced;

Q5.34 How much of the emitted carbon dioxide is absorbed by photosynthesis and other carbon sinks: how much carbon dioxide is given out by other nations; whether the global concentration of carbon dioxide is rising or falling;

Q5.35 A utilitarian framework seeks to maximise the amount of good in the world; if global warming is likely to lead to more extreme weather (storms and droughts); and to certain habitats being unable to support existing flora and fauna; it is likely to increase human unhappiness and species extinctions; a utilitarian framework would therefore probably argue that attempts should be made to slow down, halt or even reverse global warming;

Q5.36 **a** 308 ppm; **b** 365–370 ppm

Q5.37 It cannot compete with other plants that flourish in the warmer conditions;

Q5.38 Their ranges would move to higher altitude up the mountain, where it is cooler;

Q5.39 With warmer conditions species moving north will compete with the existing species; there may be nowhere for existing species to migrate to; if some species can move or become locally extinct food supply to other species may be affected; there are a wide range of other consequences, for example if temperature acts as an environmental cue for development or behaviour this could be disrupted;

Q5.40 **a** 31–32 °C

b Isoenzymes may have different temperature optimums; they could synthesise the form with the appropriate temperature optimum as temperature changes;

Q5.41 Cooler nests producing males will become less and less common; threatening the survival of the species;

Q5.42 A gene is a length of DNA with a specific sequence of nucleotide bases which codes for the amino acid sequence of a polypeptide; a gene may have more than one form; these different forms of the gene are called alleles; they have different nucleotide base sequences and produce slightly different polypeptides; the alleles that a person has make up their genotype; the characteristic caused by the genotype is known as the phenotype; most eukaryotic organisms have two copies of every gene, two alleles; in an individual, the two alleles may be the same or different; some alleles are dominant and are always expressed; others are recessive and will only be expressed if an individual has two of these alleles; some characteristics, such as height, are controlled by several genes and are said to be polygenic; alleles are the units of selection which can increase or decrease in frequency in a population, resulting in evolution; new alleles can arise in a population through spontaneous changes in the DNA, known as mutations;

Q5.43 It allows the reviewers freedom to be critical without causing offence; it prevents the author from influencing the reviewers; [Authors sometimes thank anonymous reviewers at the end of their papers for helpful criticism.]

Q5.44 Publishing online is much quicker and can lead to a rapid exchange of ideas; however, it is possible for poorly conducted experiments to be put in the public domain; this can mislead people and may delay or misdirect future scientific research;

Q5.45 It allows scientists to review new findings and question those presenting them; it allows scientists to communicate with colleagues, discussing current ideas, sparking off new ideas and future projects; it may enable funding bodies to assess the outcome of the research they are paying for;

Q5.46 The conclusion may not be very reliable because the difference between 1.6% and 2.3% is small; any errors in measuring the temperature at which the hybrid DNA denatures will have a large effect on the results;

Q5.47 Some mussels have an allele that enables them to thicken their shell in response to the chemicals released by the Asian crabs; these mussels have a selective advantage, as they are less likely to be cracked open by the crabs; they survive and pass on the allele for thicker shells to their offspring;

Q5.48 Both populations of mussels have evolved the ability to thicken their shell in response to chemicals released by the green crab;

Q5.49 Some populations may have females who use spring temperatures as an environmental cue for egg laying; some populations may have alleles that allow them to lay eggs in shorter day lengths;

Q5.50 **a** Many biologists would argue that they are different subspecies but belong to the same species;

b There are arguments for and against the control of ruddy ducks; control involves shooting the ducks; if you are in favour of conserving biodiversity you may favour control; if you are in favour of minimising animal suffering you may be against control;

Q5.51 **a** Photosynthesis; **b** Biomass; **c** Respiration; **d** Decay; **e** Combustion;

Q5.52 Because it merely causes carbon to be released through combustion instead of through respiration (decay); the fuel source has recently absorbed the carbon dioxide which is now being released;

Topic 6

Q6.1 Check for any forms of identification; e.g. diary, bank card, driving licence, etc.;

Q6.2 Person 2;

Q6.3 agta; 6;

Q6.4 Person A;

Q6.5 **a** 2; **b** 2;

Q6.6 The size of the DNA fragments; shorter fragments go further;

Q6.7

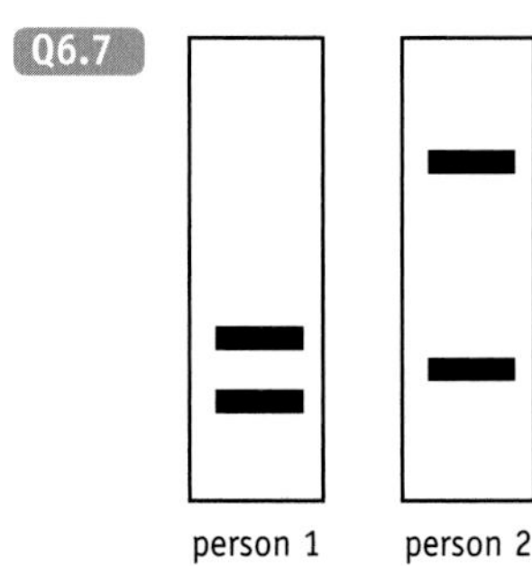

Q6.8 **a** Poppy; for each satellite the chick has inherited one copy of the repeat sequence from Poppy; one copy of A could come from Patsy but not B or C;

b Both children share some bands with each parent so the father is their biological parent; [The mother's bands are marked in red and the father's in blue.]

c They could take a tissue sample from the roots near the foundations, and then take samples from all the suspect trees in the gardens; they would then get the samples DNA profiled at a lab and match the root profile to the correct tree; as long as they were insured, their insurance company should pay for any damage to be put right!

d Test unknown sample for a number of STRs using PCR and gel electrophoresis; compare the results with a database of fragment lengths (number of repeats) for these STRs to determine which rice varieties are in the sample; different rice varieties will have a different profile for these STRs;

e The DNA is cut up using a restriction enzyme, and the fragments separated using gel electrophoresis; a DNA probe with a base sequence complementary to the allele associated with the disease is used; if the allele is present the probe will bind to the fragment on the gel;

Q6.9 The more satellites used, the more confident one can be about the conclusions drawn;

Q6.10 At elevated external temperatures such as are found in the tropics there will be little cooling of the body;

Q6.11 Larger mass; sitting upright; more clothing; warmer temperature; less air movement; low humidity;

Q6.12 The body is likely to have cooled to the surrounding temperature; the body was found in May; when the surrounding temperature would be about 18 °C; so we cannot tell if the person died in the last 24 hours;

Q6.13 **a** Speeds up; raises the body temperature so rate of enzyme action increases;

b Speeds up; allows entry of bacteria;

c Slows down; destroys the bacteria and denatures enzymes involved in decomposition;

Q6.14 Between 36 and 72 hours ago;

Q6.15 The temperatures for the local site are compared with those from the nearest weather station; if they are consistently higher or lower the temperatures recorded for the time preceding discovery of the body can be adjusted accordingly;

Q6.16 Look at the characteristic features of the insect and if not immediately recognisable use a key to identify the species;

Q6.17 6 days;

Q6.18 Approximately 30 hours;

Q6.19 Eggs laid between 24 and 28 June; earliest date of death 22 June assuming that the eggs were laid within two days of death;

Q6.20 **a** Higher temperatures would speed up succession; more rapid development of the organisms present; due to faster enzyme action; conditions on the body change more quickly allowing entry of the next species in the succession; not true if temperature greater than about 40 °C;

b Humidity; soil conditions; pH; chemical pollution;

Q6.21 **a** Atherosclerosis;

b Genetic predisposition; diet high in cholesterol and LDL lipids; smoking; high alcohol consumption;

c Aneurysms are prone to rupture (when they reach about 6–7 cm in size); the resulting blood loss from the circulatory system and the resultant shock can be fatal;

Q6.22 **a**; **b**; **c**; **d**; **f**;

Q6.23 **a** virus; **b** bacterium; **c** virus; **d** bacterium; **e** virus;

Q6.24 Some people do not consider viruses to be truly living organisms because they are not capable of independent reproduction or energy use; however, they do have characteristics of living organisms; because they can reproduce; pass genetic information to the next generation; and evolve;

Q6.25 The *Mycobacterium* is 22 times larger than the bacteriophage;

Mycobacterium length: 27 mm

$$\text{Actual length} = \frac{27\text{ mm}}{5200}$$

$$= 0.00519\text{ mm or } 5.19\ \mu\text{m}$$

Bacteriophage length: 4 mm

$$\text{Actual length} = \frac{4\text{ mm}}{17\,000}$$

$$= 0.00024\text{ mm or } 0.24\ \mu\text{m}$$

Q6.26 Barriers; between maternal and foetal blood;

Q6.27 By hydrolysing the polysaccharide;

Q6.28 Increased blood flow to the infected area makes it red; it is hot due to increased metabolic activity; it is swollen due to the leakage of fluid from capillaries into the tissue;

Q6.29 A;

Q6.30 Antibody production starts sooner; the speed of production is faster; and a higher concentration is produced; so the immune system can deal with the microbe before symptoms appear;

Q6.31 In 1915 the risk of contracting TB was 1 in 389 or 0.0026%; in 2005 the risk had decreased to 1 in 6999 or 0.00014%;

Q6.32 The conditions inside tubercules are anaerobic so there is no oxygen available for the bacteria to use in respiration;

Q6.33 **a** Less gas exchange; the lung surface area is reduced so gas exchange is less efficient;

b Breathing rate increases; to make up for the loss of gas exchange surface;

Q6.34 Metabolism is disrupted as enzymes denature; membranes are disrupted;

Q6.35 The person with scrofula would travel away from the infected area for the visit; they may experience better living conditions and improved food; this may alleviate symptoms which was interpreted as curing the disease;

Q6.36 Better diet (plenty of fruit and vegetables); reduced alcohol intake; stop smoking; better housing conditions;

Q6.37 TCCAAT;

Q6.38 **a** Passive artificial; passive natural; **b** Active artificial; active natural; **c** Passive artificial; passive natural; **d** Active artificial; active natural; **e** Active artificial; active natural; **f** Active artificial; passive artificial;

Q6.39 **a** 1991 **b** Before 1968 each measles epidemic greatly reduced the number of potential new hosts; after a couple of years enough new babies had been born to allow a new epidemic; **c** Between 1968 and 1988 the proportion of the population capable of contracting measles was falling; but herd immunity had not yet been attained;

Q6.40 Integrase inhibitors stop the DNA formed by reverse transcription integrating into the host T cell's DNA; fusion inhibitors bind to one or more of the receptors involved with HIV's entry into the host cell (CD4/gp120) and prevent entry occurring;

Q6.41 If penicillin is an enzyme it will denature and lose its antibacterial effect when heated; heating should not affect a toxic chemical; so heat penicillin and then place it on agar plates inoculated with bacteria; observe results; draw valid conclusion;

Q6.42 Option 1: *Disadvantages*: Previous patients may not have the same severity of TB as the next 100 patients that come into the hospital; previous treatments may have been so bad that anything could be better; *Advantages*: The doctor doesn't have the ethical dilemma of choosing which patients get the treatment and which don't;

Option 2: This would select a random sample but it is not a blind trial so the doctor may bias the experiment because she/he knows who is getting the treatment;

Option 3: This is really the best option if the doctor does not select the patients; in many such trials, the test is double-blind so that neither the doctor nor the patient knows whether the patient is getting the new drug or an inert substance called a placebo; this reduces the chances of the patient or the doctor exaggerating any improvements they see and deciding that these are the result or the new drug; [When the effect of streptomycin was tested by the Medical Research Council in the 1940s, they decided on option 1 for patients with tuberculous meningitis, because it was 100% fatal with the existing therapy. However, for pulmonary tuberculosis, they did a randomised trial (option 3).]

Q6.43 6;

Q6.44 Human cells do not have cell walls;

Q6.45 The waxy cell walls of *M tuberculosis* allow the bacteria to survive inside macrophages. They can also suppress the activity of T cells. HIV invades T cells, preventing the activation of macrophages, B cells and T killer cells, which weakens the immune response;

Q6.46 Avoiding the risky behaviour that resulted in HIV infection; vaccination might have protected her from TB; avoiding overcrowded hostels where there may have been a higher risk of TB infection;

Topic 7

Q7.1 A;

Q7.2 **a** Strength, flexibility/elasticity; **b** Synovial fluid; cartilage; **c** Inelastic;

Q7.3 **a** A ligament; B cartilage; C cartilage/pad of cartilage;

b Tendon; muscle; synovial membrane; synovial fluid; fibrous capsule;

Q7.4 It would take too long; to move proteins synthesised from mRNA from a single nucleus; to reach the furthest parts of the cell;

Q7.5 4;

Q7.6 The central band disappears; because the actin has moved and there is now no point at which there is only myosin on its own;

Q7.7 Better insulation reduces heat loss; so less energy is used to maintain core body temperature;

Q7.8 No oxygen is used;

Q7.9 Oxidation-reduction reactions occur as the electrons pass along the electron transport chain; the final electron acceptor is oxygen; phosphate is added to ADP to form ATP;

Q7.10 **a** 10;

b **i** 9; 3 reduced NAD and 1 reduced FAD for each turn of Krebs cycle;

ii 28; 2 reduced NAD in glycolysis, 8 reduced NAD and 2 reduced FAD from link reaction and Krebs cycle;

Q7.11 It prevents the cell overheating; and allows the controlled release of energy in small useful quantities;

Q7.12 To maintain rapid blood flow through muscles to supply oxygen; and remove lactate;

Q7.13 **a** Glycolysis; and ATP/PC; **b** Aerobic;

Q7.14 **a** B; **b** A; **c** C;

Q7.15 The untrained athlete; larger shaded area on Figure 7.38; they take up and transport oxygen more slowly so take longer to reach maximum oxygen uptake; their period of anaerobic respiration is longer;

Q7.16 It converts fats and proteins to glycogen; for storage in muscles and liver;

Q7.17 **a** Cardiac output = SV × HR; = 75 × 70; 5250 cm^3 per minute or 5.25 dm^3 per minute;

b SV = cardiac output/HR; = $\frac{5250}{50}$; 105 cm^3;

c SV = cardiac output/HR; = $\frac{5250}{33}$; 159 cm^3;

Q7.18 There are five squares between the QRS complexes, thus each heartbeat lasts one second; which gives a heart rate of 60 beats per minute; or the heart rate can be calculated by dividing 300 by 5 = 60 beats per minute;

Q7.19 **a** No clear pattern of P wave, QRS complexes and T waves; irregular timing and strength of waves;

b Very high heart rate; abnormal T wave;

Q7.20 Between the sinoatrial node and the atrioventricular node; or at the Purkyne fibres;

Q7.21 Congenital heart problem (non-hereditary condition existing at birth); extreme exhaustion; extreme dehydration; use of certain illegal drugs;

Q7.22 **a** If pressing on the neck causes increased blood pressure in the carotid artery; blood pressure sensors in the carotid artery would signal to the cardiovascular control centre; which in turn would stimulate the vagus nerve; reducing heart rate and thus pulse measurement;

b The wrist; or groin;

Q7.23 Anticipatory rise due to the effect of adrenaline on the heart; increases oxygen supply to the muscles in preparation for the activity about to occur;

Q7.24 Increased blood flow to active skeletal muscles; reduced blood flow to non-essential organs such as the digestive system; to increase supply of oxygen and glucose to respiring muscle cells;

Q7.25 6 dm^3 min^{-1};

Q7.26 **a** 0.7 dm^3;

b 2.7 dm^3;

c Tidal volume × rate of breathing; = 0.7 × 16 = 11.2 dm^3 per minute;

d Rate of oxygen consumption = volume of oxygen used (dm^3)/time (s); = 0.5/20; = 0.025 dm^3 s^{-1} or 25 cm^3 s^{-1};

Q7.27 The depth and rate of breathing increase so there is a greater volume of air inhaled and mixed with the residual air in the lungs; so concentration of oxygen increases; the higher concentration makes the diffusion gradient between the alveolar air and the blood steeper; increasing the speed of gas exchange; an advantage given the raised metabolic rate;

Q7.28 Stretch receptors signal the start of movement; allowing ventilation to increase before there is a build-up of the waste products of respiration;

Q7.29 The increased concentration of oxygen in the blood; detected by the chemoreceptors; slows breathing rate and decreases depth of breathing;

Q7.30 **a** Aerobic respiration occurs within the mitochondria; large numbers allow slow twitch fibres to have a greater capacity for aerobic respiration;

b Calcium ions released from the sarcoplasmic reticulum initiate muscle contraction; more sarcoplasmic reticulum allows rapid, repeated contraction of the muscle;

c Myoglobin stores oxygen within the cells for use in aerobic respiration;

d Slow twitch via aerobic respiration; fast twitch using anaerobic glycolysis reactions;

e Fast twitch; poor supply of oxygen to the fibre; uses anaerobic respiration; rapid build-up of lactate;

f Fast twitch fibre;

Q7.31 **a** Fast twitch; **b** Slow twitch;

Q7.32 Large percentage of fast twitch fibres, low mitochondrial density and high enzymes associated with anaerobic respiration indicate muscles have high capacity for glycolysis; the study demonstrates that these muscles of cheetahs are adapted for anaerobic exercise;

Q7.33 **a** 37–8 °C; **b** Low temperatures lead to low metabolic rates as the enzyme-controlled reactions slow; high temperatures increase the rate of metabolic reactions initially; but then it declines as the higher temperature denatures the enzymes;

Q7.34 Chemoreceptors detect any changes away from the norm blood carbon dioxide, pH and temperature; the ventilation centre is the control mechanism that sends impulses to the intercostal muscles; these effectors act to return conditions to the norm;

Q7.35 Hair erector muscles; muscles in the walls of the arterioles in the skin; skeletal muscles; liver cells;

Q7.36 **a** Increases heat energy loss by conduction; and evaporation;

b Less energy is lost by evaporation so it is harder for athletes to keep their body temperatures down to a safe level;

Q7.37 Evaporation from gas exchange surfaces; lowering hairs to increase energy loss by convection, conduction and radiation;

Q7.38 **a** Conduction; **b** Some energy is transferred to the body cells as a waste product of normal metabolism; shivering increases resting metabolism 3- to 5-fold; transferring additional energy to the body cells; nerve impulses to the arterioles in the skin cause vasoconstriction; resulting in restricted blood flow through the skin; this reduces energy loss by radiation, conduction and convection; hair raising is pretty ineffectual in humans, particularly for cross-Channel swimmers who coat themselves in protective grease (petroleum jelly); when exercising, respiration in the muscles increases, transferring additional energy to body cells;

Q7.39 **a** Reduced risk of upper respiratory tract infections with moderate amounts of or moderate intensity exercise; increased risk of infection with large amounts of or high intensity exercise;

b At low to moderate intensity of training there is a negative correlation, i.e. as intensity increases the risk of URT infection decreases; as intensity increases from moderate to high intensity training there is a positive correlation with risk of URT infection, i.e. as training intensity increases the risk of infection increases; there is a correlation between the two variables but this does not necessarily mean that the change in one is responsible for the change in the other;

Q7.40 Apoptosis is likely to destroy the infected cell and any viruses inside it; cell lysis would split open the cell releasing any viruses inside;

Q7.41 **a** B cells; **b** Inflammation; phagocytosis; antimicrobial proteins; **c** Few T helper cells so less cytokines produced; T killer cells will not be activated by the cytokines;

Q7.42 Moderate exercise increases the number of natural killer cells; intense exercise reduces the number and activity of natural killer cells, phagocytes, B cells and T helper cells;

Q7.43 **a** Tendons connect muscles to bones;

b Only one operation is required; there should be no immune response to a tendon from the same person whereas a tendon from someone else could be rejected;

c The tendon will be less elastic than a ligament, so she is likely to be advised to exercise her knee gently to stretch the tendon so that she regains full movement in her joint; she must avoid overstretching and damaging the tendon;

d Muscles usually have more than one attachment to bones (e.g. the triceps in the upper arm has three attachments), so that removal of one tendon will still leave other tendons attaching the muscle to the bone; the remaining tendons may grow stronger with use so that the missing tendon does not cause any weakness;

Q7.44 They are lipid based and so dissolve;

Q7.45 The hormone binds to a complementary receptor to bring about a response; only the target cells have the complementary receptors;

Q7.46 It would increase the number of red blood cells and hence the amount of haemoglobin; improving the blood's oxygen-carrying capacity; enhancing oxygen delivery to muscle tissue and hence improving aerobic capacity;

Q7.47 Unspecialised stem cells in the bone marrow divide and differentiate to form red blood cells; EPO binds to receptors on target cells in the bone marrow; altering gene expression that will lead to formation of the red blood cells; this includes transcription of the enzymes involved in synthesis of haemoglobin.

Q7.48 Blood clots in arteries or veins;

Q7.49 Sprint events rely on anaerobic respiration so performance is not dependent on the athlete's aerobic capacity;

Q7.50 To help determine if athletes are taking EPO as a performance-enhancing drug;

Topic 8

Q8.1 They produce rapid responses; important for protection and survival;

Q8.2 Other neurones must be involved; relay neurones within the central nervous system; and motor neurones to muscles in the arm; which are under conscious control;

Q8.3 **a** The one on the left; **b** c. 60%

Q8.4 Radial;

Q8.5 **a** Rods and cones in the retina; **b** Optic nerve; **c** Brain; **d** Iris muscle;

Q8.6 To prevent damage to the retina from high-intensity light; in dim light it ensures maximum light reaches the retina;

Q8.7 To protect the eye from sudden flashes of bright light;

Q8.8 Three (per eye);

Q8.9 The channel's opening is dependent on changes in voltage;

Q8.10 No (unless ATP was added); the polarisation of the membrane is maintained by the concentration gradients achieved by the action of energy-requiring sodium-potassium pumps; membrane integrity is lost in a dead axon;

Q8.11 A new action potential will only be generated at the leading edge of the previous one; because the membrane behind it will be recovering/incapable of transmitting an impulse; the membrane has to be repolarised and return to resting potential before another action potential can be generated;

Q8.12 Presynaptic neurone is on the left; it contains synaptic vesicles;

Q8.13 Ion channels open in the postsynaptic membrane; ions move through the channels increasing the polarisation of the membrane; more excitatory synapses would be required to depolarise the membrane;

Q8.14 **a** A larger insect landing on your arm will stimulate more receptors, so there is more likely to be an action potential due to spatial summation;

b A large insect crawling along your arm will stimulate lots of receptors over a longer period of time; this spatial and temporal summation is more likely to produce a response;

Q8.15 Pupil constriction: excitation of synapses to circular muscles, inhibition of synapses to radial muscles; Pupil dilation: excitation of synapses to radial muscles, inhibition of synapses to circular muscles;

Q8.16 Binds to receptor on target cell surface or within target cell; directly or indirectly via a second messenger molecule the hormone affects gene expression;

Q8.17 Site of production: produced in the testes by males and in small amounts by the adrenal glands in both males and females;

Method of transport: in the blood;

Location of target cells: throughout the body including male sex organs, skin and muscle cells, any cells involved in the development of the secondary sexual characteristics;

Effect on the target cells: binds to androgen receptors on target cells, modifying gene expression to alter the development of the cell; for example, increasing anabolic reactions such as protein synthesis in muscle cells, increasing the size and strength of the muscle;

Q8.18 B – shows that something in the tip is influencing the growth and curving of the shoot towards the light; C – shows that the tip is producing some sort of chemical that can diffuse into and out of an agar block; D – shows that in unidirectional light more IAA passes down the shaded side of the shoot;

Q8.19 The response in uniform light will be the same as occurs in the dark;

Q8.20 The same experiment completed with an agar block that has not been placed under the cut shoot; this will check that it is not something in the agar itself that is producing the response;

Q8.21 Might expect photoreceptors to be on the surface of the retina, but they form a deeper layer/light has to travel through the other layers including blood vessels before reaching the photoreceptors;

Q8.22 **a** **i** Active transport; **ii** Diffusion;

b One that lets any positive ions through, such as Na^+ and Ca^{2+};

c Because sodium ions are being actively transported out; and their re-entry through ion channels is prevented; increasing the potential difference across the membrane;

Q8.23 Both are light-sensitive pigment molecules made up of a protein and non-protein component; in both cases they are photoreceptors that change shape (rhodopsin breaks down, phytochrome molecules change from one isomer to another) allowing the organism to detect light;

Q8.24 **a** P_{fr}; **b** In the dark no red light is absorbed so no P_{fr} is formed that would stimulate germination; any P_{fr} present is converted back to P_r which inhibits germination;

Q8.25 Long days have a short period of uninterrupted darkness (short night); it is not long enough to convert all the P_{fr} back into P_r; the P_{fr} present stimulates flowering;

Q8.26 **a** From the diagram the short-day plant needs about 6 hours of light; the long-day plant requires 18 hours of light; **b** It would be better to call them short-night and long-night plants because it is the period of darkness that controls the flowering;

Q8.27 Although warmer days should improve plant growth the chrysanthemums are short-day plants and the additional lighting will mean the plants do not get the critical period of darkness needed to initiate flowering; not enough time to convert all the flowering-inhibiting P_{fr} back into P_r;

Q8.28 Flowering is not affected by day length;

Q8.29 **a** The plant grown in the dark is much taller than the one grown in the light, and it has longer internodes (length of stem between leaves); the plant grown in the dark is not quite as dark green as the one grown in the light; [Note that in total darkness for extended periods the leaves would be yellow, lacking chlorophyll.] **b** The formation of a longer stem would help the plant grow into the light; once in the light, the resources diverted into stem growth would be used in chlorophyll formation;

Q8.30 **a** P_{fr} inhibits the shade tolerance response; in the shade the P_{fr} is converted back into P_r so the shade tolerance response occurs;

b Increased growth with longer internodes to seek out light, the same response that occurs in the dark;

Q8.31 The region of the brain concerned with vision processing is the occipital lobe which sits at the back of the cortex and is closest to the back of the head (thus a blow to this area would cause a disturbance in vision);

Q8.32 Frontal lobe; parietal lobe; motor cortex; cerebellum;

Q8.33 Parietal lobe/basal ganglia;

Q8.34 Thalamus routes sensory information to the occipital lobe (visual cortex) which processes information from the eyes; the frontal lobe and parietal lobe may also be involved in interpreting the information;

Q8.35 There are many conditions that can cause brain damage, including lack of oxygen, carbon monoxide poisoning, toxic chemical exposures, infectious diseases, tumours, strokes and genetic conditions;

Q8.36 Cerebellum; motor cortex in frontal lobe;

Q8.37 **a** **i** fMRI; **ii** MRI; **b** The posterior hippocampus is involved in remembering detailed mental maps;

Q8.38 Visual stimulation with both light and patterns;

Q8.39 Because kittens are born blind, early deprivation (under three weeks) would have no effect; by three months connections to the brain have been made, and deprivation has no effect/the critical period has ended; the critical period is at about four weeks of age so lack of stimulation from the kitten's environment at this time severely affects visual development;

Q8.40 For some birds their song seems to be innate, due to nature and not nurture; for others both nature and nurture seem to be required;

Q8.41 1; 5; 3; 2; 4 or 1; 5; 3; 4; 2 or 1; 3; 4; 5; 2 or 1; 3; 5; 4; 2

Q8.42 Synapses from the optic nerve axons to the visual cortex will have been weakened or eliminated; binocular cells in the visual cortex may only receive sensory information from one eye; so they will not have two views to compare, making stereoscopic vision difficult or impossible;

Q8.43 The fact that there is a correlation between two variables does not necessarily mean that there is a causal relationship;

Q8.44 Familiarity with the relative sizes of the elephant and the antelope; the elephant in the picture is smaller than the antelope and man so it is in the background; overlap; one hill partly hiding another tells us that it is closer;

Q8.45 Some apparent cross-cultural differences in perception may occur because people cannot report perceptual differences; there may, for instance, be a language difficulty; researchers from Western societies chose the tests; there has been a strong emphasis on two-dimensional visual illusions which may favour subjects from the carpentered world;

Q8.46 They will show symptoms of distress or fear, and may refuse or cry;

Q8.47 **a** The baby has not experienced this before so cannot have learned it;

b Some perceptual development has certainly taken place since birth; the experiment requires the baby to crawl and this isn't possible for several months;

Q8.48 The fact that these animals had had little time for learning suggests that the behaviour is innate;

Q8.49 Kenge was not used to vast open spaces and would have had little opportunity to look far into the distance; with no experience of seeing distant objects he would have had little experience of certain environmental depth cues such as *size constancy* – the tendency to perceive the same object as always being the same size, however far away it is; as a result he saw the buffalo simply as very small animals;

Q8.50 Strawberry; pencil;

Q8.51 Parietal lobe; temporal lobe; hippocampus;

Q8.52 It suggests they are involved in forming new long-term memories but are not involved in recalling long-term memories;

Q8.53 **a** The neurones are not myelinated; **b** They have large diameter axons;

 **Q8.54**

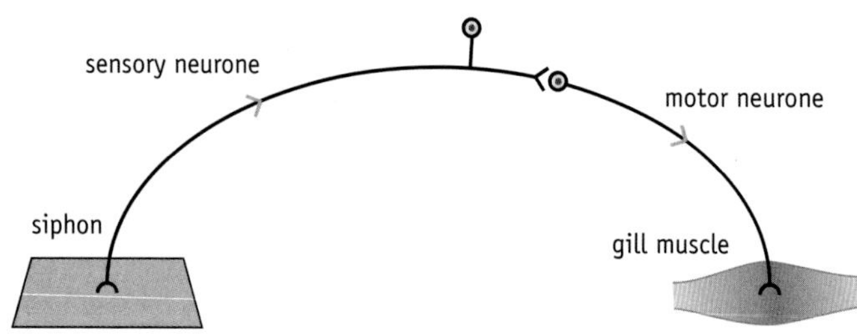

Q8.55 Breast cancer is common in women; over time hundreds of thousands of women die from it; each woman suffers more than each mouse; friends and relatives of a woman who dies young also suffer emotional pain;

Q8.56 Non-peptide compounds will not be digested as they pass through the gut so can be given to the patient orally;

Q8.57 **a** Polygenic inheritance; **b** Alzheimer's disease, schizophrenia, diabetes, coronary heart disease and some cancers;

Q8.58 Carriers of the inherited disease gene may not ever develop the disease; carriers of the inherited disease gene may develop the disease but it my not affect their ability to do their job or alter their insurance risk;

Q8.59 Eukaryotes contain a membrane bound nucleus;

Q8.60 CF patients experience lots of lung infections, therefore there are always high levels of neutrophils in their lungs releasing more elastase than the naturally occurring AAT can cope with; giving additional AAT could help break down the excess elastase reducing the inflammatory response and lung damage;

Q8.61 Clinical trials phase I, II and III;

Q8.62 Chickens' lifecycles are short so a flock of transgenic chickens can be bred quickly; the chickens are easy to keep and the eggs easy to collect making it possible to produce the proteins in bulk relatively cheaply;

Index

Picture credits

The publisher would like to thank the following for their kind permission to reproduce their photographs:

(Key: b-bottom; c-centre; l-left; r-right; t-top)

www.dreamstime.com: 49r, 140c, 229t; Action Plus Sports Images: Glyn Kirk 158t, 159t, 161t; Alamy Images: blickwinkel 10tr; Clark Brennan 190tr; Daniel Gotshall/Visuals Unlimited 251t; JUPITERIMAGES/ Polka Dot 163; Marc Schlossman 200; Neil Harwick 48; Nigel Cattlin 220; OJO Images Ltd 140; Wolfgang Kaehler 66; Ardea: Asgeir Helgestad 14t; Bob Gibbons 12b; D. Parer & E. Paer-Cook 55b; Duncan Usher 10b; John Mason 62t; M.Watson 2t; Blackthorn Arable Ltd: 230, 272; Corbis: Charles Gullung/zefa/ 196; Didrik Johnck 175; Michael St. Maur Sheil 13; Nicolas Asfouri/AFP 174; Vic Sievey/ Eye Ubiquitous 246t; Dr Dmitri Mauquoy : 32t; Mary Evans Picture Library: 192; FLPA Images of Nature: B. Borrell Casals 28; Bob Gibbons 62c; Ian Rose 9; Nigel Cattlin 95tr, 228b, 229b; Rinie van Muers 25; S & D & K Maslowski 62l; Stefan Auth/Imagebroker 138b; Suzi Eszterhas/Minden Pictures 138t; Getty Images: AFP 43; Agence Zoom 193; Jamie McDonald 185; Stephen Munday/ALLSPORT 139t; Robert Harding World Imagery: Patrick Dieudonne 12cl; iStockphoto: 67, 249bl; Nature Picture Library: Jose B. Puiz 11t; NHPA Ltd / Photoshot Holdings: Brian Hawkes 10tl; Laurie Campbell 7t; M.I. Walker 14c; Normand Glennard: Nicola Wiberforce 171t; OSF: Jorge Sierra 62r; PA Photos: 178, 197; Phil Walter/EMPICS Sprot 190tl; Photolibrary.com: Amanaimages/Clover 258b; Photoshot Holdings Limited: AllCanadaPhotos 53r; Reuters: Carlos Barria 139c; Science & Society Picture Library: Science Museum 270; Science Photo Library Ltd: Adam Hart-Davis 205l, 205r; Alfred Pasieka 93cr; Andree Syred 32b; Andrew Lambert 150; Andrew McClenaghan 130t; b. Murton / Southampton Oceanography Centre 15; Biology Media 144; Biophoto Associates 176; Brad Nelson / Custom Medical Stock Photo 188l; BSIP, Beranger 92; BSIP, Vem 115; CDC 117; Centre for Bioimaging, Rothamsted Research 95tl; Charles D. Winters 16; Clive Freeman, The Royal Institution 98; CNRI 153t, 188r, 232; Costantino Margiotta 70; Darwin Dale 149br (b); David Parker 83, 262t; Doug Plummer 167; Dr Gary Gaugler 133; Dr Jeremy Burgess 87; Dr Kari Lounatmaa 19, 100; Dr Klaus Boller 129; Dr P. Marazzi 99t, 190b; Edward Kingsman 179; Eric Grave 143b; Eye of Science 186; George Bernard 75; Health Protection Agency 80; J.C. Revy 208; James King-Holmes 74; Jerry Mason 79bl; John Burbidge 112; John Durham 132; Juergen Berger 94r; Kent Wood 149c; Living Art Enterprises, LLC 237; Louise Lockley / CSIRO 79t (fig6); Makoto Iwafuji / Eurelios 266; Martin Dohrn 73t; Martin M. Rotker 128b; Martyn F. Chillmaid 222; Maximilian Stock Ltd 131b; Microfield Scientific Ltd 50; Omikron 251b; Pascal Goetgheluck 254; Peter Menzel 109b; Prof. K. Seddon & Dr. T. Evans, Queen's University Belfast 149bl, 151 (a); Prof. P. Motta / Dept. Of Anatomy / University "La Sapienza", Rome 225; Richard Kirby, David Spears Ltd 34cl; Scimat 160c; scott camazine 189; scott sinklier / agstockusa 268; sidney moulds 34cr; simon fraser 173; sinclair stammers 269b; sovereign, ism 238; Steve Gschmeissner 202; Susumu Nishinaga 249; Tek Image 262b; Thomas Deerinck, NCMIR 148, 215; Vaughan Fleming 231; Volker Steger 89; Western Ophthalmic Hospital 241; Zephyr 236; Spectrum Photofile: Ottmar Bierwagen/www.photographersdirect.com 59; The Davis Enterprise: Florence Low 244; TopFoto: Berkeley MR ©Elizabeth Crews/The Image Works 248; Wellcome Trust Medical Photographic Library: Welcome Library, London 96

Picture Research by: Charlotte Lippmann

Acknowledgements

The authors and publishers would like to thank the following for permission to use copyright material: Figure 5.1 p2 reproduced courtesy of the Geophysical Fluid Dynamics Laboratory of the National Oceanic and Atmospheric Administration; Figure 5.34 p15 and Figure 5.37 p16 reproduced from Blake, Dise, N., Jones, B., Murphy, P. and Taylor, P. (1998) S103 *Discovering Science.* Block 2 *A Temperate Earth?* The Open University, Milton Keynes; Table 5.4 p39 reproduced with permission of International Panel on Climate Change; Figure 5.41A, Figure 5.41B p36 reproduced courtesy of the UK Meteorological Office; Figure 5.44 p40 © Nature Publishing Group, from *Nature*, Vol 329, November 1987 reproduced with permission; Table 5.5 p40 reproduced courtesy of the Department for Environment, Food and Rural Affairs; Figure 5.47 p46 reproduced courtesy of the UK Meteorological Office; Figure 5.49 p49 © Nature Publishing Group, from *Nature*, Vol 399, 30 June 1999 reproduced with permission; Table 5.7 p64 reproduced courtesy of the Department for Business, Enterprise and Regulatory Reform; Figure 6.18, p87, Table 6.5, p90 and Table 6.6, p91 adapted from Kenneth GV Smith (1986), *A Manual of Forensic Entomology*, London: Trustees of the British Museum (Natural History); Figure 6.21, p89 adapted from article 'On maggots and murder; forensic entomology' by Martin Hall on the Natural History Museum website (http://www.nhm.ac.uk); M Lee Goff quote, p90 reprinted by permission of the publisher from *A Fly for the Prosecution: How Insect Evidence Helps Solve Crimes* by M Lee Goff, p14, Cambridge, Mass: Harvard University Press © 2000 by the President and Fellows of Harvard College; Figure 6.36, p101 adapted from Figure 43.4, p842 in N Campbell, J Reece and L Mitchell (1999) *Biology*, 5th Edn © 1999 by Benjamin/Cummings, an imprint of Pearson Education Ltd; Table 6.7, p110 adapted from Fact sheet 104 'Tuberculosis', revised April 2005, on the World Health Organisation website (http://www.who.int); Table 6.8 reproduced courtesy of the Health Protection Agency, Centre for Infections; Figure 6.47, p116 reproduced by kind permission of the Joint United Nations Programme on HIV/AIDS (http://www.unaids.org); Figure 6.59 reproduced courtesy of the Office for National Statistics and Health Protection Agency (www.hpa.org.uk); David Livermore quote, p136 reproduced with permission from article 'Setback in superbug battle: Patient shows resistance to new class of antibiotic', Sarah Boseley, *The Guardian*, 7 December 2002 © Guardian Newspapers Ltd 2002; Figure 7.38, p161 and Figure 7.60, p182 adapted from W McArdle, F Katch and V Katch (2005) *Essentials of Exercise Physiology*, Philadelphia: Lippincott, Williams and Wilkins; Figure 7.59 p178 reproduced courtesy of NASA; Figure 7.64, p185 from LT Mackinnon (2000) 'Special features for the Olympics: effects of exercise on the immune system; overtraining effects on immunity and performance in athletes', *Immunology and Cell Biology* **78**: 502–9. All Crown Copyright material is reproduced with the permission of the Controller of HMSO and the Queen's Printer for Scotland.